Advances in Thermoresponsive Polymers

Editor

Mattia Sponchioni

MDPI • Basel • Beijing • Wuhan • Barcelona • Belgrade • Manchester • Tokyo • Cluj • Tianjin

Editor
Mattia Sponchioni
Department of Chemistry,
Materials and Chemical
Engineering "Giulio Natta"
Politecnico di Milano
Milano
Italy

Editorial Office
MDPI
St. Alban-Anlage 66
4052 Basel, Switzerland

This is a reprint of articles from the Special Issue published online in the open access journal *Polymers* (ISSN 2073-4360) (available at: www.mdpi.com/journal/polymers/special_issues/adv_thermoresponsive_polymer).

For citation purposes, cite each article independently as indicated on the article page online and as indicated below:

LastName, A.A.; LastName, B.B.; LastName, C.C. Article Title. *Journal Name* **Year**, *Volume Number*, Page Range.

ISBN 978-3-0365-4012-2 (Hbk)
ISBN 978-3-0365-4011-5 (PDF)

Contents

About the Editor . . . . **vii**

Preface to "Advances in Thermoresponsive Polymers" . . . . **ix**

Nicolò Manfredini, Marco Tomasoni, Mattia Sponchioni and Davide Moscatelli
Influence of the Polymer Microstructure over the Phase Separation of Thermo-Responsive Nanoparticles
Reprinted from: *Polymers* **2021**, *13*, 1032, doi:10.3390/polym13071032 . . . . **1**

Ekaterina M. Zubanova, Sergei V. Kostjuk, Peter S. Timashev, Yury A. Rochev, Alexander I. Kokorin and Mikhail Ya. Melnikov et al.
Inhomogeneities in PNIPAM Aqueous Solutions: The Inside View by Spin Probe EPR Spectroscopy
Reprinted from: *Polymers* **2021**, *13*, 3829, doi:10.3390/polym13213829 . . . . **13**

Lorenzo Marsili, Michele Dal Bo, Giorgio Eisele, Ivan Donati, Federico Berti and Giuseppe Toffoli
Characterization of Thermoresponsive Poly-N-Vinylcaprolactam Polymers for Biological Applications
Reprinted from: *Polymers* **2021**, *13*, 2639, doi:10.3390/polym13162639 . . . . **29**

Wenli Gao, Zhidan Wang, Fei Song, Yu Fu, Qingrong Wu and Shouxin Liu
Temperature/Reduction Dual Response Nanogel Is Formed by In Situ Stereocomplexation of Poly (Lactic Acid)
Reprinted from: *Polymers* **2021**, *13*, 3492, doi:10.3390/polym13203492 . . . . **45**

Chao Geng, Shixue Wang and Hongda Wang
Recent Advances in Thermoresponsive OEGylated Poly(amino acid)s
Reprinted from: *Polymers* **2021**, *13*, 1813, doi:10.3390/polym13111813 . . . . **61**

Pavel I. Semenyuk, Lidia P. Kurochkina, Lauri Mäkinen, Vladimir I. Muronetz and Sami Hietala
Thermocontrolled Reversible Enzyme Complexation-Inactivation-Protection by Poly(*N*-acryloyl glycinamide)
Reprinted from: *Polymers* **2021**, *13*, 3601, doi:10.3390/polym13203601 . . . . **81**

György Kasza, Tímea Stumphauser, Márk Bisztrán, Györgyi Szarka, Imre Hegedüs and Endre Nagy et al.
Thermoresponsive Poly(*N*,*N*-diethylacrylamide-*co*-glycidyl methacrylate) Copolymers and Its Catalytically Active -Chymotrypsin Bioconjugate with Enhanced Enzyme Stability
Reprinted from: *Polymers* **2021**, *13*, 987, doi:10.3390/polym13060987 . . . . **91**

Vincent Huynh, Natalie Ifraimov and Ryan G. Wylie
Modulating the Thermoresponse of Polymer-Protein Conjugates with Hydrogels for Controlled Release
Reprinted from: *Polymers* **2021**, *13*, 2772, doi:10.3390/polym13162772 . . . . **109**

Aggeliki Stamou, Hermis Iatrou and Constantinos Tsitsilianis
NIPAm-Based Modification of Poly(L-lysine): A pH-Dependent LCST-Type Thermo-Responsive Biodegradable Polymer
Reprinted from: *Polymers* **2022**, *14*, 802, doi:10.3390/polym14040802 . . . . **121**

Aziz Ben-Miled, Afshin Nabiyan, Katrin Wondraczek, Felix H. Schacher and Lothar Wondraczek
Controlling Growth of Poly (Triethylene Glycol Acrylate-*Co*-Spiropyran Acrylate) Copolymer Liquid Films on a Hydrophilic Surface by Light and Temperature
Reprinted from: *Polymers* **2021**, *13*, 1633, doi:10.3390/polym13101633 . . . . **133**

Tatyana Kirila, Alina Amirova, Alexey Blokhin, Andrey Tenkovtsev and Alexander Filippov
Features of Solution Behavior of Polymer Stars with Arms of Poly-2-alkyl-2-oxazolines Copolymers Grafted to the Upper Rim of Calix[8]arene
Reprinted from: *Polymers* **2021**, *13*, 2507, doi:10.3390/polym13152507 . . . . **149**

Tatyana Kirila, Anna Smirnova, Vladimir Aseyev, Andrey Tenkovtsev, Heikki Tenhu and Alexander Filippov
Self-Organization in Dilute Aqueous Solutions of Thermoresponsive Star-Shaped Six-Arm Poly-2-Alkyl-2-Oxazines and Poly-2-Alkyl-2-Oxazolines
Reprinted from: *Polymers* **2021**, *13*, 1429, doi:10.3390/polym13091429 . . . . **171**

About the Editor

Mattia Sponchioni

Mattia Sponchioni received his MSc cum laude in Chemical Engineering in 2015 and his PhD cum laude in Industrial Chemistry and Chemical Engineering in 2018, both at Politecnico di Milano. The focus of his PhD research was on the development of thermoresponsive polymers for biomedical applications. After his PhD, Mattia Sponchioni spent one year as a post-doc at ETH Zürich in 2019, working on the development of continuous countercurrent chromatographic processes for the purification of biopharmaceuticals. From 2020 to present, he has been an assistant professor at the Department of Chemistry, Materials and Chemical Engineering "Giulio Natta"at Politecnico di Milano. Here, Mattia Sponchioni leads a biomanufacturing group, active in the development of continuous perfusion processes for the production (upstream) of mRNA and bioplastics via fermentation and in the conversion of the traditional single-column batch chromatographic purifications (downstream) into continuous countercurrent processes based on two twin columns. The vision of this group, inspired by the principles of Industry 4.0, is towards process intensification, to be achieved through both continuous processes and integrated manufacturing, as a way not only to increase the productivity but also the sustainability of biomanufacturing. Within this framework, Mattia Sponchioni is a contact for the EU H2020 Marie-Curie innovative training network "Continuous Downstream Processing of Bioproducts".

Preface to "Advances in Thermoresponsive Polymers"

Stimuli-responsive polymers are leading to a breakthrough in many fields, from biomedicine to oil and gas, to advanced separations. The possibility introduced by these materials to dynamically respond to environmental changes, mimicking the complex behavior of naturally occurring macromolecules that regulate fundamental living functions by responding to external stimuli, allows accessing complex applications that could not even be imagined before. Among these materials, thermoresponsive polymers, able to respond to temperature changes with a sharp and often reversible phase separation, are attracting growing attention from the literature, as confirmed by the exponential increase in the scientific publications about this topic in the last twenty years. This is due to the spontaneous occurrence of thermal gradients and to the possibility of applying temperature stimuli quite easily and inexpensively, which makes thermoresponsive polymers extremely appealing in medical and industrial settings. Given the interest towards these materials, in this work entitled "Advances in Thermoresponsive Polymers", the recent advances in the field are reported. The aim is to provide a clear picture of the frontiers reached in the understanding of the mechanistic behavior associated with the temperature-induced phase separation, the influence of the polymer structure in regulating the macroscopic behavior of these materials and the latest applications for which thermoresponsive polymers show great potential.

Mattia Sponchioni

Editor

polymers MDPI

Article

Influence of the Polymer Microstructure over the Phase Separation of Thermo-Responsive Nanoparticles

Nicolò Manfredini, Marco Tomasoni, Mattia Sponchioni * and Davide Moscatelli

Department of Chemistry, Materials and Chemical Engineering "Giulio Natta", Politecnico di Milano, via Mancinelli 7, 20131 Milan, Italy; nicolo.manfredini@polimi.it (N.M.); marco3.tomasoni@mail.polimi.it (M.T.); davide.moscatelli@polimi.it (D.M.)

* Correspondence: mattia.sponchioni@polimi.it

Abstract: Thermo-responsive nanoparticles (NPs), i.e., colloids with a sharp and often reversible phase separation in response to thermal stimuli, are coming to the forefront due to their dynamic behavior, useful in applications ranging from biomedicine to advanced separations and smart optics. What is guiding the macroscopic behavior of these systems above their critical temperature is mainly the microstructure of the polymer chains of which these NPs are comprised. Therefore, a comprehensive understanding of the influence of the polymer properties over the thermal response is highly required to reproducibly target a specific behavior. In this study, we synthesized thermo-responsive NPs with different size, polymeric microstructure and hydrophilic-lipophilic balance (HLB) and investigated the role of these properties over their phase separation. We first synthesized four different thermo-responsive oligomers via Reversible Addition-Fragmentation Chain Transfer (RAFT) Polymerization of poly(ethylene glycol)methyl ether methacrylate. Then, exploiting the RAFT living character, we chain-extended these oligomers with butyl methacrylate obtaining a library of NPs. Finally, we investigated the NP thermo-responsive behavior, their physical state above the cloud point (Tcp) as well as their reversibility once the stimulus is removed. We concluded that the solid content plays a minor role compared to the relative length of the two blocks forming the polymer chains. In particular, the longer the stabilizer, the more favored the formation of a gel. At the same time, the reversibility is mainly achieved at high HLB, independently from the absolute lengths of the block copolymers.

Keywords: polymeric nanoparticles; emulsion polymerization; RAFT; thermo-responsive polymers; smart materials; LCST; phase diagram; phase separation

Citation: Manfredini, N.; Tomasoni, M.; Sponchioni, M.; Moscatelli, D. Influence of the Polymer Microstructure over the Phase Separation of Thermo-Responsive Nanoparticles. *Polymers* **2021**, *13*, 1032. https://doi.org/10.3390/polym13071032

Academic Editors: Alina Sionkowska and Sebastien Lecommandoux

Received: 9 March 2021
Accepted: 23 March 2021
Published: 26 March 2021

1. Introduction

Polymeric nanoparticles (NPs) have been extensively studied since the middle of the previous century for their appealing and controllable interfacial and composition properties [1–3]. These features determined their consolidation in several fields, such as coating [4], painting [5], textile [6], or cosmetics [7].

Nowadays, these colloids are mainly produced via free radical emulsion polymerization (FRPe) [8]. This technique allows the synthesis of concentrated (up to 50–60% *w/w*) NP suspensions, commonly referred to as latexes, directly in water, avoiding the use of any organic solvent [9,10]. Although economically viable and brought to a multi-ton scale to satisfy the ever-increasing market demand for polymer latexes, one of the main drawbacks of FRPe is the limited control over the polymer microstructure and hence over the NP physico-chemical properties [11,12].

In the last decades, this limitation was progressively overcome by the advent of controlled radical polymerizations, which gave new lymph to the field [13–15]. In particular, the high control over the polymer chain microstructure achievable with these polymerization techniques paved the way to highly engineered colloids [16–22], contributing to the spreading of the NP technology towards previously unexplored applications.

In a similar direction, such controlled radical polymerizations were exploited to tune the response of a particularly appealing class of NPs, the so-called stimuli-responsive NPs, i.e., colloids with a sharp and often reversible phase separation behavior in response to external stimuli. Among the different stimuli that can be exploited to trigger such behavior (e.g., pH [23], temperature [24–28], CO_2 [29,30], redox potential [31]), the temperature is surely one of the most interesting, promising and versatile. In fact, the possibility of applying thermal stimuli in a controlled way, coupled with the natural occurrence of thermal gradients in living organisms, made thermo-responsive materials attractive for several applications.

In the literature, two opposite behaviors are reported for thermo-responsive polymers. With reference to the phase diagram, namely the diagram of temperature vs. polymer volume fraction, for a solvent/polymer binary mixture, the thermo-responsive polymers can be miscible with the solvent in any proportion below their binodal curve, while phase separating in a polymer-rich phase above it. The minimum of this curve is called lower critical solution temperature (LCST), while each other point is commonly referred to as cloud point (Tcp), since the formation of polymer-rich droplets above this critical temperature often determines the cloudiness of the system. Oppositely, the binodal curve can form a maximum, known as upper critical solution temperature (UCST), with the thermo-responsive polymer being miscible with the solvent above this curve, while phase separating below it [24,32,33]. Despite examples of both kinds of systems have been reported, the easier capability of modulating the Tcp and the almost insensitivity of their phase behavior to environmental properties rendered polymers with a LCST the most studied. These have been successfully investigated in many different applications ranging from pharmaceutical [17,24,34–36] to oil and gas [37] industries.

What clearly appears from these studies is that the peculiar microstructure of the chains constituting the final NPs has a tremendous impact on the macroscopic thermo-responsive behavior of these colloids, including their reversibility (i.e., the capacity to recover the original physico-chemical properties when the stimulus is removed) [38]. At the same time, guidelines for the selection of the polymer chain requisites necessary to access a desired macroscopic behavior are lacking.

In this work, with the aim of rationalizing this concept, we considered NPs constituted by amphiphilic block copolymers comprising a thermo-responsive chain and a hydrophobic portion and systematically investigated the impact of the length of each chain on the final thermo-responsive behavior. In particular, we first synthesized four different thermo-responsive macromolecular chain transfer agents (macro CTAs) via reversible addition-fragmentation chain transfer (RAFT) polymerization. This technique allows to set a priori the average degree of polymerization (n) of the oligomers by simply changing the thermo-responsive monomer (in our case, oligo(ethylene glycol)$_4$ methyl ether methacrylate, hereinafter EG_4) to chain transfer agent (CTA) molar ratio. The choice of this thermo-responsive monomer instead of the more common N-isopropylacrylamide (NIPAM) was driven by the higher biocompatibility and poor hysteresis of EG_4 that contributed to intensify the research towards this compound [17]. Then, exploiting the living nature of the RAFT polymerization, we chain-extended these macro CTAs with butyl methacrylate (BMA) and targeted different chain lengths (p). The adoption of the RAFT emulsion polymerization allowed us to obtain the formation of NPs with a lipophilic core and a thermo-responsive shell (nEG_4-pBMA) directly in water. These colloids are expected to be stable as long as the temperature is kept below the Tcp. In fact, in this configuration the thermo-responsive chains should be well hydrated providing steric stabilization to the colloids. On the contrary, as soon as the temperature is raised above the Tcp, a sharp coil-to-globule transition of the EG_4 moieties is expected. With the thermo-responsive chains collapsing on the NP surface, the colloidal stability is lost, which leads to the NP aggregation, as schematically reported in Scheme 1. For each combination of n and p, we evaluated the physical state (precipitate or gel in Scheme 1) of the separated phase as well as the reversibility of these aggregates once the temperature is lowered again below the Tcp and concluded on the factors governing such macroscopic behavior.

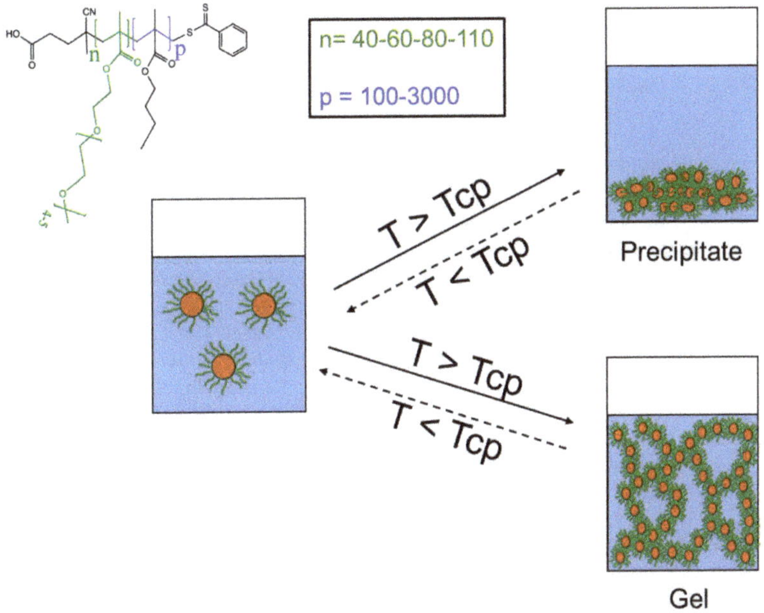

Scheme 1. Schematic representation of the expected thermo-responsive behavior for NPs stabilized by a thermo-responsive polymer.

2. Materials and Methods

2.1. Materials

4-Cyano-4-(phenyl-carbonothioylthio) pentanoic acid (CPA, MW = 279.38, 99%, Sigma Aldrich, Saint Louis, MO, USA), 2,2′-azobis(2-methylpropionamidine)dihydrochloride (V-50, MW = 271.19, 98%, Acros Organics, Geel, Belgium), poly(ethylene glycol)methylether methacrylate (EG_4, MW = ca300, Sigma Aldrich, Saint Louis, MO, USA), 4-4′-azobis (4-cyanovaleric acid) (ACVA, MW = 280.28, Sigma Aldrich, Saint Louis, MO, USA), butyl methacrylate (BMA, ≥99%, MW = 142.20, Fluka Chemika, Buchs, Switzerland), diethyl ether (99.7%, MW = 74.12, Sigma Aldrich, Saint Louis, MO, USA), deuterated chloroform ($CDCl_3$, 99.8%, MW = 120.38, Sigma Aldrich, Saint Louis, MO, USA), ethanol (EtOH, MW = 46.07, ≥99.8%, Sigma Aldrich, Saint Louis, MO, USA), and tetrahydrofuran (THF, ≥99.9%, MW = 72.11, Sigma Aldrich, Saint Louis, MO, USA) were used as received.

2.2. Synthesis of the Thermo-Responsive Macro CTAs

Four different thermo-responsive macro CTAs were synthesized via RAFT polymerization of EG_4 using ACVA as initiator and CPA as CTA. The ACVA to CPA molar ratio (IA) was kept equal to 1/3 for all the syntheses while the EG_4 to CPA molar ratio (n) was varied in order to obtain different degrees of polymerization (i.e., n = 40, 60, 80, and 100). The polymerizations were carried out in ethanol at a monomer concentration equal to 20% *w/w*. Briefly, to synthesize the macro CTA with n = 80, hereinafter $80EG_4$, 8 g of EG_4 (26.6 mmol), 0.093 g of CPA (0.33 mmol, i.e., EG_4/CPA = 80 mol/mol), and 0.031 g of ACVA (0.11 mmol, i.e., IA = 1/3 mol/mol) were placed in a round bottom flask equipped with a magnetic stirrer. Then, 32.5 g of ethanol were added and the solution was purged with nitrogen for 20 min. The reaction was let occurring at 65 °C for 24 h under stirring at 300 rpm. The solution was then purified through three consecutive precipitations in diethyl ether in order to remove all the impurities and the unreacted monomer. At the end of the purification process, the solution was centrifuged for 10 min at 5000 rpm to separate the two phases and the polymer pellet was recovered and dried inside a vacuum drying oven (VACUCELL) at 35 °C overnight. Finally, the polymer was stored at −20 °C. Aliquots of the samples were dissolved in deuterated chloroform ($CDCl_3$) at a concentration equal to 14 mg/mL and analyzed via nuclear magnetic resonance (^{1}H-NMR) on a Bruker Ultrashield 400 MHz spectrometer (Bruker, Billerica, MA, USA) with 64 scans (see reaction scheme and spectrum in Figure S1). The monomer conversion and average degree of polymerization, *n*, were calculated according to Equations (S1) and (S2), respectively, and the results are summarized in Table S1.

In order to calculate the number-average molecular weight (*Mn*) and dispersity (*Đ*), the samples were dissolved in tetrahydrofuran (THF) at a concentration of 4 mg/mL, filtered through a 0.45 μm polytetrafluoroethylene (PTFE) membrane and injected in a Jasco LC-2000Plus (Jasco, Easton, MD, USA) gel permeation chromatograph (GPC). The analysis was performed at a flow rate of 1 mL/min and at a temperature equal to 35 °C. Three styrene/divinyl benzene columns in series (300 × 8 mm^2, particle size 5 μm, pore size 1000, 10^5 and 10^6 A, respectively) were used for the separation and the signal collected with a refractive index (RI) detector. A pre-column (50 × 8 mm^2) was added before the first column. The instrument calibration was performed with polystyrene standards.

2.3. Synthesis of Amphiphilic Block Copolymers Self-Assembled into NPs

The thermo-responsive macro CTAs were chain-extended with butyl methacrylate via RAFT emulsion polymerization in order to produce amphiphilic block copolymers self-assembled into NPs. These block copolymers, hereinafter nEG_4-pBMA, were synthesized setting the initiator to macro CTA molar ratio to 1/3 and varying the BMA to macro CTA molar ratio (*p*) in order to obtain copolymers with different degrees of polymerization. The polymerization was performed in water at a concentration equal to 20% *w/w*.

For example, in the synthesis of $80EG_4$-500BMA (i.e., molar ratio between BMA and $80EG_4$ equal to 500), 1 g of BMA (7 mmol), 0.363 g of $80EG_4$ (0.015 mmol, i.e., BMA/$80EG_4$ = 500 mol/mol), and 1.27 mg of V-50 (0.005 mmol, i.e., V-50/$80EG_4$ = 1/3 mol/mol) were mixed in 3.64 g of distilled water. The suspension was purged with nitrogen for 20 min at room temperature to avoid side reactions, mixed using a vortex and placed in a pre-heated oil bath at 50 °C under magnetic stirring for 24 h.

At the end of the reaction, the NP suspensions were placed in a cellulose membrane (Spectra/Por) with a molecular weight cut-off of 3.5 kDa and were dialyzed against distilled water for 48 h.

The reaction conversion was first evaluated via thermogravimetric analysis performed on an Ohaus MB35 moisture analyzer [6] and then confirmed via ^{1}H-NMR (Figure S2, Equation (S3)) following the same procedure described in Section 2.2. From the ^{1}H-NMR spectrum, the actual *p* was calculated according to Equation (S4). The block copolymer *Mn* and *Đ* (Table S2) were measured via GPC according to the procedure described in Section 2.2.

The NP size (reported in terms of Z-average, indicated as Dn) and polydispersity index (PDI) were measured via dynamic light scattering (DLS) on a Zetasizer Nano ZS (Malvern, UK) at a scattering angle of 173° (Table 1). Each sample was diluted before the analysis to a concentration equal to 1% *w/w* and analyzed in triplicate with 13 runs per measurement.

Table 1. Size (Dn), polydispersity index (PDI) and temperature of cloud point (Tcp) of the NP synthesized.

Sample	Dn [nm]	PDI [-]	Tcp [°C]	HLB [-]
$40EG_4$-100BMA	58.9 ± 0.3	0.15 ± 0.01	64	9.1
$40EG_4$-200BMA	72.6 ± 0.7	0.11 ± 0.01	65	5.9
$40EG_4$-300BMA	135.9 ± 0.7	0.12 ± 0.01	63	4.5
$40EG_4$-500BMA	194.9 ± 3.1	0.08 ± 0.01	63	2.8
$40EG_4$-1000BMA	412.5 ± 3.5	0.09 ± 0.01	64	1.6
$40EG_4$-3000BMA	1251.8 ± 94.1	0.07 ± 0.01	62	0.6
$60EG_4$-150BMA	48.6 ± 0.2	0.15 ± 0.03	65	9.3
$60EG_4$-275BMA	73.3 ± 0.5	0.05 ± 0.02	65	6.2
$60EG_4$-400BMA	125.2 ± 0.8	0.03 ± 0.02	64	4.8
$60EG_4$-725BMA	192.7 ± 0.1	0.06 ± 0.02	64	2.9
$80EG_4$-450BMA	110.3 ± 2.4	0.22 ± 0.01	64	5.9
$80EG_4$-650BMA	160.6 ± 2.1	0.05 ± 0.01	63	4.2
$80EG_4$-850BMA	177.3 ± 1.8	0.04 ± 0.08	63	3.5
$100EG_4$-275BMA	57.7 ± 2.1	0.05 ± 0.10	66	9.2
$100EG_4$-600BMA	102.1± 0.9	0.03 ± 0.01	64	5.9
$100EG_4$-850BMA	126.6 ± 2.5	0.03 ± 0.03	65	4.5
$100EG_4$-1200BMA	168.9 ± 4.1	0.03 ± 0.04	64	3.1

The block copolymer hydrophilic-lipophilic balance (HLB, Table 1) was calculated according to Equation (1):

$$HLB = 20\frac{M_h}{M_t} \quad (1)$$

where M_h is the molecular mass of the hydrophilic portion of the molecule (i.e., the thermo-responsive portion in our case), and M_t is the molecular mass of the whole molecule. This value is a number comprised within 0–20 with a value of 0 corresponding to a completely lipophilic molecule and a value of 20 corresponding to a completely hydrophilic molecule.

2.4. *Study of the Thermo-Responsive Properties*

The NP cloud point (Tcp) was calculated via DLS analyzing the inflection points of Dn vs. temperature curves [17]. The temperature was ramped from 25 °C up to 70 °C at intervals of 1 °C, leaving 120 s of equilibration time before each measurement.

In order to evaluate the NP reversibility, all the samples were diluted at 20%, 10%, 5%, and 1% *w/w* and placed in a pre-heated oven at 80 °C for 1 h. At the end of the heating procedure, the NPs were removed and left to equilibrate at room temperature for an additional hour.

The samples showing a reversibility (i.e., turned again cloudy) were analyzed via DLS in order to compare their actual size with the original one (i.e., before the heating procedure).

3. Results and Discussion

Thermo-responsive amphiphilic block copolymers with a well-defined microstructure and self-assembled into NPs were synthesized via two consecutive RAFT polymerizations.

We first synthesized four thermo-responsive macro CTAs (nEG_4, with n equal to 40, 60, 80, and 100, respectively) through RAFT solution polymerization of EG_4 in ethanol. High monomer conversion (>94%), poor $Đ$ (<1.2) and controllable Mn were obtained (see Table S1 in the Supporting Information). In fact, one peculiarity of the RAFT process is the possibility of setting *a priori* the desired molecular weight by simply changing the monomer to CTA molar ratio (n) as shown in Equation (2) [39].

$$M_{n_{m\mathrm{CTA}}} = M_{\mathrm{CTA}} + nXM_{\mathrm{EG}_4} \quad (2)$$

being X the monomer conversion, $M_{n_{m\mathrm{CTA}}}$, M_{CPA} and M_{EG_4} the molecular weights of the produced macro CTA (i.e., number-average molecular weight), CTA and EG_4 respectively.

This Equation is validated by the linear increase in M_n reported in Figure 1a for different values of n. At the same time, the RAFT polymerization ensures that the actual n of the product, as measured through ^{1}H-NMR, closely matches the target value, as shown in the comparative plot reported in Figure S3a, thus granting a good control over the polymer average chain length.

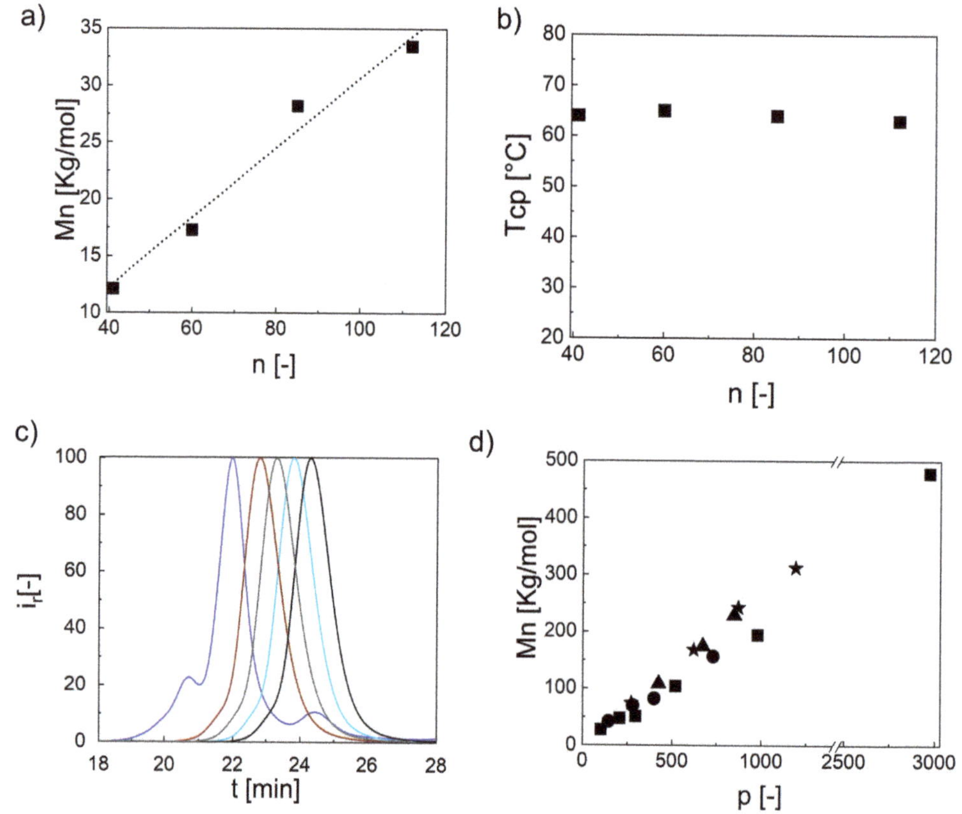

Figure 1. (**a**) Number average molecular weight as function of the degree of polymerization of the thermo-responsive block (**b**) Tcp as a function of the degree of polymerization of the thermo-responsive block (**c**) Normalized chromatograms of the block copolymers synthesized from 100EG$_4$ having p equal to 0 (black), 275 (light blue), 600 (grey), 850 (red), and 1200 (blue), respectively. i_r represents the normalized intensity of the refractive index detector. (**d**) Number-average molecular weight as a function of the degree of polymerization of the lipophilic block for polymers having n equal to 40 (square), 60 (circle), 80 (triangle), and 100 (star), respectively.

It is worth noticing that this increase in the polymer molecular weight is not followed by a change in its Tcp, which changed in a very narrow range between 62 and 66 °C, as demonstrated in Figure 1b. The phase separation at this temperature may find application in the oil and gas filed, where thermo-responsive polymers can be applied as viscosity modifiers or to enhance the emulsification/separation of oil in water during the so-called enhanced oil recovery [37,40,41]. However, we would like to point out that with this work we are not aiming at a particular application. Rather, EG$_4$ was selected as a representative of the oligo(ethylene glycol)methyl ether methacrylates, with a controllable phase separation. The Tcp of these polymers can be readily adjusted to the desired value by choosing the appropriate oligo(ethylene glycol) chain length or by copolymerization with hydrophilic (i.e., Tcp increases) or hydrophobic (i.e., Tcp decreases) monomers [24]. The poor sensitivity of the Tcp to the molecular weight for this class of polymers, quite expected for LCST-type polymers at least for these low values of *n*, allows to obtain a portfolio of macro CTAs with comparable Tcp and different *Mn*.

These macro CTAs were then chain-extended with BMA via RAFT emulsion polymerization to obtain amphiphilic block copolymers self-assembled into NPs. Different degrees of polymerization for the polyBMA portion (*p*) were targeted. Again, the RAFT polymerization ensured to achieve values of *p* very close to the target (Figure S3b). In addition, all the copolymers synthesized presented high monomer conversion (>95%) and poor Đ (<1.4, Table S2) as expected from a controlled radical polymerization. The possibility of forming well-defined block copolymers is further demonstrated by the high blocking efficiency. In fact, in most of the samples, the macro CTA showed high conversion during the polymerization as its characteristic peak is strongly reduced in intensity in the GPC chromatogram of the block copolymers. This is shown as an example in Figure 1c for the 100EG_4-*p*BMA samples. The peak of the macro CTA at 24.5 min is seen as a shoulder in the chromatogram of the 100EG_4-1200BMA, while it is not separated and resolved in all the other samples, suggesting high conversion. Indeed, for these samples, low molecular weight tails that could hide the signal from the macro CTA were recorded. However, their intensity is quite small compared to that of the main copolymer distribution, suggesting that in the worst scenario the residual macro CTA concentration is low as well.

This two-step RAFT polymerization strategy provides two degrees of freedom (*n* and *p*) enabling to independently control the partition of the thermo-responsive and hydrophobic units and the copolymer *Mn*, according to Equation (3) (Figure 1d):

$$M_n = M_{n_{m\mathrm{CTA}}} + pXM_{\mathrm{BMA}} = M_{\mathrm{CTA}} + nM_{\mathrm{EG}_4} + pXM_{\mathrm{BMA}} \tag{3}$$

with M_{BMA} and M_n being the BMA and final block-copolymer number-average molecular weight, respectively. In particular, through the tuning of *n* and *p* it is possible to produce copolymers with the same *Mn* and different hydrophilic-to-lipophilic balance (HLB). To rationalize this concept, we calculated for all the samples (Table 1) the corresponding HLB (Equation (1)), which was found to play a relevant role in dictating the macroscopic thermo-responsive behavior, as discussed in the following. The HLB could be varied in the range 1–10 and, therefore, block copolymers from very hydrophobic to slightly hydrophilic could be produced with this approach.

Another advantage of using the RAFT emulsion polymerization is the simultaneous synthesis of the copolymer and its self-assembly into NPs at high solid content. The high control in the block copolymer synthesis allows the formation of colloids with tunable size (Dn) and poor polydispersity indexes (PDI) as clearly shown in Table 1.

In particular, the two-step approach adopted in this work makes the NP size a function of *n* and *p* (Equation (4)) [39], thus providing the opportunity of decoupling Dn from the polymer molecular weight and in turn of obtaining NPs with comparable size but different microstructure:

$$D_n = \frac{6\,p\,M_{\mathrm{BMA}}}{A\,N\,\rho} = \frac{6\,p\,M_{\mathrm{BMA}}}{(\alpha + n\beta)\,N\,\rho} \tag{4}$$

with M_{BMA} being the molecular weight of BMA, ρ the density of the lipophilic chains, *N* the Avogadro number, *A* the area on the NP surface covered by a single polymeric chain, α and β two constants characteristic of the polymer. In particular, *A* is a function of the molecular weight of the hydrophilic portion of the block copolymer when amphiphilic stabilizers as $n\mathrm{EG}_4$ are used at a concentration higher than their critical micellar concentration (CMC). In this case, the lipophilic backbone is expected to dispose tangentially to the NP surface with the hydrophilic moieties exposed to the bulk. In a similar conformation, the higher *n* (and in turn *Mn* of the stabilizer), the higher the portion of the surface that it is able to cover and, in turn, the lower the average NP size [39]. The NP synthesized in this work are in line with this prediction and their size, at given *p*, decreases with *n* as clearly visible in Figure 2a. In fact, the higher *n* the higher the surface area covered by a single chain with a consequent reduction in the NP size (Equation (4)). Moreover, a linear increase in the NP size with *p*, at a given *n*, is obtained independently from the stabilizer used (Figure 2a).

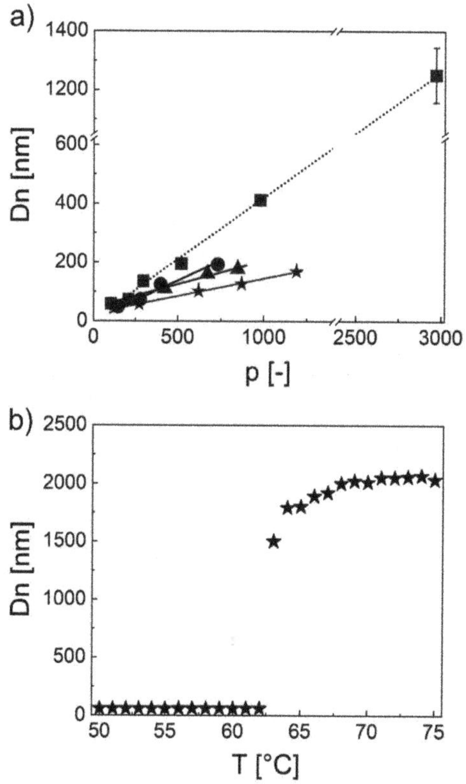

Figure 2. (**a**) NP size (Dn) as a function of *p* in the case of *n* equal to 40 (square), 60 (circle), 80 (triangle), and 100 (star), respectively. (**b**) NP size as function of external temperature in the case of $80EG_4$-650BMA.

It is worth mentioning that a broad range of Dn can be accessed with this strategy, from around 50 nm up to 1 μm (i.e., $40EG_4$-3000BMA). This is a remarkable result for colloids produced via RAFT emulsion polymerization, which typically leads to NPs with an upper size limit of 500 nm [42].

The high control achievable with this two-step RAFT polymerization strategy allows the formation of a wide portfolio of colloids with different size and HLB suitable for a systematic analysis of the impact of these parameters over the macroscopic thermo-responsive behavior.

First, we measured the Tcp of the different samples via DLS, according to the procedure described in Section 2.4. All the NPs show a similar trend with a sharp change in their size from nm to few μm once the Tcp is reached (Figure 2b). It is worth mentioning that the absolute value of the size reached above the Tcp should be only considered as an indication, since at this size scale the gravity causes the sedimentation of the larger aggregates already within the time of the analysis. What is reliable is the dynamic of the phase separation, which is due to the loss of steric stability. In fact, once the Tcp is overcome, the thermo-responsive shell collapses on the NP core leading to aggregation (Scheme 1). It is worth noticing that the Tcp is independent from *p* for all the *n* considered (Table 1). In fact, the high conversions obtained in the synthesis of the stabilizer coupled with the high compartmentalization achievable with the RAFT emulsion polymerization permits to obtain a library of NPs with a transition temperature that is dependent only on the thermo-responsive monomer chosen [17–25].

After having demonstrated that all of the samples exhibit a thermo-responsive behavior, their physical state above the Tcp as well as their reversibility once the thermal stimulus is removed were systematically investigated and related to the three degrees of freedom characterizing the system, i.e., solid content, n and p.

We explored the thermo-responsive behavior at four different concentrations, namely 1, 5, 10, and 20% *w/w* according to the procedure described in Section 2.4. Interestingly, no significant effects of NP concentration over their reversibility were noticed for all of the NPs analyzed, as visible in Figure S4. Thus, samples with a reversible transition at given n and p, are so at any of the concentrations tested. Moreover, only a modest effect on the NP physical state above the Tcp was observed with a tendency to the formation of gel when the concentration is increased. Then, the effect of n and p was investigated in details at a fixed concentration of 20% *w/w*. We divided the phase diagram into four classes depending on the NP reversibility as well as the physical state of the aggregate above the Tcp. In particular, we individuated two possible physical states, namely precipitate and gel (Figure S5). In the former, the destabilization led to the NP aggregation and precipitation of a bulky polymer phase, while in the case of the gel, the destabilization involved the whole mass of the sample, leading to a self-standing monolith. Both of these states can be reversible or irreversible depending on the capability of the NPs to recover their original size once the thermal stimulus is removed, as shown for example in Figure S6 for the sample $60EG_4$-150BMA. The four different behaviors are indicated in the phase diagram of Figure 3a as reversible precipitate, irreversible precipitate, irreversible gel, and reversible gel, respectively. In general, we observed that an increase in p always led to a progressive loss in the reversibility of the phase separation, independently from the n considered. This behavior can be justified considering that an increase in p is associated with an increase in the NP lipophilicity, leading to stronger hydrophobic interactions between the aggregated NPs. This hypothesis is supported by the evidence that by counter-balancing the increase in lipophilicity of the block copolymer with an increase in the thermo-responsive portion (i.e., n), the range of p leading to reversible transitions is expanded. In addition, n seems to play a relevant role in the physical state of the phase separated system, favoring the formation of hydrogels. This may be attributed to the decrease in the NP size, and then to an increase in their number, associated with an increase in n, when the solid content is fixed. This higher number of particles, together with the partial shell interpenetration could lead to a percolating system, in which at a critical NP distance the system stops flowing forming a self-standing gel [43].

Since, as shown, a change in p and n is always accomplished with a variation in the NP size making the understanding of the real impact of these parameters difficult, we re-analyzed the samples as a function of their HLB (Equation (1)). Interestingly, it was found that NPs with comparable HLB present similar size (Figure S7). Moreover, the decreasing trend of Dn with the HLB observed for all the n considered is in line with Equation (4). Grouping NPs with similar size allows to understand the impact of the absolute length of both blocks on the NP reversibility. In fact, as visible from Figure 3b, for HLB lower than 4.5, an irreversible aggregation is obtained independently from the n considered. However, the systems stabilized by shorter n (40 and 60 in Table S1) are more prone to the reversibility (achieved at lower HLB compared to the samples with $n = 80$ and 100). This result suggests that the longer the stabilizer, the lower the NP reversibility when size and lipophilicity are fixed, which may be due to a certain interaction between the thermo-responsive shells. Hence, a fine tuning of the block copolymer microstructure allows to decouple the NP size and HLB from their thermo-responsive response.

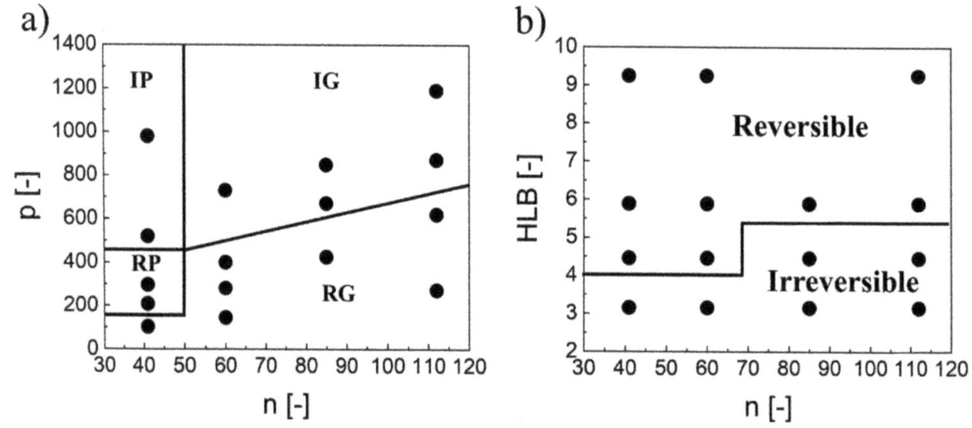

Figure 3. (**a**) Phase diagram reporting the physical state of the NP suspension above its Tcp at a fixed concentration of 20% *w/w* being RP, reversible precipitate, IP, irreversible precipitate, IG, irreversible gel, and RG, reversible gel, respectively. (**b**) Phase diagram reporting the reversibility of the phase separation at a given concentration of 20% *w/w* and as a function of the HLB of the block copolymer and *n*.

4. Conclusions

In this work, we correlated the macroscopic behavior of thermo-responsive NPs with the microstructure of the block copolymer they are made up of. The use of two sequential RAFT polymerization steps allows the synthesis of colloids with high blocking efficiency and controllable properties including size, molecular weight, HLB and microstructure. Moreover, exploiting the two degrees of freedom that this procedure allows, it was possible to decouple all the NP properties in order to better clarify the role of each parameter over the final macroscopic behavior.

First, it was found that the solid content does not play a significative role on the properties of these systems. In general, a decrease in the nanoparticle size is accomplished with a better reversibility of the system and a major tendency to form gel. However, this behavior is impacted also from the absolute length of both the portions (*n* and *p*) of the block copolymers that form the NP. In fact, NPs with comparable HLB and size recover their reversibility more easily when the stabilizing block is shorter.

Supplementary Materials: The following are available online at https://www.mdpi.com/article/10.3390/polym13071032/s1, Figure S1: ^{1}H-NMR spectrum of 60EG$_4$; Figure S2: ^{1}H-NMR spectrum of 60EG$_4$-400BMA; Figure S3: (a) thermo-responsive polymer degree of polymerization calculated via ^{1}H-NMR (n_{NMR}) as a function of the EG$_4$/CTA mole ratio (n_{target}). (b) hydrophobic block degree of polymerization calculated via ^{1}H-NMR (pNMR) as function of the BMA/macro CTA mole ratio (ptarget); Figure S4: Phase diagram reporting the reversibility of the phase separation as a function of the NP concentration and p in the case of *n* equal to 40 (a), 60 (b), 80 (c) and 100 (d), being RP = Reversible Precipitate, IP = Irreversible Precipitate, IG = Irreversible Gel and RG = Reversible Gel, respectively; Figure S5: NP forming a precipitate (left) or a hydrogel (right) once the temperature is increased above the Tcp; Figure S6: NP size as function of temperature during heating (triangle) and cooling (star) in the case of 60EG$_4$-150BMA; Figure S7: NP size as function of block copolymers HLB in the case of *n* equal to 40 (square), 60 (circle), 80 (triangle) and 100 (star); Table S1: conversion (X), degree of polymerization (*n*), number-average molecular weight (Mn) and dispersity (Đ) of the macro CTAs synthesized; Table S2: conversion (X), degree of polymerization (*n*, *p*), number-average molecular weight (Mn) and dispersity (Đ) of the block copolymers synthesized; Equation (S1): EG$_4$ conversion; Equation (S2): Degree of polymerization for the macro CTA from the corresponding ^{1}H-NMR spectrum; Equation (S3): BMA conversion; Equation (S4): Degree of polymerization for the polyBMA from the corresponding ^{1}H-NMR spectrum.

Author Contributions: Funding acquisition: D.M.; investigation: N.M.; methodology: M.S.; supervision: D.M.; validation: M.T.; writing—original draft: N.M.; writing—review and editing: M.S. All authors have read and agreed to the published version of the manuscript.

Funding: This research received no external funding.

Institutional Review Board Statement: Not applicable.

Informed Consent Statement: Not applicable.

Data Availability Statement: The data presented in this study are available on request from the corresponding author.

Acknowledgments: The authors are grateful to Roberto Panebianco for the help with the experimental part.

Conflicts of Interest: The authors declare no conflict of interests for this publication.

References

1. Bennet, D.; Kim, S. Polymer Nanoparticles for Smart Drug Delivery. *Appl. Nanotechnol. Drug Deliv.* **2014**. [CrossRef]
2. Birrenbach, G.; Speiser, P.P. Polymerized Micelles and Their Use as Adjuvants in Immunology. *J. Pharm. Sci.* **1976**, *65*, 1763–1766. [CrossRef] [PubMed]
3. Kreuter, J. Nanoparticles and Nanocapsules-New Dosage Forms in the Nanometer Size Range. *Pharm. Acta Helv.* **1978**, *53*, 33–39. [PubMed]
4. Devonport, I.W.; Even, R.C.; Hermes, A.R.; Lorah, D.P.; Tanzer, J.D.; VanDyk, A.K. Aqueous Composition Containing Polymeric Nanoparticles. U.S. Patent No. US7071261B2, 4 July 2006.
5. Serhan, M.; Sprowls, M.; Jackemeyer, D.; Long, M.; Perez, I.D.; Maret, W.; Tao, N.; Forzani, E. Industrial Applications of Nanoparticles. *AIChE Annu. Meet. Conf. Proc.* **2019**. [CrossRef]
6. Manfredini, N.; Ilare, J.; Invernizzi, M.; Polvara, E.; Contreras Mejia, D.; Sironi, S.; Moscatelli, D.; Sponchioni, M. Polymer Nanoparticles for the Release of Fragrances: How the Physicochemical Properties Influence the Adsorption on Textile and the Delivery of Limonene. *Ind. Eng. Chem. Res.* **2020**, *59*, 12766–12773. [CrossRef]
7. Patel, A.; Prajapati, P.; Boghra, R. Overview on Application of Nanoparticles in Cosmetics. *Asian J. Pharm. Sci. Clin. Res.* **2011**, *1*, 40–55.
8. Nasir, A.; Kausar, A.; Younus, A. A Review on Preparation, Properties and Applications of Polymeric Nanoparticle-Based Materials. *Polym. Plast. Technol. Eng.* **2015**, *54*, 325–341. [CrossRef]
9. Asua, J.M. Emulsion Polymerization: From Fundamental Mechanisms to Process Developments. *J. Polym. Sci. Part A Polym. Chem.* **2004**, *42*, 1025–1041. [CrossRef]
10. Chern, C.S. Emulsion Polymerization Mechanisms and Kinetics. *Prog. Polym. Sci.* **2006**, *31*, 443–486. [CrossRef]
11. Monteiro, M.J.; Cunningham, M.F. Polymer Nanoparticles via Living Radical Polymerization in Aqueous Dispersions: Design and Applications. *Macromolecules* **2012**, *45*, 4939–4957. [CrossRef]
12. Chem, R. Reaction Chemistry & Engineering Progress in Reactor Engineering of Controlled Radical Polymerization: A Comprehensive Review. *React. Chem. Eng.* **2016**, *1*, 23–59. [CrossRef]
13. Nicolas, J.; Guillaneuf, Y.; Lefay, C.; Bertin, D.; Gigmes, D.; Charleux, B. Nitroxide-Mediated Polymerization. *Prog. Polym. Sci.* **2013**, *38*, 63–235. [CrossRef]
14. Matyjaszewski, K. Atom Transfer Radical Polymerization (ATRP): Current Status and Future Perspectives. *Macromolecules* **2012**, *45*, 4015–4039. [CrossRef]
15. Chiefari, J.; Chong, Y.K.; Ercole, F.; Krstina, J.; Jeffery, J.; Le, T.P.T.; Mayadunne, R.T.A.; Meijs, G.F.; Moad, C.L.; Moad, G.; et al. Living Free-Radical Polymerization by Reversible Addition—Fragmentation Chain Transfer: The RAFT Process. *Macromolecules* **1998**, *31*, 5559–5562. [CrossRef]
16. Capasso Palmiero, U.; Maraldi, M.; Manfredini, N.; Moscatelli, D. Zwitterionic Polyester-Based Nanoparticles with Tunable Size, Polymer Molecular Weight, and Degradation Time. *Biomacromolecules* **2018**, *19*, 1314–1323. [CrossRef] [PubMed]
17. Sponchioni, M.; Ferrari, R.; Morosi, L.; Moscatelli, D. Influence of the Polymer Structure over Self-Assembly and Thermo-Responsive Properties: The Case of PEG-b-PCL Grafted Copolymers via a Combination of RAFT and ROP. *J. Polym. Sci. Part A Polym. Chem.* **2016**, *54*, 2919–2931. [CrossRef]
18. Sponchioni, M.; Palmiero, U.C.; Moscatelli, D. HPMA-PEG Surfmers and Their Use in Stabilizing Fully Biodegradable Polymer Nanoparticles. *Macromol. Chem. Phys.* **2017**, *218*, 1–12. [CrossRef]
19. Sponchioni, M.; Morosi, L.; Lupi, M.; Capasso Palmiero, U. Poly(HPMA)-Based Copolymers with Biodegradable Side Chains Able to Self Assemble into Nanoparticles. *RSC Adv.* **2017**, *7*, 50981–50992. [CrossRef]
20. Zanoni, A.; Gardoni, G.; Sponchioni, M.; Moscatelli, D. Valorisation of Glycerol and CO_2 to Produce Biodegradable Polymer Nanoparticles with a High Percentage of Bio-Based Components. *J. CO_2 Util.* **2020**, *40*, 101192. [CrossRef]
21. Zanetti, M.; Carniel, T.K.; Dalcanton, F.; dos Anjos, R.S.; Gracher Riella, H.; de Araújo, P.H.H.; de Oliveira, D.; Antônio Fiori, M. Use of Encapsulated Natural Compounds as Antimicrobial Additives in Food Packaging: A Brief Review. *Trends Food Sci. Technol.* **2018**, *81*, 51–60. [CrossRef]

22. Kumari, A.; Yadav, S.K.; Yadav, S.C. Biodegradable Polymeric Nanoparticles Based Drug Delivery Systems. *Colloids Surf. B Biointerfaces* **2010**, *75*, 1–18. [CrossRef]
23. Wang, F.; Luo, Y.; Li, B.G.; Zhu, S. Synthesis and Redispersibility of Poly(Styrene- Block—N -Butyl Acrylate) Core-Shell Latexes by Emulsion Polymerization with Raft Agent-Surfactant Design. *Macromolecules* **2015**, *48*, 1313–1319. [CrossRef]
24. Sponchioni, M.; Capasso Palmiero, U.; Moscatelli, D. Thermo-Responsive Polymers: Applications of Smart Materials in Drug Delivery and Tissue Engineering. *Mater. Sci. Eng. C* **2019**, *102*, 589–605. [CrossRef] [PubMed]
25. Sponchioni, M.; Rodrigues Bassam, P.; Moscatelli, D.; Arosio, P.; Capasso Palmiero, U. Biodegradable Zwitterionic Nanoparticles with Tunable UCST-Type Phase Separation under Physiological Conditions. *Nanoscale* **2019**, *11*, 16582–16591. [CrossRef]
26. Sponchioni, M.; O'Brien, C.T.; Borchers, C.; Wang, E.; Rivolta, M.N.; Penfold, N.J.W.; Canton, I.; Armes, S.P. Probing the Mechanism for Hydrogel-Based Stasis Induction in Human Pluripotent Stem Cells: Is the Chemical Functionality of the Hydrogel Important? *Chem. Sci.* **2020**, *11*, 232–240. [CrossRef]
27. Sponchioni, M.; Manfredini, N.; Zanoni, A.; Scibona, E.; Morbidelli, M.; Moscatelli, D. Readily Adsorbable Thermoresponsive Polymers for the Preparation of Smart Cell-Culturing Surfaces on Site. *ACS Biomater. Sci. Eng.* **2020**, *6*, 5337–5345. [CrossRef] [PubMed]
28. Sponchioni, M.; Capasso Palmiero, U.; Manfredini, N.; Moscatelli, D. RAFT Copolymerization of Oppositely Charged Monomers and Its Use to Tailor the Composition of Nonfouling Polyampholytes with an UCST Behaviour. *React. Chem. Eng.* **2019**, *4*, 436–446. [CrossRef]
29. Zhang, Q.; Yu, G.; Wang, W.J.; Yuan, H.; Li, B.G.; Zhu, S. Switchable Block Copolymer Surfactants for Preparation of Reversibly Coagulatable and Redispersible Poly(Methyl Methacrylate) Latexes. *Macromolecules* **2013**, *46*, 1261–1267. [CrossRef]
30. Jessop, P.G.; Mercer, S.M.; Heldebrant, D.J. CO_2-Triggered Switchable Solvents, Surfactants, and Other Materials. *Energy Environ. Sci.* **2012**, *5*, 7240–7253. [CrossRef]
31. Li, Y.; Liu, L.; Liu, X.; Chen, S.; Fang, Y. Reversibly Responsive Microemulsion Triggered by Redox Reactions. *J. Colloid Interface Sci.* **2019**, *540*, 51–58. [CrossRef] [PubMed]
32. Halperin, A.; Kröger, M.; Winnik, F.M. Poly(N-Isopropylacrylamide) Phase Diagrams: Fifty Years of Research. *Angew. Chem. Int. Ed.* **2015**, *54*, 15342–15367. [CrossRef] [PubMed]
33. Zhang, Q.; Weber, C.; Schubert, U.S.; Hoogenboom, R. Thermoresponsive Polymers with Lower Critical Solution Temperature: From Fundamental Aspects and Measuring Techniques to Recommended Turbidimetry Conditions. *Mater. Horiz.* **2017**, *4*, 109–116. [CrossRef]
34. Ward, M.A.; Georgiou, T.K. Thermoresponsive Polymers for Biomedical Applications. *Polymers* **2011**, *3*, 1215–1242. [CrossRef]
35. Sanoj Rejinold, N.; Muthunarayanan, M.; Divyarani, V.V.; Sreerekha, P.R.; Chennazhi, K.P.; Nair, S.V.; Tamura, H.; Jayakumar, R. Curcumin-Loaded Biocompatible Thermoresponsive Polymeric Nanoparticles for Cancer Drug Delivery. *J. Colloid Interface Sci.* **2011**, *360*, 39–51. [CrossRef] [PubMed]
36. Sponchioni, M. Polymeric Nanoparticles for Controlled Drug Delivery. In *Nanomaterials for Theranostics and Tissue Engineering*; Rossi, F., Rainer, A.B.T.-N., Eds.; Elsevier: Amsterdam, The Netherlands, 2020; pp. 1–28. [CrossRef]
37. Cao, P.-F.; Mangadlao, J.D.; Advincula, R.C. Stimuli-Responsive Polymers and Their Potential Applications in Oil-Gas Industry. *Polym. Rev.* **2015**, *55*, 706–733. [CrossRef]
38. Winninger, J.; Iurea, D.M.; Atanase, L.I.; Salhi, S.; Delaite, C.; Riess, G. Micellization of Novel Biocompatible Thermo-Sensitive Graft Copolymers Based on Poly(ε-Caprolactone), Poly(N-Vinylcaprolactam) and Poly(N-Vinylpyrrolidone). *Eur. Polym. J.* **2019**, *119*, 74–82. [CrossRef]
39. Palmiero, U.C.; Agostini, A.; Gatti, S.; Sponchioni, M.; Valenti, V.; Brunel, L.; Moscatelli, D. RAFT Macro-Surfmers and Their Use in the Ab Initio RAFT Emulsion Polymerization To Decouple Nanoparticle Size and Polymer Molecular Weight. *Macromolecules* **2016**, *49*, 8387–8396. [CrossRef]
40. Maddinelli, G.; Bartosek, M.; Carminati, S.; Moghadasi, L.; Mandredini, N.; Moscatelli, D. Design of Thermoresponsive Polymers to Selective Permeability Reduction in Porous Media. In Proceedings of the Abu Dhabi International Petroleum Exhibition and Conference, Abu Dhabi, United Arab Emirates, 9–12 November 2020.
41. Boulif, N.; Sebakhy, K.O.; Joosten, H.; Raffa, P. Design and Synthesis of Novel Di- and Triblock Amphiphilic Polyelectrolytes: Improving Salt-Induced Viscosity Reduction of Water Solutions for Potential Application in Enhanced Oil Recovery. *J. Appl. Polym. Sci.* **2021**, *138*, 50366. [CrossRef]
42. Parker, B.R.; Derry, M.J.; Ning, Y.; Armes, S.P. Exploring the Upper Size Limit for Sterically Stabilized Diblock Copolymer Nanoparticles Prepared by Polymerization-Induced Self- Assembly in Non-Polar Media. *Langmuir* **2020**, *36*, 3730–3736. [CrossRef]
43. Wu, H.; Xie, J.; Morbidelli, M. Soft Matter Kinetics of Colloidal Gelation and Scaling of the Gelation point. *Soft Matter* **2013**, *9*, 4437–4443. [CrossRef]

polymers

MDPI

Article

Inhomogeneities in PNIPAM Aqueous Solutions: The Inside View by Spin Probe EPR Spectroscopy

Ekaterina M. Zubanova [1,*], Sergei V. Kostjuk [2,3,4], Peter S. Timashev [1,3,5], Yury A. Rochev [3,6], Alexander I. Kokorin [5,7], Mikhail Ya. Melnikov [1] and Elena N. Golubeva [1]

1 Faculty of Chemistry, Lomonosov Moscow State University, 1-3, Leninskie gory, 119991 Moscow, Russia; timashev.peter@gmail.com (P.S.T.); melnikov46@mail.ru (M.Y.M.); legol@mail.ru (E.N.G.)
2 Research Institute for Physical Chemical Problems of the Belarusian State University, 14, Leningradskaya str., 220030 Minsk, Belarus; kostjuks@bsu.by
3 Institute for Regenerative Medicine, Sechenov First Moscow State Medical University, 8-2, Trubetskaya str., 119992 Moscow, Russia; yury.rochev@nuigalway.ie
4 Faculty of Chemistry, Belarusian State University, 220006 Minsk, Belarus
5 N.N. Semenov Federal Research Center for Chemical Physics, Russian Academy of Sciences, Kosygin st. 4, 119991 Moscow, Russia; alex-kokorin@yandex.ru
6 National University of Ireland Galway, University Road, H91 TK33 Galway, Ireland
7 Plekhanov Russian University of Economics, Stremyannyi per., 117997 Moscow, Russia
* Correspondence: kate_zub@mail.ru

Abstract: Coil to globule transition in poly(N-isopropylacrylamide) aqueous solutions was studied using spin probe continuous-wave electronic paramagnetic resonance (CW EPR) spectroscopy with an amphiphilic TEMPO radical as a guest molecule. Using Cu(II) ions as the "quencher" for fast-moving radicals in the liquid phase allowed obtaining the individual spectra of TEMPO radicals in polymer globule and observing inhomogeneities in solutions before globule collapsing. EPR spectra simulations confirm the formation of molten globules at the first step with further collapsing and water molecules coming out of the globule, making it denser.

Keywords: thermoresponsive polymers; electronic paramagnetic resonance; spin probe; nitroxides; coil to globule

Citation: Zubanova, E.M.; Kostjuk, S.V.; Timashev, P.S.; Rochev, Y.A.; Kokorin, A.I.; Melnikov, M.Y.; Golubeva, E.N. Inhomogeneities in PNIPAM Aqueous Solutions: The Inside View by Spin Probe EPR Spectroscopy. *Polymers* **2021**, *13*, 3829. https://doi.org/10.3390/polym13213829

Academic Editor: Mattia Sponchioni

Received: 9 October 2021
Accepted: 2 November 2021
Published: 5 November 2021

1. Introduction

Thermoresponsive polymers are of great interest because they undergo coil-to-globule transitions of single-polymer chains in polar solvents near the lower critical solution temperature (LCST) [1–3]. This peculiarity gives them potential for biomedical and pharmaceutical applications such as drug and gene delivery, tissue engineering, cell expansion, sensors, microarrays, and imaging [4–9]. Phase transition in the thermoresponsive polymers solutions passes due to molecular interaction and the cohesion of solvent molecules with hydrophilic fragments in the polymer chains [10,11]. Hydrophilic groups form hydrogen bonds with water molecules at low temperatures, resulting in good solubility in aqueous solutions. Increasing the temperature leads to the degradation of the hydrogen bonds system and the formation of an intramolecular interaction between polymers chains, further collapsing the polymer globule. Poly(N-isopropylacrylamide) (PNIPAM) is one of the most studied thermoresponsive polymers, which has LCST in a physiologically relevant temperature range of ≈32 °C in aqueous solutions [1]. Macroscopic methods (turbidimetry, DSC, etc.) usually fix sharp and reversible changes of PNIPAM properties in the vicinity of LCST [11,12]. In addition, the formation of small, even nanoscopic inhomogeneities of polymer gels or films in different solvents before LCST is proved by continuous-wave electron paramagnetic resonance spectroscopy (CW EPR) [13–15]. The EPR spectra of paramagnetic molecules (spin probes) are sensitive to the microenvironment and can give valuable information concerning collapse processes at the molecular level [14,16,17]. Stable nitroxide radicals containing a >N–O•·group have been the most popular spin probes during the last 55 years [17,18]. Hyperfine interaction of the unpaired electron spin with the magnetic moment of 14N (S = 1) nucleus leads to splitting of the EPR signal of nitroxides to three lines. Local polarity, viscosity, and the ability of a media to form hydrogen bonds influence the electron density distribution in the spin probe, affecting the shape of the EPR signal and spin-Hamiltonian parameters (g-tensor and hyperfine splitting (hfs) tensor), which can be estimated by modeling the EPR spectra. The small amphiphilic radical 2,2,6,6-tetramethylpiperidin-1-yl)oxyl (TEMPO) (ca. 6.7 Å in diameter) is a powerful tool to detect and control the formation of polymer globules [14,19–21]. It is known that the micropolarity in polymeric globules of thermoresponsive polymers is significantly lower than the polarity of water and is close to that of chloroform [18]. A lower polarity and higher viscosity of the globules [15] result in changes of spin Hamiltonian parameters and the line widths of TEMPO spectra compared with those in aqueous solution, reflecting a decrease of the amplitudes of the spectral lines belonging to the radicals in the solution and the appearance of broader components of the TEMPO spectrum corresponding to probes localized in the globules. This effect was used by Junk et al. [19] to reveal the formation of heterogeneities in thin photocrosslinked films of PNIPAM notably earlier and later the LCST detected by macroscopic methods. However, the spin probe technique was not applied to study coil-to-globule transitions and the nature of inhomogeneities in PNIPAM aqueous solutions up to now. In the present paper, we applied CW EPR to study the structure and features of formation of nano- and/or micro heterogeneities of PNIPAM with two different polydispersities and the dynamics of spin probes inside the globules and in aqueous solutions upon heating from room temperature.

2. Materials and Methods

2.1. Substances

The stable radical (2,2,6,6-tetramethylpiperidin-1-yl)oxyl (TEMPO) and copper(II) chloride dihydrate $CuCl_2 \cdot 2H_2O$ purchased from Sigma-Aldrich were used without further purification. N-Isopropylacrylamide (NIPAM) (99%, Acros, Geel, Belgium) was recrystallized from solution in *n*-hexane, dried in vacuum, and then stored under argon atmosphere. 2,2′-Azobis(2-methylpropionitrile) (AIBN) (98%, Sigma-Aldrich, Burlington, MA, USA) was recrystallized from ethanol and dried in vacuum at 20 °C. Benzene (anhydrous, 99.8%, Sigma-Aldrich, USA) and *n*-hexane (reagent grade, Ekos-1) were used as received.

2.2. Polymer Synthesis

Two poly(N-isopropylacrylamide) (PNIPAM) samples I and II were synthesized via conventional radical and RAFT polymerization, respectively. The sample I (number-average molar mass M_n = 175.5 kDa, polydispersity index Đ = 4.3; yield: 97%) was prepared by free radical polymerization in benzene at 60 °C for 24 h using azobisisobutyronitrile (AIBN) as an initiator according to the procedure reported in [20]. Then, the reaction mixture was precipitated in n-hexane. Then, the obtained polymer was purified by dissolving in acetone followed by precipitation in n-hexane at least three times, and the product was dried at 45 °C in a vacuum oven. Sample II (M_n = 107.6 kDa, Đ = 2.05; yield: 95%) was prepared by RAFT-mediated radical polymerization. The polymerization was carried out in benzene at 60 °C for 24 h using 2-(dodecylthiocarbonothioylthio)-2-methylpropionic acid as the RAFT agent. The polymerization was conducted in an argon atmosphere in a Schlenk reactor equipped with a magnetic stir bar. The reactor was charged by N-isopropylacrylamide (0.502 g, 4.44 mmol), vacuumed, and filled with argon. Then, 0.22 mL of 0.02 M benzene solution of RAF-agent, 1.11 mM of 0.02 M benzene solution of AIBN, and 2.87 mL of benzene were added into the reactor. The mixture was bubbled with argon for 30 min, and then, the reactor was placed into an oil bath heated to 60 °C. After 24 h, the reactor was opened and frozen with liquid nitrogen. The polymer purification was performed similarly to the purification of sample I (see above).

2.3. *Polymers Characterization*

2.3.1. H NMR Spectroscopy

^{1}H NMR (500 MHz) spectra were recorded in $CDCl_3$ at 25 °C on a Bruker AC-500 (Bruker, Karlsruhe, Germany) spectrometer calibrated relative to the residual solvent resonance. The ^{1}H NMR spectra of sample I and sample II are presented in Figure A1 (see Appendix A).

2.3.2. Spectrophotometry

The temperature transition was monitored by UV-Vis spectrophotometry using a Victor Nivo instrument (Perkin Elmer, Waltham, MA, USA) at a wavelength of 405 nm. The studied polymer solution was poured into a 96 (48)-well plate, and the absorbance at each point was registered. The range of measurement was 4–40 °C. The cloud points were determined at 10% of the transmission reduction.

2.3.3. Differential Scanning Calorimetry (DSC)

The DSC studies were carried out using a NETZSCH STA 449 F3 synchronous thermal analyzer (Selb, Germany) in a helium atmosphere at gas flow rates of 70 mL/min (main) and 50 mL/min (protective). An aluminum crucible was used with a solution weight of 30–45 mg. The calibration of the temperature and sensitivity of the device was carried out using standard samples of adamantane, indium, and distilled water. The aqueous solution of the polymer sample (5 wt %) was heated and cooled at 2 K/min. TGA measurements were performed with the same device using solid PNIPAM samples at heating rate of 20 °C min^{-1} under nitrogen flow.

2.3.4. Size Exclusion Chromatography

Size exclusion chromatography was performed using an Ultimate 3000 Thermo Scientific chromatographic complex (Thermo Fisher Scientific, Waltham, MA, USA) equipped with PLgel precolumn guard (Agilent, Santa Clara, CA, USA, size 7.5 × 50 mm, particle size 5 μm) and PLgel MIXED-C column (Agilent, size 7.5 × 300 mm, particle size 5 μm) thermostated at 50 °C. The elution was performed in the isocratic mode with DMF containing 0.10 M LiBr at a flow rate of 1.0 mL min^{-1}. SEC traces were recorded as mentioned above, and polymethylmethacrylate standards (ReadyCal Kit, PSS GmbH) with Mw/Mn $\leq$ 1.05 were used to calculate Mw/Mn.

2.4. *EPR Samples Preparation*

In all cases, ≈0.5 mM TEMPO and 10 wt % PNIPAM freshwater solutions were prepared, dissolving the required predefined amounts of TEMPO radical and PNIPAM polymer. In the first step, the dissolution of TEMPO was performed at room temperature using ultrasonication. Then, PNIPAM was added, and the mixture was aged at 4 °C for 24 h until the complete dissolution of the polymer. The solutions were put into glass tubes with 2 mm inner diameter; then, the tubes were sealed to prevent water evaporation.

2.5. *EPR Measurements*

EPR spectra were recorded using X-band spectrometer Bruker EMX-500 (Bruker, Karlsruhe, Germany). The temperature of the samples was varied in the range of 300–353 K using the flow of nitrogen gas. The thermostatic device from Bruker was used; the accuracy of the temperature setting was about ±0.5 K. Each sample was left at the particular temperature for precisely 5 min for equilibration before recording. At 305 K, the waiting time was 60 min to equilibrate the samples. Typical parameters of the spectra recording were a microwave power of 0.8 mW, modulation amplitude of 0.04 mT, and a sweep width of 8 mT. «Quenching» of fast spin probes in water was performed by adding 10 mg of $CuCl_2 \cdot 2H_2O$ to 0.5 mL PNIPAM solution, as recommended in [21].

2.6. Data Analysis

The integration of EPR spectra and the amplitude measuring were performed by the EsrD program developed at the Chemistry Department of Lomonosov Moscow State University [16]. For LCST measurements, the amplitude value of TEMPO EPR spectra was normalized to the Q-factor [22] at each temperature and to the amplitude of the left component of the spectra at 300 K. All spectra simulations were performed using homemade scripts for the MATLAB program employing an Easyspin (v. 5.2.28) toolbox [23]. Spectra of TEMPO radical in solutions (type *A*) were simulated using a model based on fast motion implemented as a 'garlic' function in Easyspin. Simulations of the spectra of radicals in polymer globule (type ***B***) were made in a slow-motion regime using a 'chili' function from the Easyspin program. The slow-motion model 'chili' is based on the Schneider–Freed theory [24], solving equations for slow tumbling nitroxides. Anisotropic values of Spin-Hamiltonian parameters (*g*-tensor and hyperfine splitting (hfs) tensor, usually denoted as *a*-tensor) were averaged to obtain g_{iso} and a_{iso} values, the rotational correlation time tensor (τ_{corr}), denoted as the average time required for the rotation of a molecule at one radian, was calculated from averaged rotational diffusion constant. Fitting errors for g_{iso} and a_{siso} were ± 0.00003 and ± 0.01 mT, respectively. More details of the spectra simulation are presented in Appendix A.

3. Results

According to DSC (Figure A2 in Appendix A) data from the heating step, the measured LCST of the polymers I and II in aqueous solutions was $\approx$305 K and did not depend on the synthesis method and molar mass distribution of polymers, which is typical for PNIPAM solutions [25]. TGA measurements showed that both PNIPAM samples are stable at the temperatures used in this work ($\leq$353 K), and the thermal decomposition (5 wt % weight loss) began at $\approx$625 K (Figure A3a). Note that the weight loss between 320 and 400 K (Figure A3a) is consistent with the presence of ca. 10–12 wt % of water in poly(N-isopropylacrylamide) samples. This is confirmed by the disappearance of this weight loss after annealing the samples at 470 K for 15 min (Figure A3b).

At temperatures below LSCT (305 K), the EPR spectrum of TEMPO in PNIPAM solution (sample I, 10 wt%) and TEMPO spectrum in water have a shape close to the isotropic fast-motion limit (Figure 1). The EPR spectrum of TEMPO in the polymer II solution and its changes during heating were very similar to those of polymer I. However, all these spectra (see Figure A4 in Appendix A) are slightly asymmetric; i.e., the signal amplitude above the baseline is bigger than that below it. This asymmetry may appear due to the following reasons. Firstly, the existence of the solvent shells with different polarities around TEMPO molecules in aqueous media revealed by the analysis of Q-band EPR spectra was postulated by Hunold et al. [26]. This fact leads to a bimodal distribution of magnetic and dynamic parameters of TEMPO probes, which manifest itself in asymmetric signals due to the superposition of the spectra. Secondly, the asymmetry of the nitroxide signal can be caused either by the intermolecular spin-exchange interaction between spin probes in solutions [27,28] or by mixing with the dispersion signal of the resonance effect [29]. Fortunately, in our case, the observed effects are too small and can be neglected to simplify the spectra modeling in aqueous solutions. According to our simulation, the TEMPO spectrum in water at room temperature corresponds to probes with a g_{iso} value equal to 2.00579 and the hfs constant a_{iso} equal to 1.73 mT. The isotropic rotational correlation time (τ_{corr}) is about 11 ps. TEMPO in 10 wt % PNIPAM solution has parameters typical for less polar media: g_{iso} equal to 2.00585 and a_{iso} is equal to 1.72 mT, wherein $\underline{\tau_{corr}}$ = 16 ps, manifesting the higher viscosity of polymer solution. A simple estimation of viscosity using the Stocks–Einstein equation from rotational correlation time shows increasing viscosity of 10 wt% PNIPAM solution below LCST in $\approx$1.5 times compared to pure water.

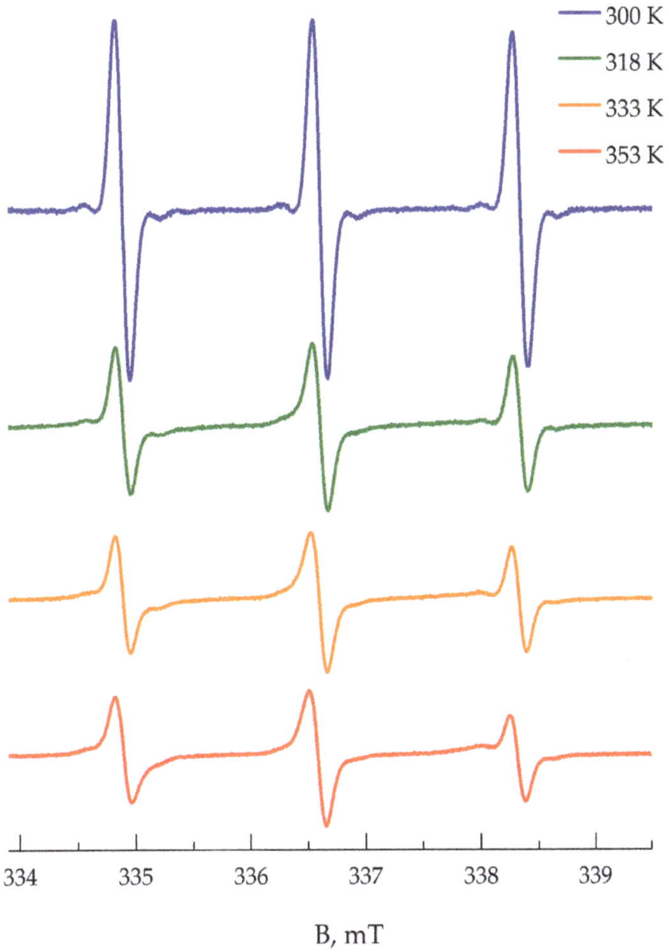

Figure 1. Normalized EPR spectra of TEMPO radical in 10 wt% PNIPAM aqueous solutions at 300–353 K registered upon heating.

At temperatures higher than 305 K, the intensity of the TEMPO spectra decreases rapidly, and the lines shape changes: the spectra become broader, and the central line turns to more asymmetric. At 320 K and higher, the components in the high-field region corresponding to the probe in the hydrophobic globule clearly manifest themselves in the spectra. Changes in the lines' shape can be empirically presented as the amplitude A of the high field component vs. temperature T (Figure 2). As seen from Figure 2, this dependence allows measuring the LCST of polymer solutions. A fast drop of the amplitude occurs due to the appearing of the more broadened and hence less intensive signal of TEMPO radicals located in globules.

At temperatures above the LCST, the TEMPO spectrum in the PNIPAM solution appears to be the sum of at least two signals. One of them belongs to the radicals in the solution (type ***A***). At the same time, the position and shape of the second signal (denoted as the spectrum of ***B***-type probes) manifest a more hydrophobic local environment and a hindered rotation of TEMPO radicals comparing to ***A*** probes. The simulation of experimental spectra as the sum of the spectra of two probes applies a considerable number of parameters, the variation of which leads to similar changes in EPR spectra. For example, a shift of the components can occur both due to changes in the hfs constant or rotational correlation time. The use of the spin-Hamiltonian and dynamic parameters of radicals *A* at different temperatures determined from the temperature dependence of TEMPO spectra in an aqueous solution makes it possible to slightly reduce the number of parameters to be varied in the modeling. Nevertheless, the parameters of probes ***B*** cannot be quantitatively determined from the simulation of experimental spectra due to their simultaneous influence on the shape of the spectral line and overlapping signals of probes ***B*** with the component corresponding to the hyperfine splitting on ^{13}C nuclei of TEMPO, the so-called satellite in the probes *A* spectrum. The spectrum of ***B*** probes can be elucidated by subtracting the spectrum of centers *A* from the total spectrum or as the spectra of polymer films swollen in the probe solution [19]. Nevertheless, the subtracting does not allow obtaining a fully "pure" single-particle spectrum and may lead to artifacts in the line shape.

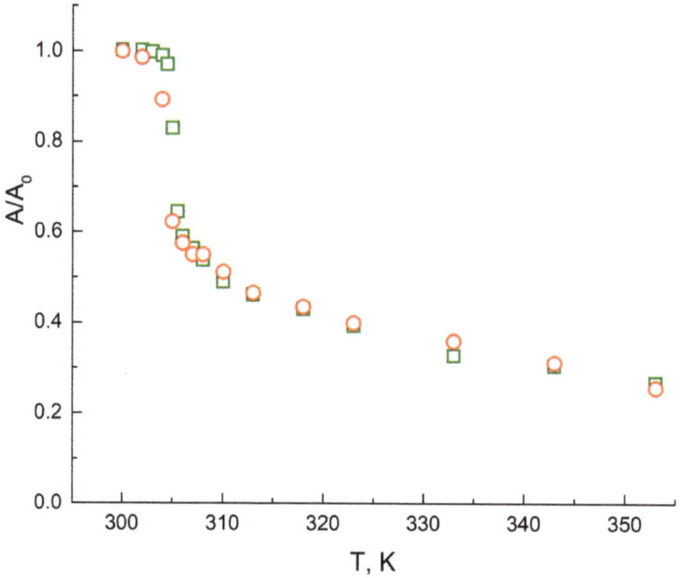

Figure 2. The normalized amplitude of the high-field component of TEMPO EPR signal in PNIPAM solution (green squares correspond to polymer I, red circles correspond to polymer II) vs. temperature.

The perspective method to get the individual spectrum of radicals inside a polymer globule is "quenching" the signal of rapidly rotating radicals *A* due to the spin-exchange broadening of EPR lines [21]. A Heisenberg exchange interaction occurs between unpaired electrons as a result of the collision of two paramagnetic particles, leading to an exchange of spin states between partners. In turn, the exchange of spin states between partners leads to a change in the width of the EPR line of the paramagnetic particle up to its confluence with the baseline. Such broadening may appear in the presence of some inert paramagnetic particles, e.g., Cu(II) ions, not chemically interacting with components of solutions. Figure A5 presents the change of the EPR spectra of TEMPO radicals in aqueous solutions in the presence of Cu(II) ions. Adding small amounts of the paramagnetic ions firstly leads to minor broadening of the signal of motile radicals in the solution. Further increasing of Cu(II) concentration continues to broaden the TEMPO signal up to the baseline. The optimal concentration of Cu(II) ions is about 0.2 M, and this amount was used for "quenching" the signal of radicals *A* in PNIPAM (polymer I) solutions.

Due to the spin exchange between Cu(II) ions and TEMPO, the experimental spectrum transfers practically to a base line at the temperature range of 295–301 K. At 302 K, a new slight signal of slow-moving radicals appears, and its intensity and double integral increase with temperature rising (see Figure A6, Appendix A). We suppose that the new signal corresponds to TEMPO radicals located in inhomogeneities of the polymer solution and not contacted with the quencher. Over 305 K, polymer globules start collapsing, and the spectra become more intensive due to an increase in the fraction of the radicals captured by polymer globules. Heating up to 323 K leads to a further increase in the TEMPO signal, but the lines' shape does not change noticeably. At 333–353 K, the components of the spectra narrow, apparently, because of the increasing mobility of TEMPO radicals. The number of spin probes that depends linearly on the double integral of EPR spectra stays constant at 333–353 K. We assume that the spectrum at 353 K shown in Figure 3 matches the TEMPO spectra of the radicals in dense PNIPAM globules (probes ***B***). According to our simulation results, these spectra correspond to the anisotropic rotation of TEMPO radicals with g_{iso} and a_{iso} equal to 2.00615 and 1.60 mT, respectively. Such value of the isotropic hfs constant matches to the local polarity close to that of the chloroform or low-molecular alcohols [30]. Rotational movements of radicals are anisotropic and more hindered along the x-axis: $D_x = 7.6 \times 10^6\ s^{-1} < D_z = 8.7 \times 10^7\ s^{-1} < D_y = 5.9 \times 10^8\ s^{-1}$, where $D_{x,y,z}$ are the components of the rotational diffusion tensor. The obtained magnetic and dynamic parameters of TEMPO radicals in PNIPAM globules (probes B) were used in our simulations of the total spectra series (Figure 1).

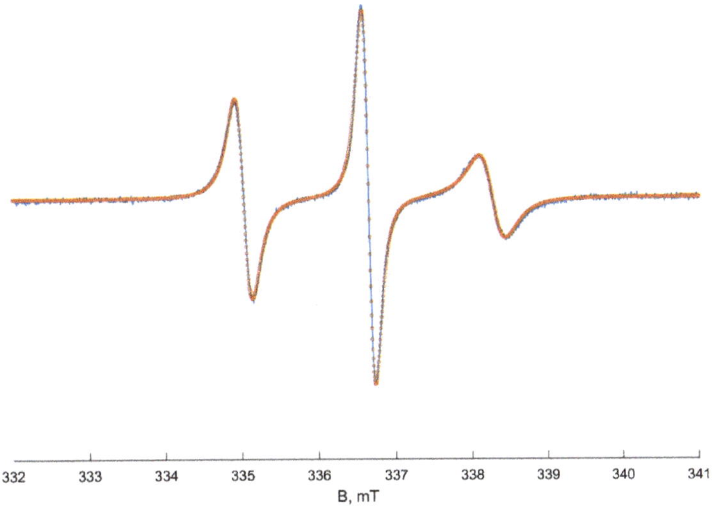

Figure 3. EPR spectrum of TEMPO radical in PNIPAM solution at 353 K in the presence of Cu(II) ions. Blue line—experimental spectrum, red dots—simulated spectrum.

All spectra of the samples I and II at temperatures above 305 K were excellently fitted as a sum of the spectra of two types: "hydrophilic, fast" ***A*** probes and "hydrophobic, slow" ***B*** probes (Appendix A, Figure A7). The root-mean-square deviation obtained from simulations of all fitted spectra was less than 1%. The complete simulation results obtained from EPR spectra fitting at several temperatures are collected in Appendix A in Table A1. The difference between the two samples comes out only as the content of the ***B*** probes at 305 K: 26% in sample I and 44% in sample II. This difference may occur due to various polydispersities of polymers I (Đ = 4.3) and II (Đ = 2.05). Actually, the content of macromolecules with similar molecular mass is bigger in polymer II, leading to a more rapid globule collapse and the capturing of bigger content of the TEMPO radical. The simulation results of the EPR spectra of samples I and II registered higher than 305 K are similar, so only the simulation results obtained for sample II will be discussed further.

Figure 4 illustrates the fitted EPR spectrum recorded at 333 K as a superposition of two individual spectra. The spectrum of probes A looks like a "fast limit" spectrum of TEMPO radicals dissolved in water. Its EPR parameters become less polar with heating due to decreasing the polarity of liquids (for example, the dielectric constant ε of water changes from 78 to 60 at temperatures 298 and 350 K, respectively) [31]. A hyperfine coupling constant also decreases from 1.73 to 1.71 mT, and giso increases from 2.00588 to 2.00598. The rotation correlation time of A probes decreases with heating from 10 to 1 ps due to the diminishing water viscosity [32]. The magnetic and dynamic parameters of type A probes after LCST do not depend on PNIPAM concentration and are similar to those for the pure aqueous solution without PNIPAM.

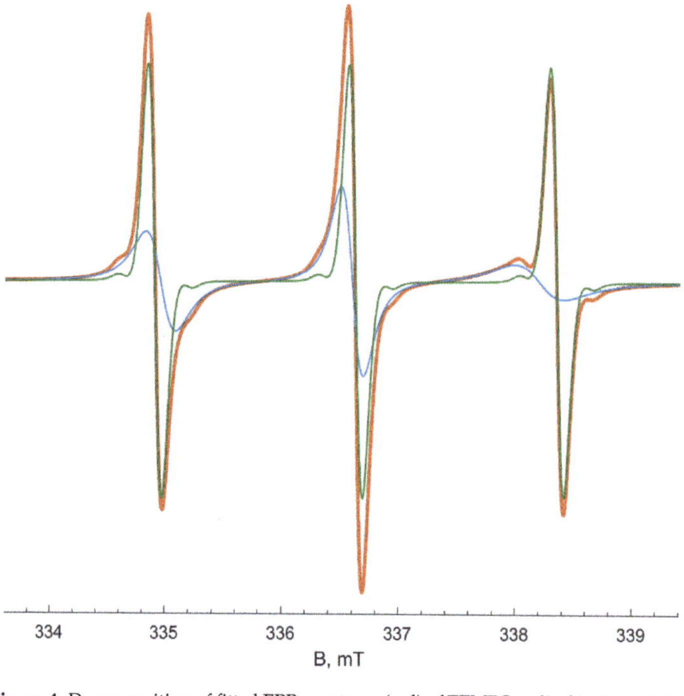

Figure 4. Decomposition of fitted EPR spectrum (red) of TEMPO radical in 10 wt % PNIPAM solution at 333 K into two signals: type ***A*** radicals (green) and type ***B*** radicals (blue).

The spectrum of ***B*** probes (radicals located in hydrophobic globules) is more broadened and looks like a «slow-motion» signal. The EPR spectrum of type ***B*** at 353 K coincides with the experimental spectrum of the TEMPO radical in the polymer globule obtained by the "quenching" experiment (Figure 3).

Figure 5 shows changes of a_{iso} and g_{iso} of radicals ***A*** and ***B*** as a function of temperature. The complete simulation results obtained from EPR spectra fitting at several temperatures are collected in Table A1 of Appendix A. At 305–313K, the g_{iso} values corresponding to ***B*** particles are similar to those of probe ***A*** in aqueous solution, but after 313 K, they rise to 2.00615 at 353 K. In contrast, a_{iso} values corresponding to ***B*** probes do not change at 305–353 K being equal to 1.60 mT. The average rotation correlation time of particles ***B*** is 1300 ps at 305–318 K. After 318 K, a two-fold decrease of τ_{corr} to 620 ps resulting in narrowing of the line width of the EPR signal of the ***B*** probes is observed, signaling that TEMPO molecules in the globule become more mobile. The mole fraction of type ***B*** probes rises upon heating to 353 K to 70%.

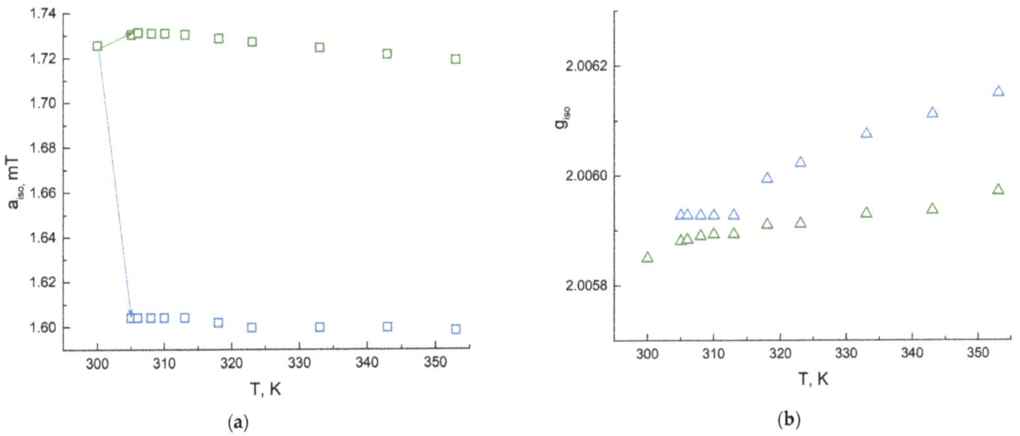

Figure 5. Change of a_{iso} (**a**) and g_{iso} (**b**) of TEMPO radical in PNIPAM solution with heating according to EPR spectra simulation. Green squares and triangles correspond to ***A*** particles, blue ones correspond to ***B*** particles.

4. Discussion

All the applied methods: DSC, spectrophotometry, and the EPR spin probe technique determine the same value of LCST (305 K) in aqueous PNIPAM solutions of polymers I and II, which is consistent with the literature data [33]. Coil-to-globule transition on PNIPAM aqueous solutions studied by the EPR spin probe method occurs in a narrow temperature range of 8–10 K. Thus, in solutions, hydrophilic bonds between water molecules and PNIPAM individual chains quickly degrade, and the rapid formation of polymer globules takes place. To the contrary, Junk et al. also fixed LCST at 305 K in PNIPAM-based photocrosslinked films swollen in water, but the temperature range of the collapse as measured by spin-probe CW EPR spectroscopy was substantially broadened to 40–50 K [19]. Therefore, only one type of radical (***A*** probes) prevails in the aqueous solution of PNIPAM below LSCT (305 K). Nevertheless, at 302–304 K, some part of TEMPO molecules is captured into polymer globules, transferring to ***B*** probes and testifying to the generation of heterogeneities that begin to collapse at 305 K, when the coil-to-globule transition takes place.

We suppose that the observed increase in g-factor values of ***B*** particles in the globules with increasing temperature occurs due to the breakdown of hydrogen bonds between water molecules and the N–O$^{\bullet}$ group of TEMPO [34,35]. At higher temperatures, these complexes decompose, and g_{iso} rises to lower polar values corresponding to individual TEMPO radicals. In addition, water molecules may come out of the globules while heating, making the media surrounding TEMPO radicals more hydrophobic. Such evolution from molten globule to tight globule was previously observed by Wang et al. [36] by laser scattering for PNIPAM with narrow mass distribution, and it may also lead to the change of g_{iso} value to be less polar.

The changes of content in *A* probes obtained from spectra simulation are similar to the normalized amplitude changes of the high-field component of the full TEMPO spectrum in PNIPAM solutions (see Figure 2 and Table A1). Therefore, the amplitude changes may be used not only for LCST determination but for estimation of the content of probe molecules $A(\chi_A)$ in the aqueous solution after the globule collapse. Since $\chi_A + \chi_B = 1$, the molar fraction of ***B*** probes could be valued, too. The content of probes in the collapsed globules relates to an affinity of the probe to the polymer and amphiphilicity of the collapsed globules. Thus, the measuring of amplitudes of EPR spectra of probe molecules in thermoresponsive polymers aqueous solutions gives the opportunity to determine and predict its properties using the spin-probe technique without making challenging spectra simulations.

5. Conclusions

The CW EPR method with the spin-probe technique and "quenching" approach was applied for studying coil-to-globule phase transition in PNIPAM aqueous solution upon heating at 298–353 K. The broadening of fast-moving radicals in water by spin exchange with Cu(II) ions allows obtaining individual spectra of the spin probe in polymer globule after collapsing and reliable spin-Hamiltonian parameters of both types of TEMPO probes in solutions. In addition, the "quenching" approach has shown the formation of inhomogeneities in PNIPAM aqueous solution at 2–3 degrees below the critical temperature, whereas the temperature range of the collapse in PNIPAM-based photocrosslinked films swollen in water is substantially broadened to 40–50 K. A simple analysis based on amplitude measurement of TEMPO EPR spectra registered at different temperatures gives a cloud point and allows estimating the content of probe in collapsed globule. According to EPR spectra simulations, the formation of dense globules at high temperatures through molten ones at the first step takes place in PNIPAM solutions.

Author Contributions: Conceptualization, E.N.G. and Y.A.R., investigation and formal analysis, E.M.Z. and S.V.K., writing—original draft preparation, E.M.Z., E.N.G. and S.V.K., writing—review and editing, A.I.K., Y.A.R., P.S.T., project administration, P.S.T. and M.Y.M. All authors have read and agreed to the published version of the manuscript.

Funding: This research was supported by the Russian Foundation for Basic Research (Grant 20-02-00712).

Institutional Review Board Statement: Not applicable.

Informed Consent Statement: Not applicable.

Data Availability Statement: The data presented in this study are available on request from the corresponding author.

Acknowledgments: This work was partially performed using MSU equipment provided by the M.V. Lomonosov Moscow State University Program of Development. The authors are grateful to A.V. Bogdanov for useful discussions and Evgenii Ksendzov for his help in the synthesis and characterization of poly(N-isopropylacrylamide) samples.

Conflicts of Interest: The authors declare no conflict of interest.

Appendix A

EPR Spectra Simulation Details

All CW EPR spectra simulations were performed using homemade scripts for MATLAB program employing the Easyspin (v. 5.2.28) toolbox [23]. Quality of fitting was controlled by the calculation of root-mean-square deviation for the difference spectra. Spectral and dynamic parameters were obtained by fitting simulated EPR spectra to experimental data using least-squares fitting algorithms.

The spectra of TEMPO radical (type *A*) in solutions were simulated using a model based on fast-motion implemented as a 'garlic' function in Easyspin. Line broadening was described by isotropic rotation correlation time, and by additional convolutional line broadening caused mostly by unresolved hyperfine splitting between unpaired electron and nuclear spin (usually ^{1}H). As initial values for the diagonal elements of the g-tensor and hyperfine splitting tensor on ^{14}N nuclei for type ***A*** probes, the following values were used:

$$g_x = 2.0092,\ g_y = 2.0062,\ g_z = 2.0022;$$

$$A_{xx} = 0.71\ \text{mT},\ A_{yy} = 0.70\ \text{mT},\ A_{zz} = 3.71\ \text{mT}.$$

In addition to splitting on ^{14}N nuclei in the case of type ***A*** probes, isotropic hyperfine splitting on ^{13}C isotopes of six CH_3-groups with a_{iso} = 0.54 mT was considered. To simplify the modeling procedure, we change abundances for carbon isotopes instead of adding six paramagnetic nuclei.

Simulations of the spectra of radicals in polymer globule (type ***B***) were made in a slow-motion regime using the 'chili' function from the Easyspin program. The slow-motion model 'chili' is based on the Schneider–Freed theory [24], solving equations for slow-tumbling nitroxides. Line broadening was described by the anisotropic rotation correlation time and additional Gaussian line broadening. To start fitting, the following diagonal values of g-tensor and hyperfine interaction a-tensor were used:

$$g_x = 2.0098,\ g_y = 2.0062,\ g_z = 2.0022;$$

$$A_{xx} = 0.71\ \text{mT},\ A_{yy} = 0.70\ \text{mT},\ A_{zz} = 3.35\ \text{mT}.$$

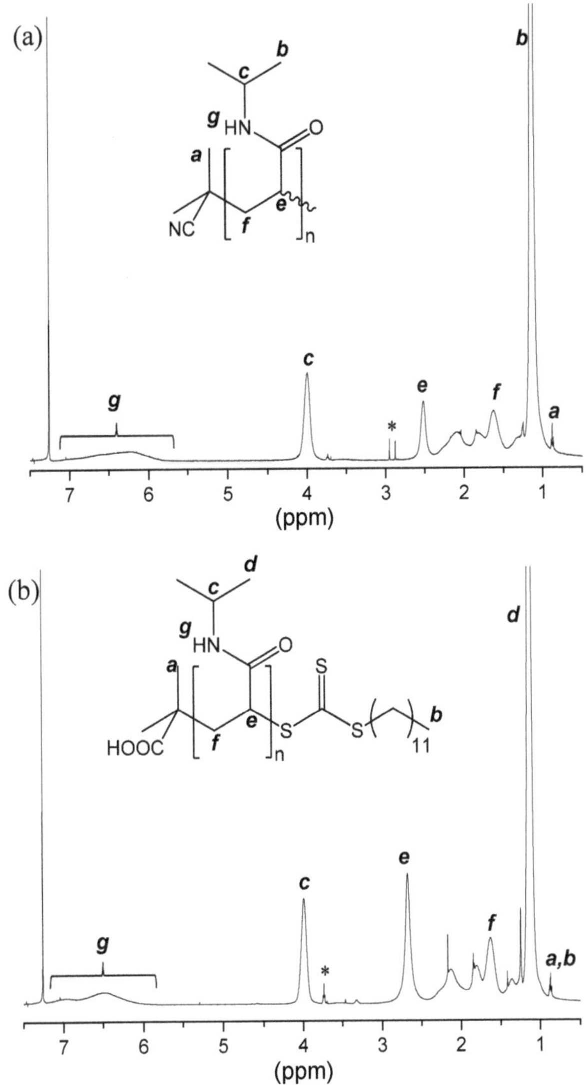

Figure A1. ^{1}H NMR spectra of (**a**) sample I and (**b**) sample II. The traces of solvents are marked by asterisks.

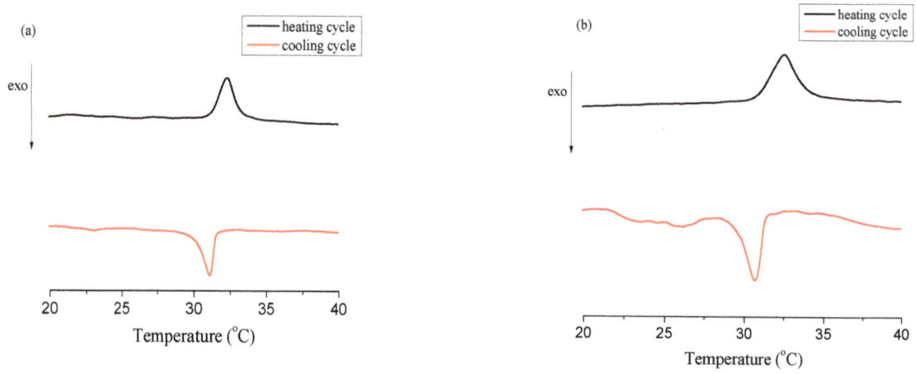

Figure A2. DSC thermograms of 5 wt % aqueous solutions of poly(N-isopropylacrylamide): (**a**) sample I and (**b**) sample II.

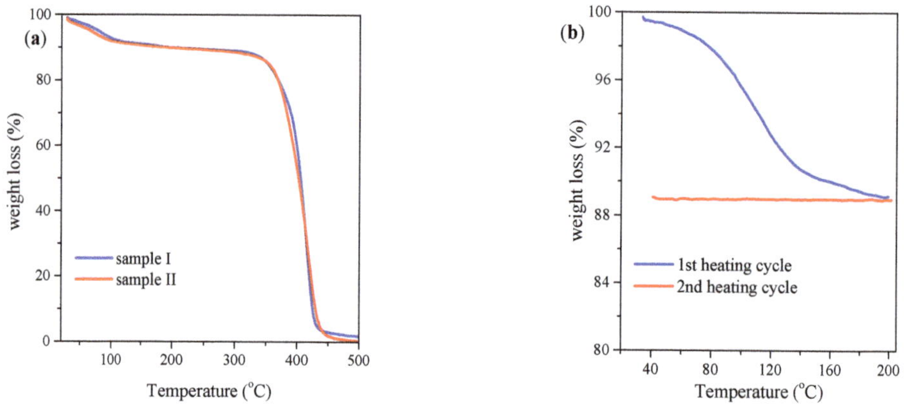

Figure A3. TGA curves of (**a**) sample I and sample II studies in this work and (**b**) sample II before and after annealing at 470 R for 15 min.

The initial values of magnetic parameters were taken from [37] and corrected using experimental isotropic g_{iso} and a_{iso} of TEMPO radical in different solvents [30]. During the simulation, only g_x and A_{zz} components were varied, being more sensitive to the environment. Isotropic g_{iso} and a_{iso} were calculated as an average value of diagonal elements:

$$g_{iso} = \frac{1}{3}\left(g_x + g_y + g_z\right)$$

$$a_{iso} = \frac{1}{3}\left(A_{xx} + A_{yy} + A_{zz}\right)$$

The average rotational correlation time τ_{corr} was calculated using the following equations from diagonal elements of diffusion tensor $D_r = [D_{xx}\ D_{yy}\ D_{zz}]$:

$$D_r = \frac{1}{6\tau_{corr}}$$

$$\tau_{corr} = \frac{1}{6\sqrt[3]{D_{xx}D_{yy}D_{zz}}}$$

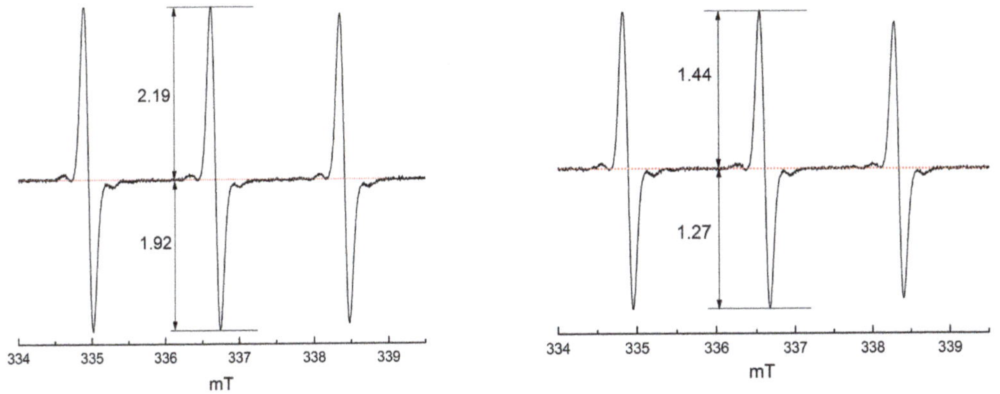

(a) (b)

Figure A4. EPR spectra of TEMPO radical in water (0.3 mM) (**a**) and 10% PNIPAM solution (0.5 mM) (**b**) at 298 K. Arrows show amplitude value of spectra above and below the baseline. Red dot line shows baseline.

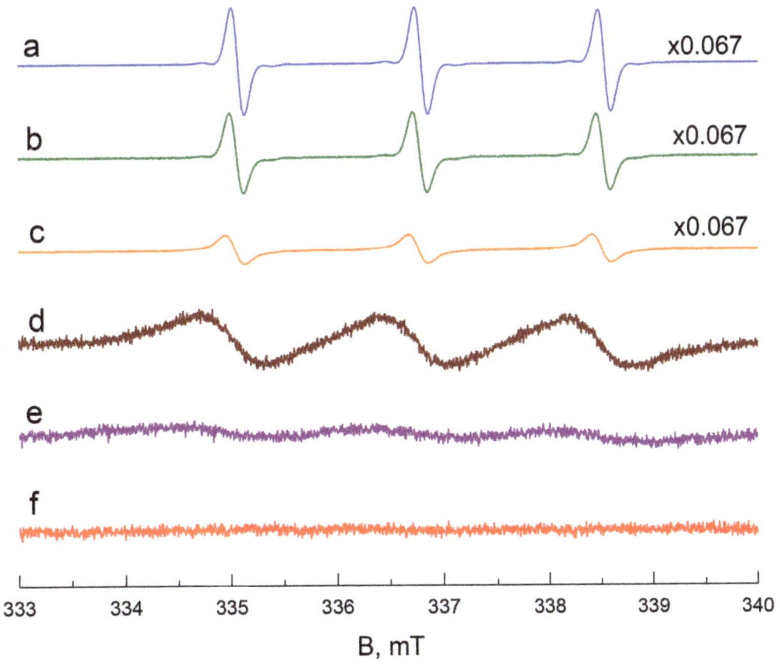

Figure A5. EPR spectra of TEMPO radical in water in the presence of Cu(II) ions: (**a**)—[Cu(II)] = 0, [TEMPO] = 0.5 mM; (**b**)—[Cu(II) = 0.003 M, [TEMPO] = 0.33 mM; (**c**)—[Cu(II)] = 0.01 M, [TEMPO] = 0.25 mM; (**d**)—[Cu(II)] = 0.1 M, [TEMPO] = 0.5 mM; (**e**)—[Cu(II)] = 0.2 M, [TEMPO] = 0.5 mM; (**f**)—[Cu(II)] = 0.3 M, [TEMPO] = 0.5 mM. Spectra (**a**–**c**) were divided by 15 for better representation and comparison with (**d**–**f**).

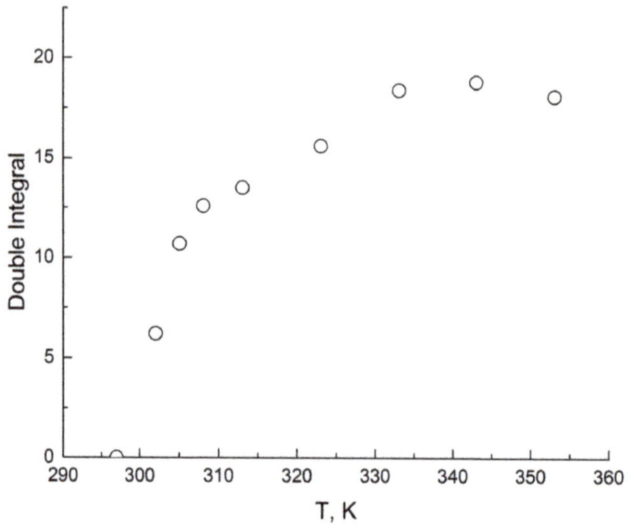

Figure A6. Change of double integral of EPR spectra of TEMPO in PNIPAM aqueous solution in the presence of Cu(II) vs. temperature.

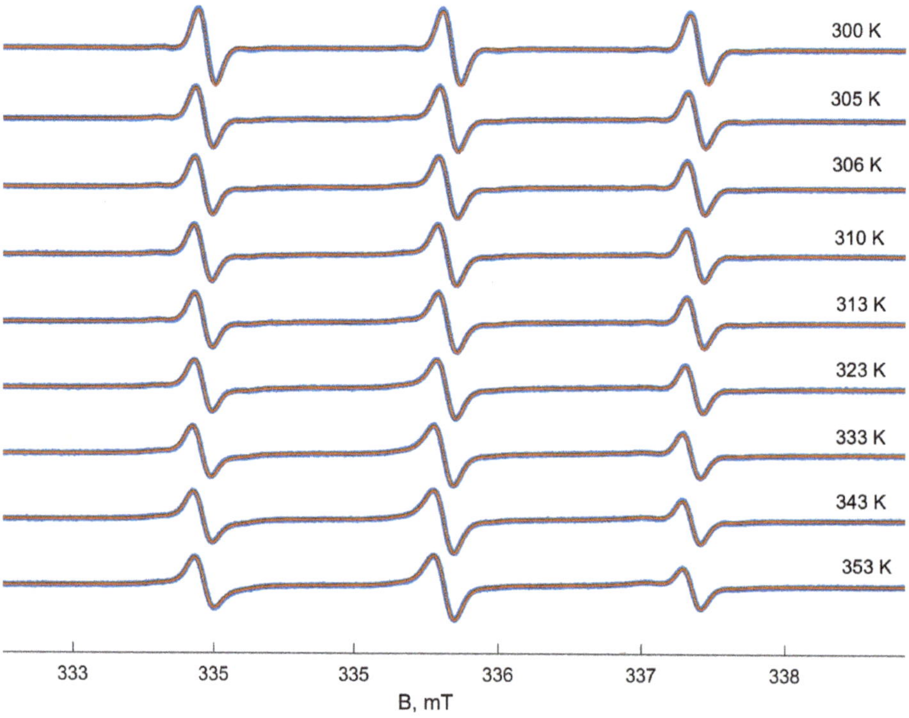

Figure A7. Simulation of spectra of system TEMPO/PNIPAM/H_2O at different temperatures. Blue lines—experimental spectra, red circles—simulation spectra.

Table A1. Selected spectral parameters from simulation TEMPO EPR spectra in 10 wt% aqueous solutions at 300–353 K.

T, K	Type A			Type B				
	g_{iso}	a_{iso}, mT	τ_{corr}, ps	g_{iso}	a_{iso}, mT	τ_{corr}, ns	χ_B, % (Polymer I)	χ_B, % (Polymer II)
300	2.00585	1.72	16	-	-	-	0	0
305	2.00588	1.73	10	2.00593	1.60	1.26	26.5	44.6
306	2.00588	1.73	8	2.00593	1.60	1.31	50.0	52.6
308	2.00589	1.73	8	2.00593	1.60	1.31	56.1	54.5
310	2.00589	1.73	5	2.00593	1.60	1.31	58.3	55.9
313	2.00589	1.73	5	2.00593	1.60	1.31	61.7	58.8
318	2.00591	1.73	2	2.00599	1.60	1.31	63.1	62.0
323	2.00591	1.73	2	2.00602	1.60	1.25	64.5	63.2
333	2.00593	1.72	2	2.00608	1.60	1.01	65.9	64.3
343	2.00594	1.72	2	2.00611	1.60	0.86	68.8	67.2
353	2.00597	1.72	2	2.00615	1.60	0.74	70.1	69.9

References

1. Heskins, M.; Guillet, J.E. Solution Properties of Poly(N-isopropylacrylamide). *J. Macromol. Sci. Part A-Chem.* **1968**, *2*, 1441–1455. [CrossRef]
2. Aseyev, V.; Tenhu, H.; Winnik, F.M. Non-Ionic Thermoresponsive Polymers in Water. *Adv. Polym. Sci.* **2011**, *242*, 29–89. [CrossRef]
3. Pasparakis, G.; Tsitsilianis, C. LCST Polymers: Thermoresponsive Nanostructured Assemblies towards Bioapplications. *Polymer* **2020**, *211*, 123146. [CrossRef]
4. Peppas, N.A.; Hilt, J.Z.; Khademhosseini, A.; Langer, R. Hydrogels in Biology and Medicine: From Molecular Principles to Bionanotechnology. *Adv. Mater.* **2006**, *18*, 1345–1360. [CrossRef]
5. Ward, M.A.; Georgiou, T.K. Thermoresponsive Polymers for Biomedical Applications. *Polymers* **2011**, *3*, 1215–1242. [CrossRef]
6. Nash, M.E.; Fan, X.; Carroll, W.M.; Gorelov, A.V.; Barry, F.P.; Shaw, G.; Rochev, Y.A. Thermoresponsive Substrates Used for the Expansion of Human Mesenchymal Stem Cells and the Preservation of Immunophenotype. *Stem Cell Rev. Rep.* **2013**, *9*, 148–157. [CrossRef]
7. Frolova, A.; Ksendzov, E.; Kostjuk, S.; Efremov, Y.; Solovieva, A.; Rochev, Y.; Timashev, P.; Kotova, S. Thin Thermoresponsive Polymer Films for Cell Culture: Elucidating an Unexpected Thermal Phase Behavior by Atomic Force Microscopy. *Langmuir* **2021**, *37*, 11386–11396. [CrossRef]
8. Cao, M.; Wang, Y.; Hu, X.; Gong, H.; Li, R.; Cox, H.; Zhang, J.; Waigh, T.A.; Xu, H.; Lu, J.R. Reversible Thermoresponsive Peptide–PNIPAM Hydrogels for Controlled Drug Delivery. *Biomacromolecules* **2019**, *20*, 3601–3610. [CrossRef]
9. Doberenz, F.; Zeng, K.; Willems, C.; Zhang, K.; Groth, T. Thermoresponsive Polymers and Their Biomedical Application in Tissue Engineering—A Review. *J. Mater. Chem. B* **2020**, *8*, 607–628. [CrossRef]
10. Lu, H.; Leng, J.; Du, S. A Phenomenological Approach for the Chemo-Responsive Shape Memory Effect in Amorphous Polymers. *Soft Matter* **2013**, *9*, 3851. [CrossRef]
11. Zhang, Q.; Weber, C.; Schubert, U.S.; Hoogenboom, R. Thermoresponsive Polymers with Lower Critical Solution Temperature: From Fundamental Aspects and Measuring Techniques to Recommended Turbidimetry Conditions. *Mater. Horiz.* **2017**, *4*, 109–116. [CrossRef]
12. Ding, Y.; Ye, X.; Zhang, G. Microcalorimetric Investigation on Aggregation and Dissolution of Poly(N-Isopropylacrylamide) Chains in Water. *Macromolecules* **2005**, *38*, 904–908. [CrossRef]
13. Antheunis, H.; van der Meer, J.-C.; de Geus, M.; Heise, A.; Koning, C.E. Autocatalytic Equation Describing the Change in Molecular Weight during Hydrolytic Degradation of Aliphatic Polyesters. *Biomacromolecules* **2010**, *11*, 1118–1124. [CrossRef]
14. Kurzbach, D.; Junk, M.J.N.; Hinderberger, D. Nanoscale Inhomogeneities in Thermoresponsive Polymers. *Macromol. Rapid Commun.* **2013**, *34*, 119–134. [CrossRef]
15. Winnik, F.M.; Ottaviani, M.F.; Bossmann, S.H.; Garcia-Garibay, M.; Turro, N.J. Consolvency of Poly(N-Isopropylacrylamide) in Mixed Water-Methanol Solutions: A Look at Spin-Labeled Polymers. *Macromolecules* **1992**, *25*, 6007–6017. [CrossRef]
16. Kokorin, A. (Ed.) *Nitroxides—Theory, Experiment and Applications*; InTech: London, UK, 2012; ISBN 978-953-51-0722-4.
17. Drescher, M.; Jeschke, G. (Eds.) *EPR Spectroscopy*; Springer: Berlin/Heidelberg, Germany, 2012; Volume 321.
18. Kurzbach, D.; Reh, M.N.; Hinderberger, D. Nanoscale Inhomogeneities in Thermoresponsive Triblock Copolymers. *ChemPhysChem* **2011**, *12*, 3566–3572. [CrossRef]

19. Junk, M.J.N.; Jonas, U.; Hinderberger, D. EPR Spectroscopy Reveals Nanoinhomogeneities in the Structure and Reactivity of Thermoresponsive Hydrogels. *Small* **2008**, *4*, 1485–1493. [CrossRef]
20. Rochev, Y.; O'Halloran, D.; Gorelova, T.; Gilcreest, V.; Selezneva, I.; Gavrilyuk, B.; Gorelov, A. Rationalising the Design of Polymeric Thermoresponsive Biomaterials. *J. Mater. Sci. Mater. Med.* **2004**, *15*, 513–517. [CrossRef] [PubMed]
21. Caragheorgheopol, A.; Schlick, S. Hydration in the Various Phases of the Triblock Copolymers EO13PO30EO13 (Pluronic L64) and EO6PO34EO6 (Pluronic L62), Based on Electron Spin Resonance Spectra of Cationic Spin Probes. *Macromolecules* **1998**, *31*, 7736–7745. [CrossRef]
22. Eaton, G.R.; Eaton, S.S.; Barr, D.P.; Weber, R.T. *Quantitative EPR*; Springer: Vienna, Austria, 2010; ISBN 978-3-211-92947-6.
23. Stoll, S.; Schweiger, A. EasySpin, a Comprehensive Software Package for Spectral Simulation and Analysis in EPR. *J. Magn. Reson.* **2006**, *178*, 42–55. [CrossRef] [PubMed]
24. Schneider, D.J.; Freed, J.H. Calculating Slow Motional Magnetic Resonance Spectra. In *Spin Labeling: Theory and Applications*; Berliner, L.J., Reuben, J., Eds.; Springer: Boston, MA, USA, 1989; pp. 1–76. ISBN 978-1-4613-0743-3.
25. Yanase, K.; Buchner, R.; Sato, T. Microscopic Insights into the Phase Transition of Poly(N-Isopropylacrylamide) in Aqueous Media: Effects of Molecular Weight and Polymer Concentration. *J. Mol. Liq.* **2020**, *302*, 112025. [CrossRef]
26. Hunold, J.; Eisermann, J.; Brehm, M.; Hinderberger, D. Characterization of Aqueous Lower-Polarity Solvation Shells Around Amphiphilic 2,2,6,6-Tetramethylpiperidine-1-Oxyl Radicals in Water. *J. Phys. Chem. B* **2020**, *124*, 8601–8609. [CrossRef]
27. Salikhov, K.M. Current State of the Spin Exchange Theory in Dilute Solutions of Paramagnetic Particles. New Paradigm of Spin Exchange and Its Manifestations in EPR Spectroscopy. *Physics-Uspekhi* **2019**, *62*, 951–975. [CrossRef]
28. Salikhov, K.M. New Paradigm of Spin Exchange and Its Manifestations in EPR Spectroscopy. *Appl. Magn. Reson.* **2020**, *51*, 297–325. [CrossRef]
29. Salikhov, K.M. Consistent Paradigm of the Spectra Decomposition into Independent Resonance Lines. *Appl. Magn. Reson.* **2016**, *47*, 1207–1227. [CrossRef]
30. Kecki, Z.; Lyczkowski, Z. KW Critical Comparison of Empirical Systems Used to Describe Solvent Properties. *J. Solution Chem.* **1986**, *15*, 413–422. [CrossRef]
31. Ellison, W.J. Permittivity of Pure Water, at Standard Atmospheric Pressure, over the Frequency Range 0–25THz and the Temperature Range 0–100 °C. *J. Phys. Chem. Ref. Data* **2007**, *36*, 1–18. [CrossRef]
32. Korson, L.; Drost-Hansen, W.; Millero, F.J. Viscosity of Water at Various Temperatures. *J. Phys. Chem.* **1969**, *73*, 34–39. [CrossRef]
33. Halperin, A.; Kröger, M.; Winnik, F.M. Poly(N-isopropylacrylamide) Phase Diagrams: Fifty Years of Research. *Angew. Chemie Int. Ed.* **2015**, *54*, 15342–15367. [CrossRef]
34. Franchi, P.; Lucarini, M.; Pedrielli, P.; Pedulli, G.F. Nitroxide Radicals as Hydrogen Bonding Acceptors. An Infrared and EPR Study. *ChemPhysChem* **2002**, *3*, 789–793. [CrossRef]
35. Rinkevicius, Z.; Murugan, N.A.; Kongsted, J.; Aidas, K.; Steindal, A.H.; Ågren, H. Density Functional Theory/Molecular Mechanics Approach for Electronic g-Tensors of Solvated Molecules. *J. Phys. Chem. B* **2011**, *115*, 4350–4358. [CrossRef] [PubMed]
36. Wu, C.; Wang, X. Globule-to-Coil Transition of a Single Homopolymer Chain in Solution. *Phys. Rev. Lett.* **1998**, *80*, 4092–4094. [CrossRef]
37. Lebedev, Y.S.; Grinberg, O.Y.; Dubinsky, A.A.; Poluektov, O.G. Investigation of Spin Labels and Probes by Millimeter Band EPR. In *Bioactive Spin Labels*; Springer: Berlin/Heidelberg, Germany, 1992; pp. 227–278.

polymers

Article

Characterization of Thermoresponsive Poly-N-Vinylcaprolactam Polymers for Biological Applications

Lorenzo Marsili [1,2,*], Michele Dal Bo [1], Giorgio Eisele [3], Ivan Donati [4], Federico Berti [2,†] and Giuseppe Toffoli [1,†]

1 Experimental and Clinical Pharmacology Unit, CRO National Cancer Institute, IRCCS, Via Franco Gallini 2, 33081 Aviano, Italy; mdalbo@cro.it (M.D.B.); gtoffoli@cro.it (G.T.)
2 Department of Chemical and Pharmaceutical Sciences, University of Trieste, Via Licio Giorgieri 1, 34127 Trieste, Italy; fberti@units.it
3 Centro Alta Tecnologia "Istituto di Ricerche Chimiche e Biochimiche G. Ronzoni" Srl, via G. Colombo 81, 20133 Milan, Italy; eisele@cat-ronzoni.it
4 Department of Life Sciences, University of Trieste, Via Licio Giorgieri 5, 34127 Trieste, Italy; idonati@units.it
* Correspondence: lorenzo.marsili@phd.units.it
† Contributed as last authors.

Abstract: Poly-N-Vinylcaprolactam (PNVCL) is a thermoresponsive polymer that exhibits lower critical solution temperature (LCST) between 25 and 50 °C. Due to its alleged biocompatibility, this polymer is becoming popular for biomedical and environmental applications. PNVCL with carboxyl terminations has been widely used for the preparation of thermoresponsive copolymers, micro- and nanogels for drug delivery and oncological therapies. However, the fabrication of such specific targeting devices needs standardized and reproducible preparation methods. This requires a deep understanding of how the miscibility behavior of the polymer is affected by its structural properties and the solution environment. In this work, PNVCL-COOH polymers were prepared via free radical polymerization (FRP) in order to exhibit LCST between 33 and 42 °C. The structural properties were investigated with NMR, FT-IR and conductimetric titration and the LCST was calculated via UV-VIS and DLS. The LCST is influenced by the molecular mass, as shown by both DLS and viscosimetric values. Finally, the behavior of the polymer was described as function of its concentration and in presence of different biologically relevant environments, such as aqueous buffers, NaCl solutions and human plasma.

Keywords: poly-N-vinylcaprolactam; thermoresponsive polymers; LCST

Citation: Marsili, L.; Dal Bo, M.; Eisele, G.; Donati, I.; Berti, F.; Toffoli, G. Characterization of Thermoresponsive Poly-N-Vinylcaprolactam Polymers for Biological Applications. *Polymers* **2021**, *13*, 2639. https://doi.org/10.3390/polym13162639

Academic Editors: Mattia Sponchioni and Florian J. Stadler

Received: 16 July 2021
Accepted: 4 August 2021
Published: 8 August 2021

1. Introduction

Thermoresponsive polymers are characterized by a drastic and discontinuous change of their physical properties with temperature. Given a solvent/polymer binary mixture, the phase diagram usually exhibits a binodal curve that divides a polymer-rich zone from a zone in which the polymer and the solvent are miscible in every proportion. Accordingly, the critical solution temperature can correspond to the minimum or the maximum of the binodal curve. If the critical point corresponds to the minimum of the curve, the corresponding temperature is called lower critical solution temperature (LCST). On the contrary, the temperature corresponding to the maximum of the curve is called upper solution temperature (UCST) [1].

The ability to respond to a change in temperature makes thermoresponsive polymers a "smart" class of materials that can be applied in a broad range of applications [2]. To date, there are hundreds of thermoresponsive polymers developed for various applications in the biological field, which include tissue engineering, bioseparation, drug and gene delivery [2–4]. As a general rule, LCST-type polymers are easily solvated in water through hydrogen bonding and polar interactions [5]. Accordingly, their biological interest relies on the presence of a desirable sharp transition in aqueous systems [2]. Their utilization in the biological field requires that the LCST in aqueous solution is close to the normal human physiological temperature, which is usually comprehended between 36.5 and 37.5 °C, but can oscillate in a wider range between 33.2 and 38.2 °C [6].

PNVCL is a thermoresponsive polymer which exhibits LCST that has been applied in biomedical and environmental applications, in cosmetics and as an anticlogging agent in pipelines [3,4]. PNVCL is usually described as a biocompatible alternative to poly-N-isopropylamide (PNIPAM) [3,4,7–15] as both polymers can exhibit LCST behavior close to the physiological temperature. According to this misconception, PNVCL has been often described as a polymer with a well-defined LCST at 32 °C [7–9,16–21], even though the LCST has been observed in broad range of temperature between 25 and 50 °C [3,4]. As a matter of fact, PNVCL and PNIPAM exhibit a complete opposite miscibility behavior [3,22]. PNIPAM exhibits a LCST at about 32 °C, which is almost independent from the molecular weight (Type II) [3,4]. On the contrary, PNVCL miscibility can be described by the Flory–Huggins theory and depends on molecular weight [3,23], polymer, salt and protein concentration (Type I). Due to its molecular weight-dependent LCST, the control of both molecular weight and polydispersity in PNVCL synthesis is of crucial importance [3,4,22–27] for any application. Similarly, the determination of molecular mass must rely on appropriate methodologies that allow to correlate the synthetic procedures with the final properties of the polymer.

Although the thermoresponsivity of PNVCL was first observed in 1968 [23], the polymer is still introduced as a novelty in biological and nanoscientific publications [3,4,7–9,16–21]. The first "in vitro" evaluation of PNVCL cytotoxicity on intestinal Caco-2 and pulmonary Calu-3 cell lines was published in 2005. The test demonstrated great cell tolerance although PNVCL exhibited toxic effects above its LCST [28]. To date, the biocompatibility of PNVCL has been established in several different cell lines, including different types of human carcinomas [4,29–31]. PNVCL was also investigated as a suitable environment for cell proliferation and manipulation [32] and it is commercialized as a hair setting product under the name of Luviscol® Plus [33]. The utilization of PNVCL with carboxyl end groups has been reported for the preparation of PNVCL-based delivery systems in conjunction with biocompatible polysaccharides, such as alginates [34], chitosan [7,9,16–21,35], or dextran [3,36,37]. A common approach is to use PNVCL-COOH polymers with a 32 °C LCST as starting materials for the preparation of biocompatible copolymers and thermoresponsive particles [7–9,19,35,38]. The LCST is raised according to the degree of substitution and the molecular mass of the polymer conjugated with PNVCL. However, to ensure reproducibility, PNVCL synthesis must be optimized in order to ensure control over chemical structure, molecular mass, polydispersity and morphology [3,4]. Moreover, it is still fundamental to assess the polymer thermoresponsive behavior in biologically relevant fluids such as human plasma.

The aim of this work is to rationalize this concept, as we considered the polymerization of different PNVCL polymers with carboxyl terminations (PNVCL-COOH) that exhibit different LCSTs according to their different molecular mass, their concentration and the solution environment. Furthermore, this work provides insight on different approaches for the determination of the molecular mass and the LCST of PNVCL. The polymers were prepared by using a modified approach of the protocol reported by Prabaharan [35] and reported by many different authors [7–9,16–19]. Accordingly, we reported the preparation of PNVCL-COOH polymers exhibiting LCST between 33 and 42 °C via FRP. The hypothesis of mass-dependent LCST was first supported by the comparison of ^{13}C NMR spectra of different PNVCL-COOH, by assessing the variation of the signals related to the terminations as a function of the NVCL/AIBN ratio that were used for the synthesis. Finally, the determination of the molecular mass via viscosimetry and DLS demonstrated that the LCST of PNVCL-COOH is inversely proportional to the molecular mass. LCSTs were determined using UV-VIS and DLS at a concentration of 0.5 wt.%. The variation of LCST of PNVCL-COOH polymers was also assessed in relation to polymer concentration, the presence of NaCl and three different buffers: citrate (pH 3), acetate (pH 5) and phosphate (pH 7). Finally, we report a method for the evaluation of the polymer in human plasma. In particular, it was observed that it is possible to increase LCST by decreasing PNVCL-COOH concentration, while the presence of salts, particularly phosphates, results in a significant lowering of the LCST. The measurements performed in human plasma showed a relevant lowering of the LCST of PNVCL-COOH polymers by about 10 °C. These observations are particularly useful for physiological applications, as the polymers undergoing LCST may result in possible cytotoxic effects.

2. Materials and Methods

2.1. Materials

N-Vinyl Caprolactam (NVCL), 2,2′-Azobis(2-methylpropionitrile) (AIBN), 3-mercaptopropionic acid (MPA), were purchased from Sigma-Aldrich (St. Louis, MO, USA). N-hydroxy succinimide (NHS), Deuterium oxide (D_2O), N,N′-Dimethylformamide anhydrous (DMF), Dimethyl sulfoxide-d_6, Deuterium Chloride 37% (DCl/D_2O) were purchased from Sigma-Aldrich (St. Louis, MO, USA). Spectra/Por^{TM} Cellulose membrane tubing (1–2 kDa, MWCO) was purchased from Thermo Fisher Scientific (Waltham, MA, USA).

2.2. Synthesis of PNVCL-COOH Polymers

PNVCL-COOH was synthesized by free radical polymerization by using a modified version of the protocol reported by Prabaharan [35]. Prior to their utilization, NVCL was recrystallized in hexane and AIBN was recrystallized in ethanol, while MPA was transferred in a sealed vial that was deoxygenated with Ar. For all preparations, 5 mg of AIBN (0.304 mmol) and 28 μL of MPA (3.278 mmol) were used. Different molar ratios between NVCL and AIBN were used in order to obtain polymer with different molecular mass and LCST. The NVCL/AIBN molar ratio were, respectively, 122, 244, 305, 610, 1220 and 1690. DMF was deoxygenized with Ar for 15 min prior to the addition of the reagents (Figure 1). After the dissolution of the reagents, the reaction was carried out in sealed vials at 70 °C for 8 h. All sealed vials were dried overnight at 80 °C before the reaction. After the reaction, the solution was dialyzed in a cellulose membrane tubing (MWCO of 1–2 kDa) against distilled water for at least 2 days to remove impurities and unreacted materials. Finally, the frozen product was freeze-dried at −50 °C and 0.05 mbar and stored at 4 °C.

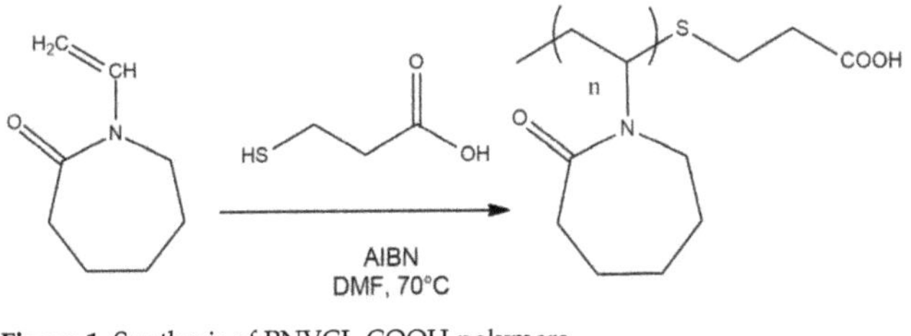

Figure 1. Synthesis of PNVCL-COOH polymers.

2.3. Structural Characterization of PNVCL-COOH Polymers

NMR spectra were acquired using a High-resolution 500 MHz Bruker NEOn500 Quadruple resonance (H/C/N/2H) equipped with a high-sensitivity TCI 5 mm CryoProbe (Bruker, Billerica, MA, USA) at 25 °C in D_2O and d_6-DMSO.

FT-IR spectra were recorded with a double-beam Perkin Elmer System 2000 Ft-IR Spectrometer (Perkin Elmer, Waltham, MA, USA) in the range of 4500–370 cm^{-1} using KBr pellets.

The number of terminal carboxyl groups was determined via conductimetric titration. The polymers were dissolved using diluted HCl and the solutions were titrated using a standardized solution of NaOH 0.1 M. The deprotonation of the -COOH end groups resulted in the formation of a small plateau in the conductimetric titration curve. The number of carboxyl groups was determined by the volume difference of added NaOH solution between the initial and the final point of the plateau (Figure S10).

2.4. UV-VIS Determination of LCST

Phase transition and absorbance measurements were carried out in a Shimazdu UV-visible spectrophotometer model UV-2450 equipped with a TCC-240A Thermoelectrically Temperature Controlled Cell Holder (Shimazdu, Kyoto, JP). Transmission data were used for the realization of the miscibility curves. The miscibility curves were fitted with a sigmoidal model in order to calculate the LCST as the inflection point by using the following equation:

$$y = Tr_{max} + \frac{(Tr_{max} - Tr_{min})}{1 + e^{\frac{(T-LCST)}{dT}}} \quad (1)$$

where Tr is the calculated transmittance at a specific wavelength at a fixed temperature.

2.5. DLS Determination of LCST

The LCSTs of PNVCL-COOH polymers were determined using a Malvern Zetasizer Nano ZS90 (Malvern Panalytical S.R.L., Malvern, UK). Samples (at a concentration of 0.5 wt.%) were incubated at different temperatures for 5 min in the temperature range between 25 and 50 °C and the cuvettes were examined to check visible turbidity. The LCST was attributed to the temperature at which a dramatic change in the shape of the autocorrelation curve was observed [39].

2.6. Determination of Molecular Mass

2.6.1. Size Exclusion Chromatography

Samples were analyzed using a Viscotek TDA 302 (HP-SEC-TDA, Viscotek, USA), using degassed 0.3 M acetate buffer (pH = 8.0) as solvent. A volume of 100 µL of sample was injected, flow rate was maintained at 0.6 mL/min and the column and detectors temperature were kept at 25 °C. Before injection, polymer solutions were filtered through a 0.45 µm cellulose nitrate disposable membrane and the eluent was filtrated through a 16–40 µm glass filter to ensure a low light scattering noise level. A polyethyleneoxide standard (M_W = 22,411, $[\eta]$ = 0.384 dL/g, M_w/M_n = 1.03) was used to normalize the viscometer and the light scattering detectors.

2.6.2. Dynamic Light Scattering

The molecular mass of the polymers was estimated using a Malvern Zetasizer Nano ZS90 (Malvern Panalytical S.R.L., Malvern, UK). All measurements were performed at 25 °C at a concentration of 0.5 wt.% in milliQ water. Each sample was analyzed three times and provided a measurement of an average hydrodynamic diameter (D_h) corresponding to the position of a peak in size distribution located between 5 and 30 nm. D_h standard deviation is referred to the position of the peaks and does not provide any information on peak width. For a random coil conformation, the average radius of gyration was calculated as $R_g = D_h \times 0.75$. The average molecular mass was estimated from R_g using the equation reported by Lau [40] (Equation (2)) and Eisele [41] (Equation (3)), respectively:

$$R_g = 2.94 \times 10^{-2} {M_w}^{0.54} \quad (2)$$

$$\left\langle s^2 \right\rangle = 1.77 \times 10^{-18} {M_w}^{1.15} \quad (3)$$

where $\langle s^2 \rangle$ is the mean square radius of gyration in cm^{-1}.

2.6.3. Intrinsic Viscosity Measurements

The intrinsic viscosity $[\eta]$ of PNVCL-COOH was measured at 25 °C by means of a Schott Geräte AVS/G automatic measuring apparatus and a Ubbelhode capillary viscometer, using water as a solvent. Polymer solutions and solvents were filtered prior to the analysis through 0.45 μm nitrocellulose filters (Millipore, Germany). $[\eta]$ was calculated from the polymer concentration dependence of the reduced specific viscosity, η_{sp}/c, according to the Huggins Equation (4):

$$\eta_{red} = \frac{\eta_{sp}}{c} = [\eta] + k[\eta]^2 c \quad (4)$$

where k is Huggins constant. The values of intrinsic viscosity $[\eta]$ was calculated at infinite dilution by using two calibration lines per sample, one of which was obtained by excluding the values for the most diluted sample. $[\eta]$ was calculated as an average of the values calculated by applying both equations, in order to provide a reasonable confidence interval for the calculated value. The corresponding average viscosimetric molecular weight (M_η) of PNVCL-COOH was calculated in agreement with the Mark–Houwink–Sakurada (MHS) Equation (5).

$$[\eta] = K \cdot M_\eta^\alpha \quad (5)$$

K and a parameters used for the calculation are reported by Kirsh [12,42] for the calculation of the molecular mass in different size range at 25 °C. The MHS equations that were used were, respectively:

$$[\eta] = 35 \times 10^{-3} M_\eta^{0.57} \quad (6)$$

$$[\eta] = 38.9 \times 10^{-3} M_\eta^{0.69} \quad (7)$$

3. Results and Discussion

PNVCL-COOH polymers were prepared via free radical polymerization as reported by Prabaharan [35]. The initial results ruled out the possibility that PNVCL-COOH was able to precipitate in diethyl ether in the conditions described by the original procedure. Accordingly, the procedure was modified in order to produce PNVCL-COOH polymers with higher molecular weight. In free radical polymerization, the degree of polymerization $\overline{X}_n$ is directly proportional to the square of the concentration of the monomer, according to the following equation:

$$\overline{X}_n = \alpha \frac{(k_p)^2}{2k_t v_p} [M]^2 \quad (8)$$

Consequently, the preparation of PNVCL-COOH with higher molecular weight was achieved by changing the molar ratio between initiator and monomer (M/I). For this reason, PNVCL-COOH were distinguished according to the M/I molar ratio that was used for their synthesis. The M/I was, respectively, 122 (equal to that reported by Prabaharan), 244, 305, 610, 1220 and 1690. The precipitation in diethyl ether was achieved from values of M/I above 610. The elongation of the hydrophobic portion resulted in a different hydrophilic-hydrophobic balance that allowed the precipitation of PNVCL_610, PNVCL_1220 and PNVCL_1690 in diethyl ether. FRP of PNVCL is associated with a lack of control on polymer polydispersity. This could be related to the low yield of the synthesis after diethyl ether precipitation (<20%). Accordingly, the fraction of the polymer under a critical molecular mass was unable to precipitate in the solvent. The average yield of the process was raised by purifying the polymer directly with dialysis using with membrane tubings (MWCO = 1 kDa) that were compatible with DMF solutions. In this way, PNVCL-COOH polymers were obtained without diethyl ether precipitation. The main drawback of this purification method is the inability to remove eventual traces of DMF that remain in water after dialysis. In a few polymers, it was possible to identify a minor peak at 2.9 ppm (Figure 2) that demonstrates the presence of a residual DMF in freeze-dried products. The peak had been previously mistaken for the methylene group (-C-S-CH_2) present in the terminal group [35]. This hypothesis was excluded with the utilization of HSQC-DEPT heterocorrelated spectra (see Supplementary Figure S4). All polymers exhibited different LCSTs in relation to their different molecular mass and their corresponding M/I ratio.

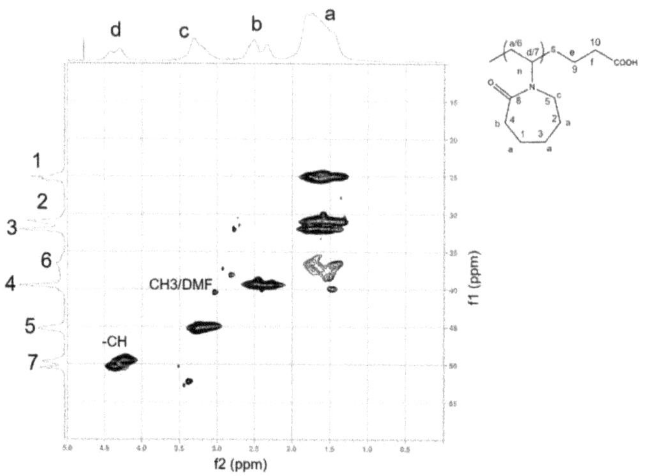

Figure 2. HSQC spectra of PNVCL_305.

3.1. Structural Characterization of PNVCL-COOH Polymers

NMR characterization was performed in D_2O and d_6-DMSO. In the ^1H spectrum of PNVCL-COOH, four main signals were observed. The formation of the polymer was confirmed by the presence of broad signals and from the disappearance of the vinylic signals at 7.36 ppm. Similarly, amide I vibrations at 1631 and 1480 cm^{-1} (C-N stretching vibration) were observed, while the characteristic signals related to the monomer (C=C, 1658 cm^{-1}, CH= and CH2=, 3000 and 3100 cm^{-1}) disappeared (see Supplementary Figure S5). All ^1H NMR PNVCL-COOH spectra exhibited peaks at 1.77 ppm (3H, -CH_2), 2.49 ppm (2H, -$COCH_2$), 3.31 (2H, -NCH_2) and 4.36 (1H, -NCH). The HSQC correlation spectrum with ^1H and ^{13}C assignments is reported in Figure 2. ^1H and ^{13}C spectra are provided in the Supplementary Materials (Figures S1 and S3). The presence of carboxyl end groups was confirmed by analyzing PNVCL_244 and PNVCL_305 in d_6-DMSO using a high number of acquisitions. The signals were identified in small peaks observed at about 12 ppm in both spectra (Figure S2). In FT-IR spectra, the carboxyl end groups were recognized from the presence of broad signals at 3450 cm^{-1} (Figure S5). The carboxyl group content was estimated via conductimetric titration and was found to be inversely dependent on the M/I ratio. The number of terminations for PNVCL_122, PNVCL_305, PNVCL_610, PNVCL_1220 and PNVCL 1690 were, respectively, 0.96 $\pm$ 0.25, 0.73 $\pm$ 0.17, 0.63 $\pm$ 0.14, 0.55 $\pm$ 0.12 and 0.45 $\pm$ 0.13 mmol/g. ^{13}C spectra confirmed the structure of the polymers based on the previous literature [10,12] (Figure 3 and Figure S2). Hetero-correlated 2D-HSQC spectra (see Supplementary Figure S4) allowed the identification of two minor signals at 1.55/29 ppm and 2.4/38 ppm that were related to the presence of the sulphur-bonded aliphatic methylene and the methylene bonded to the carboxyl termination group. The comparison between ^{13}C spectra (Figure 3) of PNVCL_122, PNVCL_244 and PNCVL_1220 in the range between 50 and 25 ppm confirmed that the signals are associated to the aliphatic portion of the termination groups. As shown, the signal intensity decreases as M/I ratio increases. This observation is in accordance with the presence of different PNVCL-COOH polymer with increasing molecular mass and increasing M/I ratio. The signals at 2.9/37 ppm were associated to CH_3 residues of residual DMF solvent.

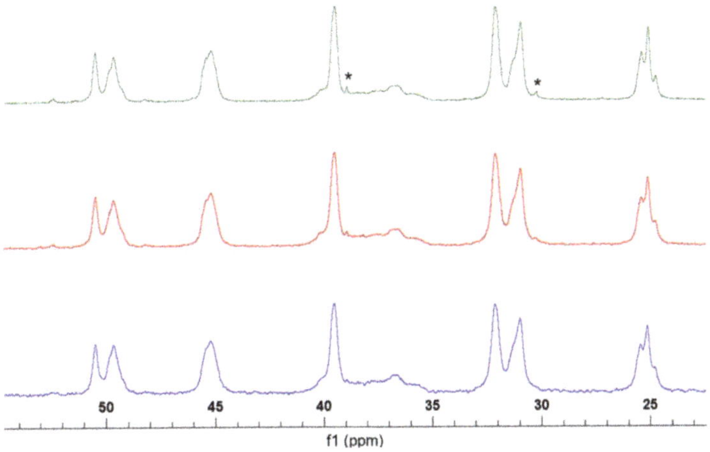

Figure 3. Comparison between ^{13}C spectra of PNVCL_122 (green), PNVCL_305 (red) and PNVCL_1220 (blue).

3.2. Spectroscopic Determination of LCST

The miscibility curves were represented by plotting the transmittance of the solutions as a function of temperature (Figure 4a). LCSTs were determined from the inflection points of the sigmoidal functions that were used to fit the miscibility curves. PNVCL-COOH polymers (0.5 wt.% solutions) exhibited LCST in a range between 33 and 42 °C. The results showed that LCST diminishes in relation to the M/I ratios that were used for the synthesis of the polymers (Table 1). This molecular mass-dependent behavior is in accordance with a "classical" type I miscibility behavior. Accordingly, the increase of the hydrophobicity of the polymer results in the reduction of the LCST. The results excluded the presence of "step" transitions, that are generally observed in highly polydisperse polymers with different LCSTs. As a result, the variation of the M/I ratio provided a reliable procedure for the control of the LCST with simple FRP.

The type I behavior of PNVCL-COOH was further validated by measuring the LCSTs in the presence of salt species. The 0.5 wt.% PNVCL-COOH solutions were prepared in citrate (pH 3), acetate (pH 5) and phosphate buffers (pH 7). All buffers were prepared at the concentration of 0.1 M to compare the effect of the different ions on the LCSTs of the polymers. The results (Figure 4c) showed a strong dependence towards the types of ions dissolved in solution, as it was previously demonstrated by other studies on PNVCL polymers [4]. The results suggested that the ionic environment has a stronger effect in relation to the pH of the solution. The effect of salts was more pronounced for short chain polymers, which have a higher LCST (Figure 4c). The most evident effect was observed by dissolving PNVCL-COOH in phosphate buffer. This could be relevant from the biological point of view, since phosphate buffers solutions (e.g., PBS) are widely used in cell treatment kits [4,28–31]. During LCST transition, PNVCL-COOH undergoes a coil-to-globule transition [3,4]. Consequently, the lowering of the LCST due to the cell culture medium could result in potential cytotoxic effect due to the conformational change of the polymer. The effect of polymer concentration was assessed by measuring the LCSTs of PNVCL-COOH solutions at the concentration of 0.5, 0.1, 0.05 wt.% (Figure 4b). Normally, the LCST is associated with the point where the transmittance goes to zero [8,35,43]. However, this model was not suitable for describing the behavior of solutions at lower concentration (<0.1 wt.%). The determination of the LCST of dilute solution (<0.1 wt.%) of PNVCL has been previously reported with DLS, static light scattering and differential scanning calorimetry [4,44]. The calculation of the LCST at lower concentration was facilitated using a sigmoidal model for the interpretation of curves (Equation (1)). Dilution resulted in higher values of LCST and slower transition. Accordingly, PNVCL-COOH concentration affects the kinetics and the energy of LCST transition. Finally, the LCSTs were measured in presence of NaCl in range of concentrations between 0 and 0.15 M (Figure 4d). Results showed that LCST decreases proportionally in relation to NaCl concentration [45]. LCST decreased by about 1 °C in a physiological solution (0.9% NaCl, or 0.15 M). Consequently, pharmaceutical applications of PNVCL-COOH polymer in a physiological solution would require a polymer with a LCST of at least > 38 °C in order to avoid cytotoxic effects due to the polymer transition.

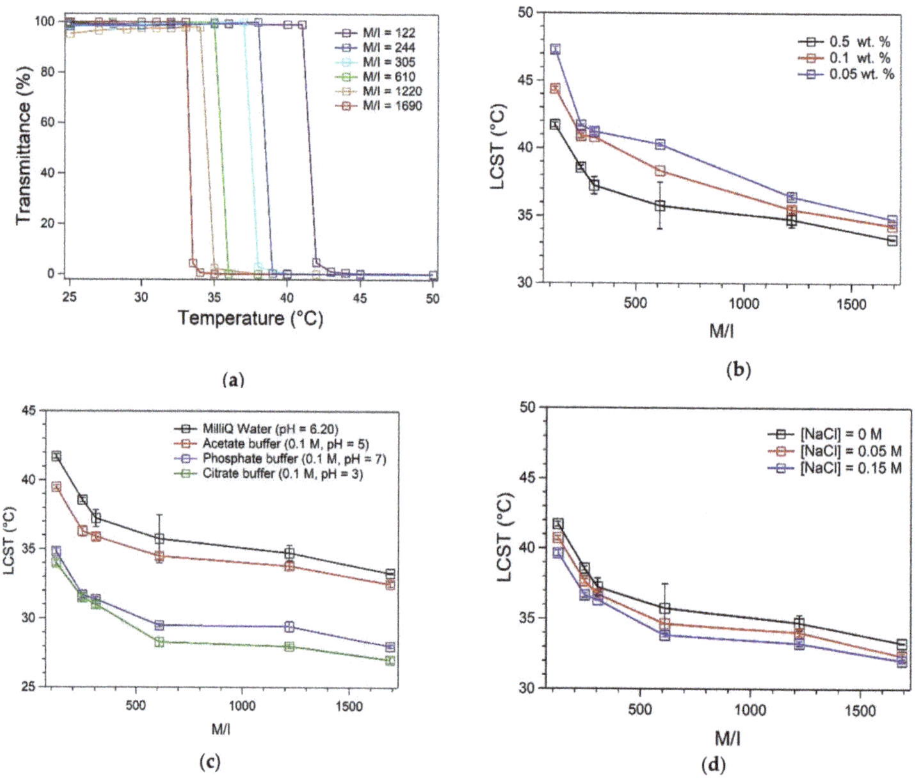

Figure 4. (**a**) Miscibility curves (transmittance vs. temperature) of PNVCL-COOH polymers at the concentration of 0.5 wt.% in milliQ water (pH = 6.20). (**b**) Variation of LCST reported as a function of the PNVCL-COOH concentration in milliQ water (pH = 6.20). (**c**) Variation of LCST of 0.5 wt.% PNVCL-COOH solutions associated to different type of buffers. (**d**) Variation of LCST of 0.5 wt.% PNVCL-COOH solutions (milliQ water, pH = 6.20) associated to the concentration of NaCl.

Table 1. LCST, average molecular weight and LCST of the synthesized PNVCL-COOH polymers.

M/I Ratio	LCST (°C)		Average Molecular Weight (kDa)			
	UV-VIS	DLS	Equation (2)	Equation (3)	Equation (6)	Equation (7)
122	41.71 ± 0.17	41	13.29 ± 0.36	29.60 ± 0.99	-	-
244	38.56 ± 0.14	37	22.60 ± 0.06	48.74 ± 0.19	19.74 ± 1.12	19.56 ± 1.18
305	37.23 ± 0.63	37	25.96 ± 0.06	55.53 ± 0.17	21.87 ± 0.88	21.65 ± 0.89
610	35.75 ± 1.74	34	52.20 ± 0.36	107.0 ± 1.00	-	-
1220	34.73 ± 0.56	33.5	79.54 ± 0.25	158.9 ± 0.70	42.87 ± 1.90	42.94 ± 1.93
1690	33.27 ± 0.03	32.5	124.1 ± 0.60	241.4 ± 1.64	-	-

3.3. Scattering Determination of LCST

DLS provided lower values of LCST in relation to UV-VIS spectroscopy. The analyses of polymers with higher M/I (PNVCL_610, PNVCL_1220, PNVCL_1690) provided a direct measurement of LCST from the displacement of the autocorrelation curve due to the transition. The analyses of the other samples did not provide a direct estimation due to the small dimensions of the PNVCL-COOH macromolecules in solution. Consequently, the LCST was estimated by checking visible turbidity. The observed differences between DLS and UV-VIS measurement were related to the polydispersity of the polymers. DLS is more sensitive to the formation of aggregates in the proximity of LCST. The formation of globular aggregates at temperature inferior to LCST is related to the fraction of PNVCL-COOH polymers with higher molecular mass. Since the scattering intensity is proportional to D_h^6, a small fraction of polymer undergoing coil-to-globule transition is able to produce variation in the autocorrelation curve shape or position (Figure S7) [39]. Accordingly, the entity of the differences between UV-VIS and DLS measurements provided an overview on the polydispersity of PNVCL-COOH polymers. According to the results, the polymer with highest polydispersity was PNVCL_610.

3.4. Determination of Molecular Mass

The molecular weight determination of polymer was carried out by SEC, DLS and viscosimetry. DLS and viscosimetry provided an estimation of the average molecular mass, while SEC did not produce any appreciable results due to the sorption of PNVCL on the column. This issue has been previously addressed for the analysis of PNVCL column in most solvents, including aqueous buffers and THF [10,46]. DLS allowed the estimation of the molecular mass according to the models described by Lau [40] and Eisele [41]. The viscosimetric molecular mass of PNVCL_244, PNVCL_305 and PNVCL_1220 mass was determined using water as a solvent at 25 °C. Samples were analyzed in dilute conditions ($\leq$ 0.2 wt.%) in order to prevent the formation of foam. Due to the low viscosity of the sample, the differences between the run times were very small. Similarly, the values of inherent viscosity (η_{inh}) were considered too low for the calculation of the molecular mass with the model described by Kraemer [47]. The comparison between DLS and viscosimetric measurements is reported in Table 1. The estimated molecular mass as a function of the M/I value is reported in Figure 5a and the variation of LCST as a function of the molecular mass is reported in Figure 5b. Both DLS and viscosimetric results are in line with a type I thermoresponsive polymer behavior. As the molecular mass increases, the LCST decreases. The utilization of K and a constant reported by Kirsh [12,42] demonstrated that the two models can be equally applied for PNVCL-COOH polymers in this molecular weight range (15–50 kDa). Molecular mass values obtained through viscosimetry were slightly lower than those obtained by interpreting DLS data with the model provided by Lau (Equation (2)), while the use of Eisele's equation led to the overestimation of the molecular mass. Interestingly, it was observed that the values calculated using Lau's equation were almost half the values obtained with Eisele's. The bigger differences between DLS and viscosimetry data are observed with increasing molecular mass (Table 1, Figure 5a,b). Accordingly, scattering methods should be considered for the estimation of the molecular mass of PNVCL polymers with lower molecular mass, that are not suitable for viscosimetric analysis due to their low viscosity. The reported relations between LCST and M_η are in accordance with the mixing behavior of aqueous PNVCL solutions reported by Meeussen [22] and Kirsh [42]. Accordingly, the small difference between the calculated LCST may be related to the different end groups of the polymers under consideration. The presence of terminations or compounds that increase the hydrophilicity of PNVCL is known to increase the LCST [27,48]. Furthermore, PNVCL-COOH polymers with a 32°C LCST have been frequently used as starting polymers for the preparation of thermoresponsive particles [7–9,19,35,38]. This LCST value has been previously associated to the behavior of PNVCL-COOH polymers with a molecular weight of 1 kDa by means of GPC-SEC measurements in THF [35] and never discussed in subsequent publications [7–9,16–19].

PNVCL_1690 shows similar properties as the LCST is close to 33°C. According to DLS measurements, PNVCL-COOH polymers require a molecular mass higher than 100 kDa in order to exhibit LCST at 32 °C. Similarly, viscosimetric analyses demonstrated that the value should be at least higher than 42 kDa, which is the value that we associated to a LCST of 34 °C. Accordingly, the importance of the molecular mass of PNVCL polymers appeared to be underestimated in studies oriented towards biological applications [7–9,16–19,35]. The molecular mass of the PNVCL has a central role in defining the thermoresponsive behavior of nanoparticles, microgels, gels and other biocompatible devices for drug and gene delivery. Consequently, it is essential to correctly characterize the molecular mass with precise, standardized, and reproducible methods.

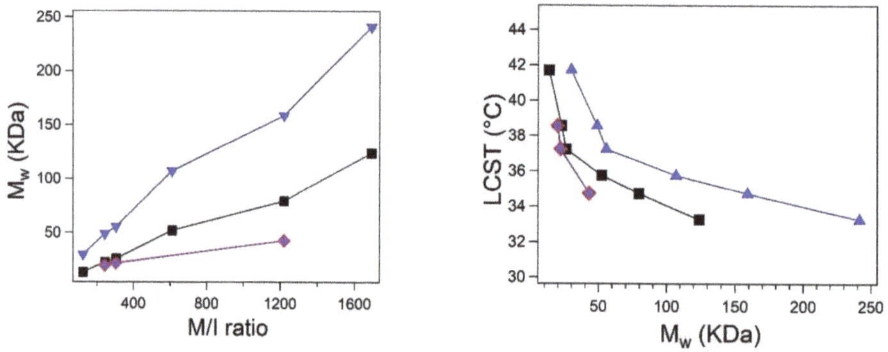

(a) (b)

Figure 5. (**a**) Average molecular weight as a function of M/I ratio. (**b**) Spectroscopic LCST reported as a function of the average molecular weight. The graphics show the comparison between the values calculated using the models reported by Lau (black), Eisele (blue), and viscosimetric masses calculated using Equations (6) (red) and (7) (violet).

3.5. Determination of LCST in Human Plasma

The critical miscibility behavior of PNVCL-COOH was observed in human plasma to evaluate possible cytotoxic effects related to the utilization of the polymer in physiological fluids. Plasma represents 55% of the blood and consists mainly of water, dissolved ions, proteins and gases. Among the main proteins, albumin, fibrinogen and globulins are present, whose functional groups give rise to the intense visible bands in the region between 200 and 600 nm [49]. In both samples analyzed, the peak at 576 nm and the shoulder at 540 nm are related to the presence of oxyhemoglobin [50] due to the hemolysis of residual erythrocytes. A first calibration of the matrix was performed on a pool of plasma realized from the union of samples taken from 12 heathy donors in a temperature range between 25 and 50 °C. The calibration excluded any spectral modification between 25 and 42 °C. Due to the complexity of the matrix, all solutions were analyzed using milliQ water as a reference. The measurements of the LCST of PNVCL_122, PNVCL_244 and PNVCL_305 were provided by the analysis of the region devoid of amino acid signals, between 600 and 800 nm (Figure 6a and Figures S8,S9). Upon heating, the absorbance signal of the spectra was shifted towards higher values due to the conformational change of the polymers. The results show that the human plasma is responsible for a significant lowering of the LCST by about 10 °C, as spectral changes were already observed at 28 °C. This is in line with the results of previous experiments, which have shown that LCST is affected by the presence of ions and proteins. By heating up the solution to 37 °C, the spectra change dramatically, and it was no longer possible to recognize any spectral information. However, the spectral properties of plasma were restored by cooling the system back to 25 °C. Accordingly, the reversibility of PNVCL-COOH transition is maintained within the plasma matrix. LCST values were calculated by fitting the transmittance vs. temperature data with a sigmoidal function (Equation (1)). Transmittance data and fitting functions are reported in Figure 6b.

The comparison between the LCST of PNVCL-COOH in milliQ water and plasma are reported in Table 2. It can be concluded that the polymer is not suitable for direct utilization within the plasma at the concentration of 0.5 wt.%. However, the reversibility of the transition could be an important starting point for future developments. According to the previously reported results, the LCST can be increased by simply decreasing the chain length by modifying the M/I ratio and by diminishing polymer concentration.

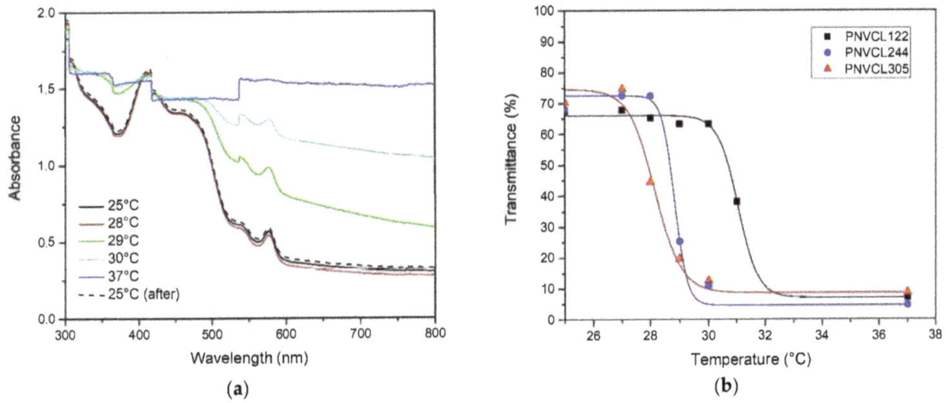

Figure 6. (**a**) Variation of UV-VIS spectrum of a solution of PNVCL_244 in human plasma (0.5 wt.%) with temperature. (**b**) Transmittance of PNVCL_122 (black squares), PNVCL_244 (blue circles) and PNVCL_305 (red triangles) solution in human plasma (0.5 wt.%) calculated at 800 nm reported as a function of temperature. The graph also shows the sigmoidal fitting functions that were used for the calculations of the LCST.

Table 2. Comparison between the LCST of PNVCL-COOH polymers (0.5 wt. %) in milliQ water and in human plasma.

M/I Ratio	LCST (°C)		
	MilliQ Water	PBS (0.1 M, pH = 7)	Human Plasma
122	41.71 ± 0.17	34.83 ± 0.21	31.04 ± 0.05
244	38.56 ± 0.14	31.68 ± 0.17	28.83 ± 0.11
305	37.23 ± 0.63	31.38 ± 0.14	28.15 ± 6.75

4. Conclusions

In this work, the preparation of PNVCL-COOH linear thermoresponsive polymers with a LCST of between 33 and 42 °C was reported. The results contradict the established thesis that PNVCL has a characteristic LCST at 32 °C and exhibits a miscibility behavior similar to PNIPAM. The utilization of different ratios between monomer and initiator (M/I) allowed to lower the LCST by increasing the molecular mass of the polymer. Accordingly, the increase in molecular mass was associated with a decrease in terminal groups content that was observed in NMR spectroscopy and conductivity titration. Molecular mass characterization was approached with different method and a comparison of the results was provided. While GPC-SEC has proven to be unreliable regarding the tendency of the polymer to adsorb on the column, DLS and viscosimetry proved to be simple and effective methods for the estimation of the molecular mass in simple aqueous solution. The behavior of the polymer was described according to its concentration and in the presence of different environments, such as buffers, NaCl solutions and human plasma. The utilization of a sigmoidal model allowed to correlate the equilibrium miscibility temperature (LCST) as the inflection point of the miscibility curve. This allowed to interpret with greater precision the behavior in diluted solutions or in complex matrices. The variability associated with the LCST values in the different solutions demonstrated the importance in reporting the LCST of PNVCL polymers in relation to their concentration and molecular mass. In addition, the study demonstrates the importance of the screening of the behavior of PNVCL polymers in biologically relevant environments (plasma, PBS, physiological solution). The LCST of PNVCL polymers for biological applications should be determined within these solutions in order to prevent cytotoxic effects due to thermo-induced conformational change.

Supplementary Materials: The following are available online at https://www.mdpi.com/article/10.3390/polym13162639/s1, Figure S1: ^{1}H spectrum of PNVCL_122; Figure S2: Carboxylic termination group signals in ^{1}H NMR spectrum of PNVCL_305 dissolved in d_6-DMSO; Figure S3: ^{13}C spectrum of PNVCL_122; Figure S4: HSQC-DEPT spectrum of PNVCL_122; Figure S5: Comparison between NVCL and PNVCL-COOH spectrum (PNVCL_305); Figure S6 Reduced viscosity reported as a function of PNVCL_244 (black squares), PNVCL_305 (red circles) and PNVCL_1220 (blue triangles) concentration. The value of η_{red} were calculated according to the Huggins method by using the calibration lines reported in the figure; Figure S7: Displacement of the correlation curve of PNVCL_1220 solution (0.5 wt.%) associated to LCST transition; Figure S8: Variation of UV-VIS spectrum of a solution of PNVCL_122 in human plasma (0.5 wt.%) with temperature; Figure S9: Variation of UV-VIS spectrum of a solution of PNVCL_305 in human plasma (0.5 wt.%) with temperature; Figure S10: Conductimetric titration of PNVCL_305.

Author Contributions: Funding acquisitions: G.T.; investigation: L.M.; methodology: L.M., F.B., G.E. and I.D.; supervision: G.T.; validation: M.D.B. and F.B; writing—original draft: L.M.; writing—review and editing: M.D.B. and F.B. All authors have read and agreed to the published version of the manuscript.

Funding: This research was funded by the Italian Ministry of Health (Ricerca Corrente).

Institutional Review Board Statement: Not Applicable.

Informed Consent Statement: Not Applicable.

Data Availability Statement: The data presented in this study are available on request from the corresponding author.

Acknowledgments: The authors are grateful to Paolo Macor for the help and the cooperation and to Josie McQuillan for her valuable assistance in revising English language.

Conflicts of Interest: The authors declare no conflict of interest in this publication.

Abbreviations

$[\eta]$	intrinsic viscosity
AIBN	2,2′-Azobis(2-methylpropionitrile)
D_2O	Deuterated water
D_h	Hydrodynamic diameter
DLS	Dynamic light scattering
DMF	N,N′-dimethylformamide
DMSO	Dimethyl sulfoxide
FRP	Free radical polymerization
FT-IR	Fourier-transform infrared
GPC-SEC	Gel Permeation Chromatography-Size exclusion chromatography
HSQC	heteronuclear single quantum coherence
HSQC-DEPT	heteronuclear single quantum coherence-distortionless enhanced polarization transfer
LCST	lower critical solution temperature
M/I	monomer to initiator ratio
M_n	Number average molecular weight
MPA	Mercaptopropionic acid
M_w	weight average molecular weight

MWCO	molecular weight cut-off
M_η	viscosimetric molecular weight
NHS	N-hydroxysuccinimide
NMR	nuclear magnetic resonance
NVCL	N-vinyl caprolactam
PBS	Phosphate-buffered saline
PNIPAM	poly-N-isopropylamide
PNVCL	Poly-N-Vinylcaprolactam
R_g	Gyration radius
THF	Tetrahydrofuran
Tr	Transmittance
UCST	upper critical solution temperature
UV-VIS	UV-Visible
η_{inh}	inherent viscosity
η_{red}	reduced viscosity
η_{sp}	specific viscosity
$\overline{X}_n$	degree of polymerization

References

1. Vihola, H. Studies on Thermosensitive Poly(N-vinylcaprolactam) Based Polymers for Pharmaceutical Applications. Ph.D. Thesis, University of Helsinki, Helsinki, Finland, 23 November 2007.
2. Gandhi, A.; Paul, A.; Sen, S.O.; Sen, K.K. Studies on thermoresponsive polymers: Phase behaviour, drug delivery and biomedical applications. *Asian J. Pharm. Sci.* **2015**, *10*, 99–107. [CrossRef]
3. Mohammed, M.N.; Bin Yusoh, K.; Shariffuddin, J.H.B.H. Poly(N-vinyl caprolactam) thermoresponsive polymer in novel drug delivery systems: A review. *Mater. Express* **2018**, *8*, 21–34. [CrossRef]
4. Cortez-Lemus, N.A.; Licea-Claverie, A. Poly(N-vinylcaprolactam), a comprehensive review on a thermoresponsive polymer becoming popular. *Prog. Polym. Sci.* **2016**, *53*, 1–51. [CrossRef]
5. Tavagnacco, L.; Zaccarelli, E.; Chiessi, E. On the molecular origin of the cooperative coil-to-globule transition of poly(: N-isopropylacrylamide) in water. *Phys. Chem. Chem. Phys.* **2018**, *20*, 9997–10010. [CrossRef]
6. Sund-Levander, M.; Forsberg, C.; Wahren, L.K. Normal oral, rectal, tympanic and axillary body temperature in adult men and women: A systematic literature review. *Scand. J. Caring Sci.* **2002**, *16*, 122–128. [CrossRef]
7. Chauhan, D.S.; Indulekha, S.; Gottipalli, R.; Reddy, B.P.K.; Chikate, T.R.; Gupta, R.; Jahagirdar, D.N.; Prasad, R.; De, A.; Srivastava, R. NIR light-triggered shrinkable thermoresponsive PNVCL nanoshells for cancer theranostics. *RSC Adv.* **2017**, *7*, 44026–44034. [CrossRef]
8. Indulekha, S.; Arunkumar, P.; Bahadur, D.; Srivastava, R. Dual responsive magnetic composite nanogels for thermo-chemotherapy. *Colloids Surfaces B Biointerfaces* **2017**, *155*, 304–313. [CrossRef]
9. Indulekha, S.; Arunkumar, P.; Bahadur, D.; Srivastava, R. Thermoresponsive polymeric gel as an on-demand transdermal drug delivery system for pain management. *Mater. Sci. Eng. C* **2016**, *62*, 113–122. [CrossRef]
10. Kozanoğlu, S.; Özdemir, T.; Usanmaz, A. Polymerization of N-vinylcaprolactam and characterization of poly(N-vinylcaprolactam). *J. Macromol. Sci. Part A Pure Appl. Chem.* **2011**, *48*, 467–477. [CrossRef]
11. Kozanoğlu, S. Polymerization and Characterization of N-Vinylcaprolactam. Master's Thesis, Middle East Technical University, Ankara, Turkey, September 2008.
12. Kirsh, Y.E.; Yanul, N.A.; Kalninsh, K.K. Structural transformations and water associate interactions in poly-N-vinylcaprolactam-water system. *Eur. Polym. J.* **1999**, *25*, 5697–5704. [CrossRef]
13. Lozinsky, V.I.; Simenel, I.A.; Kurskaya, E.A.; Kulakova, V.K.; Galaev, I.Y.; Mattiasson, B.; Grinberg, V.Y.; Grinberg, N.V.; Khokhlov, A.R. Synthesis of N-vinylcaprolactam polymers in water-containing media. *Polymer* **2000**, *41*, 6507–6518. [CrossRef]
14. Cheng, S.C.; Feng, W.; Pashikin, I.I.; Yuan, L.H.; Deng, H.C.; Zhou, Y. Radiation polymerization of thermo-sensitive poly (N-vinylcaprolactam). *Radiat. Phys. Chem.* **2002**, *63*, 517–519. [CrossRef]
15. Panja, S.; Dey, G.; Bharti, R.; Kumari, K.; Maiti, T.K.; Mandal, M.; Chattopadhyay, S. Tailor-Made Temperature-Sensitive Micelle for Targeted and On-Demand Release of Anticancer Drugs. *ACS Appl. Mater. Interfaces* **2016**, *8*, 12063–12074. [CrossRef] [PubMed]
16. Rejinold, N.S.; Thomas, R.G.; Muthiah, M.; Lee, H.J.; Jeong, Y.Y.; Park, I.K.; Jayakumar, R. Breast tumor targetable Fe_3O_4 embedded thermo-responsive nanoparticles for radiofrequency assisted drug delivery. *J. Biomed. Nanotechnol.* **2016**, *12*, 43–55. [CrossRef] [PubMed]
17. Sanoj Rejinold, N.; Thomas, R.G.; Muthiah, M.; Chennazhi, K.P.; Manzoor, K.; Park, I.K.; Jeong, Y.Y.; Jayakumar, R. Anti-cancer, pharmacokinetics and tumor localization studies of pH-, RF- and thermo-responsive nanoparticles. *Int. J. Biol. Macromol.* **2015**, *74*, 249–262. [CrossRef]
18. Rejinold, N.S.; Thomas, R.G.; Muthiah, M.; Chennazhi, K.P.; Park, I.K.; Jeong, Y.Y.; Manzoor, K.; Jayakumar, R. Radio frequency triggered curcumin delivery from thermo and pH responsive nanoparticles containing gold nanoparticles and its in vivo localization studies in an orthotopic breast tumor model. *RSC Adv.* **2014**, *4*, 39408–39427. [CrossRef]

19. Rejinold, N.S.; Chennazhi, K.P.; Nair, S.V.; Tamura, H.; Jayakumar, R. Biodegradable and thermo-sensitive chitosan-g-poly(N-vinylcaprolactam) nanoparticles as a 5-fluorouracil carrier. *Carbohydr. Polym.* **2011**, *83*, 776–786. [CrossRef]
20. Sahebi, H.; Pourmortazavi, S.M.; Zandavar, H.; Mirsadeghi, S. Chitosan grafted onto Fe_3O_4@poly(: N -vinylcaprolactam) as a new sorbent for detecting Imatinib mesylate in biosamples using UPLC-MS/MS. *Analyst* **2019**, *144*, 7336–7350. [CrossRef]
21. Banihashem, S.; Nezhati, M.N.; Panahia, H.A. Synthesis of chitosan-grafted-poly(N-vinylcaprolactam) coated on the thiolated gold nanoparticles surface for controlled release of cisplatin. *Carbohydr. Polym.* **2020**, *227*, 115333. [CrossRef]
22. Meeussen, F.; Nies, E.; Berghmans, H.; Verbrugghe, S.; Goethals, E.; Du Prez, F. Phase behaviour of poly(N-vinyl caprolactam) in water. *Polymer* **2000**, *41*, 8597–8602. [CrossRef]
23. Solomon, O.F.; Corciovei, M.; Ciută, I.; Boghină, C. Properties of solutions of poly-N-vinylcaprolactam. *J. Appl. Polym. Sci.* **1968**, *12*, 1835–1842. [CrossRef]
24. Tager, A.A.; Safronov, A.P.; Sharina, S.V.; Galaev, I.Y. Thermodynamic study of poly(N-vinyl caprolactam) hydration at temperatures close to lower critical solution temperature. *Colloid Polym. Sci.* **1993**, *271*, 868–872. [CrossRef]
25. Laukkanen, A.; Valtola, L.; Winnik, F.M.; Tenhu, H. Formation of colloidally stable phase separated poly(N-vinylcaprolactam) in water: A study by dynamic light scattering, microcalorimetry, and pressure perturbation calorimetry. *Macromolecules* **2004**, *37*, 2268–2274. [CrossRef]
26. Medeiros, S.F.; Barboza, J.C.S.; Ré, M.I.; Giudici, R.; Santos, A.M. Solution polymerization of N-vinylcaprolactam in 1,4-dioxane. Kinetic dependence on temperature, monomer, and initiator concentrations. *J. Appl. Polym. Sci.* **2010**, *118*, 229–240. [CrossRef]
27. Shao, L.; Hu, M.; Chen, L.; Xu, L.; Bi, Y. RAFT polymerization of N-vinylcaprolactam and effects of the end group on the thermal response of poly(N-vinylcaprolactam). *React. Funct. Polym.* **2012**, *72*, 407–413. [CrossRef]
28. Vihola, H.; Laukkanen, A.; Valtola, L.; Tenhu, H.; Hirvonen, J. Cytotoxicity of thermosensitive polymers poly(N-isopropylacrylamide), poly(N-vinylcaprolactam) and amphiphilically modified poly(N-vinylcaprolactam). *Biomaterials* **2005**, *26*, 3055–3064. [CrossRef] [PubMed]
29. Enomoto, Y.; Kamitakahara, H.; Takano, T.; Nakatsubo, F. Synthesis of diblock copolymers with cellulose derivatives. 3. Cellulose derivatives carrying a single pyrene group at the reducing-end and fluorescent studies of their self-assembly systems in aqueous NaOH solutions. *Cellulose* **2006**, *13*, 437–448. [CrossRef]
30. Shah, S.; Pal, A.; Gude, R.; Devi, S. Synthesis and characterization of thermo-responsive copolymeric nanoparticles of poly(methyl methacrylate-co-N-vinylcaprolactam). *Eur. Polym. J.* **2010**, *46*, 958–967. [CrossRef]
31. Ainara, I.; Jacqueline, F. N-vinylcaprolactam-based microgels for biomedical applications. *J. Polym. Sci. Part A Polym. Chem.* **2010**, *48*, 1173–1181. [CrossRef]
32. Shakya, A.K.; Holmdahl, R.; Nandakumar, K.S.; Kumar, A. Polymeric cryogels are biocompatible, and their biodegradation is independent of oxidative radicals. *J. Biomed. Mater. Res. Part A* **2014**, *102*, 3409–3418. [CrossRef]
33. BASF Luviskol ®Plus Technical Information. Available online: http://https://docplayer.net/207667766-Luviskol-plus-technical-information-nonionic-film-forming-agent-for-hair-setting-products-registered-trademark-of-basf-in-many-countries.html (accessed on 4 August 2011).
34. Madhusudana Rao, K.; Krishna Rao, K.S.V.; Sudhakar, P.; Chowdoji Rao, K.; Subha, M.C.S. Synthesis and characterization of biodegradable poly (vinyl caprolactam) grafted on to sodium alginate and its microgels for controlled release studies of an anticancer drug. *J. Appl. Pharm. Sci.* **2013**, *3*, 61–69. [CrossRef]
35. Prabaharan, M.; Grailer, J.J.; Steeber, D.A.; Gong, S. Stimuli-responsive chitosan-graft-Poly(N-vinylcaprolactam) as a promising material for controlled hydrophobic drug delivery. *Macromol. Biosci.* **2008**, *8*, 843–851. [CrossRef]
36. Feng, S.; Zheng, E.; Liu, H.; Wang, C.; Liu, F. Synthesis and characterization of poly (N-vinyl caprolactam) and its graft copolymers of dextran and dextrose. *Polymer* **2008**, *49*, 1081–1082.
37. Shi, H.Y.; Zhang, L.M. Phase-transition and aggregation characteristics of a thermoresponsive dextran derivative in aqueous solutions. *Carbohydr. Res.* **2006**, *341*, 2414–2419. [CrossRef]
38. Sanoj Rejinold, N.; Muthunarayanan, M.; Divyarani, V.V.; Sreerekha, P.R.; Chennazhi, K.P.; Nair, S.V.; Tamura, H.; Jayakumar, R. Curcumin-loaded biocompatible thermoresponsive polymeric nanoparticles for cancer drug delivery. *J. Colloid Interface Sci.* **2011**, *360*, 39–51. [CrossRef] [PubMed]
39. Stetefeld, J.; McKenna, S.A.; Patel, T.R. Dynamic light scattering: A practical guide and applications in biomedical sciences. *Biophys. Rev.* **2016**, *8*, 409–427. [CrossRef]
40. Lau, A.C.W.; Wu, C. Thermally Sensitive and Biocompatible Poly(N-vinylcaprolactam): Synthesis and Characterization of High Molar Mass Linear Chains. *Macromolecules* **1999**, *32*, 581–584. [CrossRef]
41. Eisele, M.; Burchard, W. Hydrophobic water-soluble polymers, 1. Dilute solution properties of poly(1-vinyl-2-piperidone) and poly(N-vinylcaprolactam). *Die Makromol. Chem.* **1990**, *191*, 169–184. [CrossRef]
42. Kirsh, Y.E. *Water Soluble Poly-N-Vinylamides: Synthesis and Physicochemical Properties*; John Wiley & Sons: Chicester, UK, 1998.
43. Liu, R.; Fraylich, M.; Saunders, B.R. Thermoresponsive copolymers: From fundamental studies to applications. *Colloid Polym. Sci.* **2009**, *287*, 627–643. [CrossRef]
44. Aseyev, V.; Hietala, S.; Laukkanen, A.; Nuopponen, M.; Confortini, O.; Du Prez, F.E.; Tenhu, H. Mesoglobules of thermoresponsive polymers in dilute aqueous solutions above the LCST. *Polymer* **2005**, *46*, 7118–7131. [CrossRef]

45. Mikheeva, L.M.; Grinberg, N.V.; Mashkevich, A.Y.; Grinberg, V.Y.; Thanh, L.T.M.; Makhaeva, E.E.; Khokhlov, A.R. Microcalorimetric study of thermal cooperative transitions in poly(N-vinylcaprolactam) hydrogels. *Macromolecules* **1997**, *30*, 2693–2699. [CrossRef]
46. Boyko, V.B. N-Vinylcaprolactam based Bulk and Microgels: Synthesis, Structural Formation and Characterization by Dynamic Light Scattering. Ph.D. Thesis, Technische Universität Dresden, Dresden, Germany, 29 October 2004.
47. Gooch, J.W. Kraemer Equation. In *Encyclopedic Dictionary of Polymers*; Gooch, J.W., Ed.; Springer: New York, NY, USA, 2011; ISBN 978-1-4419-6247-8.
48. Serra, A.C.; Góis, J.R.; Coelho, J.F.J.; Popov, A.V.; Costa, J.R.C. Synthesis of well-defined alkyne terminated poly(N-vinyl caprolactam) with stringent control over the LCST by RAFT. *RSC Adv.* **2016**, *6*, 16996–17007. [CrossRef]
49. Morgner, F.; Stufler, S.; Geißler, D.; Medintz, I.L.; Algar, W.R.; Susumu, K.; Stewart, M.H.; Blanco-Canosa, J.B.; Dawson, P.E.; Hildebrandt, N. Terbium to quantum dot FRET bioconjugates for clinical diagnostics: Influence of human plasma on optical and assembly properties. *Sensors* **2011**, *11*, 9667–9684. [CrossRef] [PubMed]
50. Zijlstra, W.G.; Buursma, A. Spectrophotometry of hemoglobin: Absorption spectra of bovine oxyhemoglobin, deoxyhemoglobin, carboxyhemoglobin, and methemoglobin. *Comp. Biochem. Physiol. B Biochem. Mol. Biol.* **1997**, *118*, 743–749. [CrossRef]

polymers

MDPI

Article

Temperature/Reduction Dual Response Nanogel Is Formed by In Situ Stereocomplexation of Poly (Lactic Acid)

Wenli Gao †, Zhidan Wang †, Fei Song, Yu Fu, Qingrong Wu and Shouxin Liu *

Key Laboratory of Applied Surface and Colloid Chemistry, Ministry of Education, School of Chemistry and Chemical Engineering, Shaanxi Normal University, Xi'an 710119, China; 17853465526@163.com (W.G.); 18189608643@163.com (Z.W.); 17563713257@163.com (F.S.); fy2247127430@163.com (Y.F.); wuqingrong123456@snnu.edu.cn (Q.W.)
* Correspondence: liushx@snnu.edu.cn; Tel.: +86-29-81530781
† These authors contributed equally to this work.

Abstract: A novel type of dual responsive nanogels was synthesized by physical crosslinking of polylactic acid stereocomplexation: temperature and reduction dual stimulation responsive gels were formed in situ by mixing equal amounts of PLA (Poly (Lactic Acid)) enantiomeric graft copolymer micellar solution; the properties of double stimulation response make it more targeted in the field of drug release. The structural composition of the gels was studied by proton nuclear magnetic resonance (^{1}H NMR) and Fourier transform infrared spectroscopy (FT-IR). Using transmission electron microscope (TEM) and dynamic light scattering (DLS) instruments, the differences in morphology and particle size were analyzed (indicating that nanogels have dual stimulus responses of temperature sensitivity and reduction). The Wide-Angle X-ray diffractionr (WAXD) was used to prove the stereocomplexation of PLA in the gels, the mechanical properties and gelation process of the gels were studied by rheology test. The physically cross-linked gel network generated by the self-recombination of micelles and then stereo-complexation has a more stable structure. The results show that the micelle properties, swelling properties and rheological properties of nanogels can be changed by adjusting the degree of polymerization of polylactic acid. In addition, it provides a safe and practical new method for preparing stable temperature/reduction response physical cross-linked gel.

Keywords: stereocomplexation; polylactic acid; temperature/reduction; self-recombination

Citation: Gao, W.; Wang, Z.; Song, F.; Fu, Y.; Wu, Q.; Liu, S. Temperature/Reduction Dual Response Nanogel Is Formed by In Situ Stereocomplexation of Poly (Lactic Acid). *Polymers* **2021**, *13*, 3492. https://doi.org/10.3390/polym13203492

Academic Editor: Mattia Sponchioni

Received: 30 August 2021
Accepted: 7 October 2021
Published: 12 October 2021

1. Introduction

In the field of drug delivery, there are many types of polymer-based carriers, such as nanoparticles [1], vesicles [2,3], polymer micelles [4–7], hydrogels [8–10], etc. Among these carriers, polymeric micelles are one of the most common. However, the instability of the micelle structure causes it to dissociate at a concentration lower than the critical micelle concentration, which will lead to the loss of drugs during the blood circulation and limit its further clinical application [11,12]. Scientists prepared polymers for improving the physical stability of polymer micelles through various cross-linking methods. Nanohydrogel is one of them. As one of the classifications of hydrogels, nanohydrogels are defined as a nano-scale polymer with 3D network nanoparticles, this network is formed by cross-linking polymer chains swelled in a good solvent [13,14]. Compared with commonly used nanocarriers (such as micelles), nanohydrogels contain a large amount of water in a swollen state, and stably load drugs, reducing the loss of drugs during blood circulation.

Generally speaking, hydrogels can be divided into two types according to cross-linking methods, one is a chemical crosslinking hydrogel [15], the other is a physical crosslinking hydrogel [16]. The chemical crosslinking hydrogel is usually crosslinked by chemical crosslinking agent and the force of cross-linking is the covalent bond generated by the functional groups between the polymer chains, so it has better mechanical strength and stability than physically cross-linked hydrogels [17,18]. However, the addition of a chemical crosslinking agent may lead to biodegradable hydrogels become a huge problem [19]. Physical cross-linking hydrogels mainly use intermolecular hydrogen bonding as the main cross-linking force, avoiding the formation of undegradable bonds, and has better biocompatibility, degradability and application prospect than chemical cross-linked hydrogels. The polylactic acid stereocomplex has many advantages. Due to its extraordinary biological, thermal, and mechanical properties, it shows great potential as a biological material. The group of Sytze J. Buwalda conducted research to prepare PEG-PLA star-shaped block copolymer hydrogels through physical crosslinking combined with photopolymerization. The gel is degraded by the hydrolysis of the ester groups in the PLA chain, resulting in the loss of physical and chemical crosslinks. This study shows that the injectable, photocrosslinked PEG-PLA star block copolymer hydrogel is a promising material for biomedical applications such as long-term controlled drug delivery [20].

The broad application prospects of stimulus responsive hydrogels have led to various single-response [21], double-response and multi-response hydrogels being studied and applied [22–24]. However, there are few reports on reduction/temperature dual response hydrogels. In recent years, biodegradable hydrogels with degradable bonds have become a hot topic [25,26]. Scientists have introduced degradable bonds such as hydrazine [27], enzymes [28,29], disulfide bonds [30], etc. into polymer molecules to achieve the synthesis of degradable hydrogels. Among them, disulfide bonds are favored by scientists because they can be cleaved and reformed under relatively simple and mild conditions [31,32]. In this article, we first synthesize an initiator 2-((2-Hydroxyethyl) disulfanyl) ethyl methacrylate (HSEMA) with disulfide bond through the monoesterification reaction of 2, 2′-dithiodiethanol and methacryloyl chloride, and then used it for ring-opening polymerization of poly(lactide) to form macromonomer HSEMA-PLLA and HSEMA-PDLA, the temperature-sensitive monomers 2-(2-methoxyethoxy) ethyl methacrylate (MEO_2MA) and oligo (ethylene glycol) methacrylate (OEGMA) undergo free radical polymerization with macromonomer HSEMA-PLLA or HSEMA-PDLA under the condition of AIBN as the initiator to form temperature/reduction dual responsive micelles. Mixing equal amounts of PLA enantiomeric micellar solution in situ generates a temperature reduction dual stimulus responsive gel. The physically cross-linked gel network generated by the self-recombination of micelles and then stereo-complexation provides a safe and practical new method for the preparation of temperature/reduction dual responsive physical cross-linking gel.

2. Materials and Methods

2.1. Materials

L-lactide, D-lactide (99.0%) and methacryloyl chloride (95%, M_n = 104.53 g·mol^{-1}) were purchased from Macleans (Shanghai, China). Oligo (ethylene oxide) methacrylate (OEGMA, 95%, M_n = 475 g·mol^{-1}) and 2-(2-methoxy ethoxy) ethyl methacrylate (MEO_2MA, 95%, M_n = 188.22 g·mol^{-1}) were obtained from TCI (Shanghai Development Co., Ltd., Shanghai, China). 2,2′-dithiodiethanol (90%, M_n = 154.25 g·mol^{-1}) was purchased from Alfa Aesar (Shanghai, China). 1,4-dithiothreitol (DTT, 99.0%, M_n = 154.25 g·mol^{-1}) was purchased from Rhawn (Shanghai, China), Tin(II) 2-ethylhexanoate (99%, M_n = 405.12 g·mol^{-1})was purchased from J&k (Beijing, China), 2,2′-Azoisobutyronitrile(AIBN) (Macleans, Shanghai, China) was purified by recrystallization from methanol. The solvents used are all dried and double distilled water is used in aqueous solutions.

2.2. *Synthesis of 2-((2-Hydroxyethyl) Disulfanyl) Ethyl Methacrylate (HSEMA)*

2-((2-Hydroxyethyl) disulfanyl) ethyl methacrylate (HSEMA) was synthesized by the monoesterification reaction of 2,2′-dithiodiethanol and methacryloylchloride [33,34]. The synthesis route was as follows: 2,2′-dithiodiethanol (4 mmol, 0.542 mL) and triethylamine (1 mL) were completely dissolved in 20 mL dried THF. Then methacryloyl chloride (4 mmol, 0.407 mL) was added drop-wise for 20 min in an ice bath at 0 °C, the reactants were stirred at room temperature for 24 h. The produced white solids, i.e., triethylamine salt, were removed using the method of filtration. Subsequently, the solvent was removed by rotary evaporation, and the product was purified by column chromatography.

2.3. Synthesis of Macromonomer HSEMA-PDLA

Synthesis of macromonomer HSEMA-PLLA/HSEMA-PDLA used the method of ring-opening polymerization with HSEMA as the initiator and lactide as the raw material. The synthesis route was as follows: D-lactide (6.9 mmol, 1 g), initiator HSEMA (0.53 mmol) and dry toluene (15 mL) were added to a dry and clean three-necked flask (Scheme 1). After the mixture were fully dissolved, the catalyst Tin (II) 2-ethylhexanoate ($Sn(Oct)_2$) (55 μL) was added. The reaction system was refluxed at 120 °C for 24 h under a nitrogen atmosphere, and then the crude products were precipitated and purified twice in cold anhydrous ether. Finally, the products were stored in a vacuum drying cabinet at room temperature. The synthesis of HSEMA-PLLA was similar to HSEMA-PDLA.

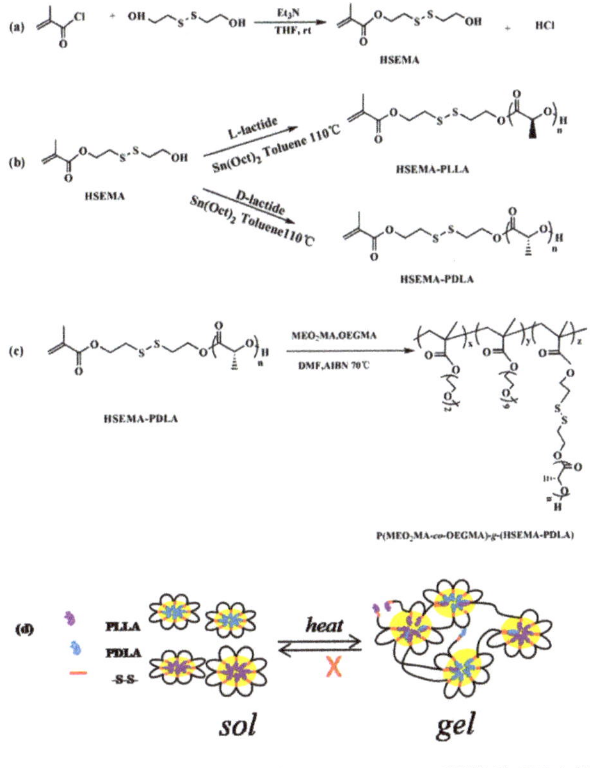

Scheme 1. Initiator HSEMA (**a**), macromonomer HSEMA-PLLA/PDLA (**b**), graft copolymer P(MEO_2MA-*co*-OEGMA)-*g*-(HSEMA-PDLA) (**c**), Schematic diagram of the synthesis and hydrogel complex mechanism (**d**).

2.4. Synthesis of Graft Copolymer P(MEO_2MA-co-OEGMA)-g-(HSEMA-PLLA)

The graft copolymer P (MEO_2MA-co-OEGMA)-g-(HSEMA-PLLA) was synthesized by free radical polymerization with 2-(2-methoxy ethoxy) ethyl methacrylate (MEO_2MA) and oligo (ethylene oxide) methacrylate (OEGMA) as temperature-sensitive monomers, HSEMA-PLLA as macromonomers and 2,2′-Azoisobutyronitrile (AIBN) as the initiator. The synthesis route was as follows: the macromonomer HSEMA-PLLA and DMF (3 mL) were added to a dry and clean Shrek tube. After HSEMA-PLLA were fully dissolved, MEO_2MA, OEGMA and initiator AIBN were added to the mixed solution. The reaction system was reacted at 70 °C for 6 h under nitrogen atmosphere. The crude products were precipitated and purified twice in ice anhydrous ether, then dialyzed at room temperature for 3 days. Finally, the products were freeze-dried and stored in a vacuum drying cabinet at room temperature.

2.5. Synthesis of Temperature/Reduction Nanogels

Copolymers of equal mass were respectively dissolved in water to obtain a uniform single enantiomeric graft copolymer solution. Then the enantiomeric graft copolymer solutions were mixed in equal volumes at 37 °C to obtain a stereocomplexed gel [35,36]. The complex mechanism was shown in Scheme 1.

2.6. Characterization of the Samples

2.6.1. Nuclear Magnetic Resonance Spectroscopy (^{1}H-NMR)

The structure of the compound was characterized using a Bruker-300MHz spectrometer (300 MHz Avance, Bruker Corporation, Karlsruhe, Germany) with $CDCl_3$ as a solvent and tetramethylsilane (TMS) as an internal standard at room temperature.

2.6.2. Fourier Transform Infrared Spectroscopy (FT-IR)

The functional group of the compound was characterized by Fourier transform infrared spectroscopy (FTIR) (Tensor 27, Bruker Corporation, Karlsruhe, Germany). Before the measurement, the sample was uniformly mixed with dry KBr, dried in a vacuum drying oven for 24 h, and finally pressed at room temperature.

2.6.3. The Wide-Angle X-ray Diffraction (WAXD)

The wide-angle X-ray diffraction (WAXD) analysis was performed on Ni-filtered Cu Kα (λ = 0.154 nm) at 25 °C on the D8 Advance diffractometer from Bruker, Karlsruhe, Germany, with a 2θ scanning rate of 2°/min and a scanning range of 5–40°.

2.6.4. Low Critical Solution Temperature (LCST)

The LCST of the gels were measured by UV-V is spectroscopy (TU-1901, Beijing Purkinje General Instrument Corporation, Beijing, China). A series of nanogel solutions (2.00 mg/mL) was arranged in a transparent sample vial. The transmittance of the nanogels was measured using the TU-1901 at a wavelength of 660 nm, photometric mode, and a temperature range of 20–50 °C.

2.6.5. The Scanning Electron Microscope (SEM) Analysis

The gels samples were immersed in the distilled water at 25 °C till swell-equilibrium. The swollen gels were quickly frozen in liquid nitrogen and then freeze-dried in a freeze dryer. Microscopic morphology of the gel was performed on a desktop scanning electron microscope (SEM, TM 3030 Hitachi, Tokyo, Japan) in low vacuum mode. Before the observation, in order to improve the conductivity of the sample, it was subjected to gold spray treatment for 80 s.

2.6.6. Dynamic Light Scattering (DLS) Analysis

Laser particle size meter (BI-90Plus, Brookhaven, New York, United States) was used to measure the particle size of nanogels under different conditions: 1 mg/mL of solution was prepared and filtered by 0.45 μm water filter. The particle size was measured at room temperature and variable temperature, respectively.

2.6.7. Swelling Kinetics of the Gel

The preparation method of the dried gel used for the swelling performance test is: 0.5 g enantiomeric copolymer was put in a sample bottle respectively, add THF right amount to dissolve completely. In an ice bath, the THF solution of the copolymer was dripped drop by drop into a round-bottomed flask containing 5 mL of secondary water.

Ultrasound at 4 °C until the solution reaches equilibrium (about 30 min), distilled THF in the solution under reduced pressure at room temperature to obtain a uniform enantiomeric graft copolymer micelle solution with a mass fraction of 10 *w*/*v*%, after mixing the same amount of enantiomeric copolymer solution and stirring for 40 min, ultrasonic for 30 min to complete the complexation, heat up in the temperature range of 25–70 °C and volatilize the solvent and obtain a completely dry PLA-based gels.

The swelling kinetics and de-swelling kinetics of the gel were explained by the weighing method. The specific method of the swelling kinetics is as follows: fully dried gels were soaked in the secondary water at 25 °C, taken out, the mass was weighed three times after a certain period of time, and the water on the surface of the gel was absorbed by filter paper. Taking the average of the data, its swelling ratio was calculated according to the following formula.

$$[\text{Swelling ratio}] = (W_t - W_d)/W_d \tag{1}$$

where W_d is the mass of dry gel (g), W_t is the mass of the gel taken out at a fixed time point during the swelling process (g).

2.6.8. Transmission Electron Microscopy (TEM) Analysis

The morphology of the gel was observed by a Field Emission Transmission Electron (JEOL JEM-2100, Tokey, Japan) with accelerating voltage 200 kV. The samples were prepared by placing a drop of 1.00 $g{\cdot}mol^{-1}$ copolymer solution on a carbon-coated copper grid (200 mesh) and then drying in the vacuum at 25 °C.

2.6.9. Determination of Disulfide Bond Reduction of Sulfhydryl in Nanohydrogel

We use the Ellman reagent to characterize the redox responsiveness of the hydrogel. Preparation of Ellman reagent is made using a buffer solution with pH = 8 as the solvent for the Ellman reagent which was specifically prepared by weighing 1.56 g of sodium phosphate and 0.0372 g of EDTA and dissolving the volume to 100 mL, and then using sodium carbonate to adjust the pH to 8. Dissolve 4 mg of 5,5′-dithiobis (2-nitrobenzoic acid) in 1 mL pH = 8 buffer solution, store in the dark, and prepare it for immediate use; the Ellman reagent was successfully prepared. Take 4 mL of the prepared 1 mg/mL nanohydrogel solution, add the reducing agent tris(2-carboxyethyl)phosphine (TCEP) hydrochloride (1 mL, 0.02 mol/L) and Ellman's reagent (0.2 mL), and let stand for 20 min [37]. At the same time, use the same method to treat without reducing as a blank comparison, the solution to be tested needs to be diluted 10 times to ensure that the concentration of sulfhydryl groups is within the measurement range. Use anultraviolet-visible spectrophotometer to measure within the measurement range of 200–500 nm.

3. Results and Discussion

The graft copolymer P (MEO2MA-co-OEGMA)-g-(HSEMA-PDLA) was synthesized by radical polymerization of 2,2-Azoisobutyronitrile (AIBN) as initiator and macromonomer HSEMA-PDLA, 2-(2-methoxy ethoxy) ethyl methacrylate (MEO2MA), oligo (ethylene oxide) methacrylate (OEGMA) (relevant data as shown in Table 1), then the same amount of enantiomeric graft copolymer micellar solution were mixed in situ to form nanogel. The amount of temperature-sensitive monomer and macromonomer used in the experiment is the same. The variable set is the degree of polymerization of the macromonomer, i.e., the amount of lactide added when synthesizing the macromonomer is different (relevant data as shown in Table 1).

Table 1. The adding data of each reaction material.

Sample	[HSEMA]:[PLLA] (n:n)	M_n [a] (Theory)	M_n [b] (^{1}H NMR)	[M]:[O] [c] (n:n)	[P(MOL)]:[P(MOD)] (n:n)
gel1	1:13	1158	1230	95:5	1:1
gel2	1:16	1374	1518	95:5	1:1
gel3	1:20	1662	1878	95:5	1:1
gel4	1:25	2022	2310	95:5	1:1

a and b are the theoretical molecular weight and ^{1}H NMR molecular weight of the macromonomer, c is molar ratio of MEO_2MA and OEGMA.

3.1. Structural Characterization of the Initiator, Macromolecular Monomers, and Branching Copolymers

The initiator 2-((2-Hydroxyethyl) disulfanyl) ethyl methacrylate (HSEMA) was mainly synthesized by the monoesterification reaction of 2,2-dithiodiethanol and methacryloyl chloride, and Et_3N acts as an acid binding agent to promote the reaction to move forward. The obtained crude product was purified by column chromatography (silica gel, eluent: ethyl acetate/petroleum ether = 1.5:1). Figure 1 shows the infrared spectrum of the initiator HSEMA. It can be seen from the figure that the peak at 3432 cm^{-1} is -OH stretching vibration, and 2944–2874 cm^{-1} is –C-H stretching vibration of -CH_2- and -CH_3. The peak at 1707 cm^{-1} is -C=O stretching vibration, at 1633 cm^{-1} is -C=C- stretching vibration, 1451–1385 cm^{-1} is -CH_3 bending vibration, and 1155 cm^{-1} is stretching vibration of –C-O-C-. Figure 2a is the proton nuclear magnetic resonance spectrum of HSEMA, as shown in the figure, ^{1}H-NMR (300 MHz, $CDCl_3$) δ ppm: 6.11, 5.58 (s, 2H, CH_2=C(CH_3)-), 4.40 (t, 2H, -OCH_2CH_2-S-S-), 3.87 (t, 2H, -CH_2OH), 2.95, 2.87 (t, 4H, -CH_2-S-S-CH_2-), 1.93 (s, 3H, CH_2=C(CH_3)-. Combining the corresponding data of the above, the synthesis of HSEMA is successful.

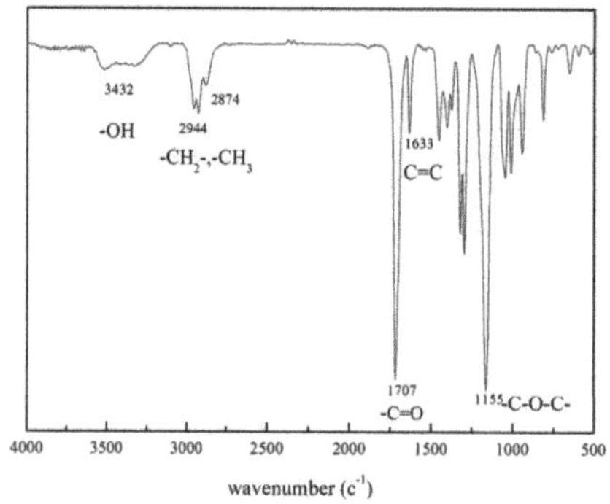

Figure 1. Fourier transform infra-red (FTIR) spectra of the initiator HSEMA.

The synthesis of macromonomer was achieved by ring-opening polymerization of lactide with HSEMA as initiator and Tin (II) 2-ethylhexanoate ($Sn(Oct)_2$) as catalyst. Figure 2b is the proton nuclear magnetic resonance spectrum of the macromonomer HSEMA-PDLA. ^{1}H NMR (300 MHz, $CDCl_3$) the corresponding chemical shift is δ ppm: 6.06, 5.52 (s, 2H, CH_2=C(CH_3)-), 5.13–5.06 (m, 1H, -(C=O)CH(CH_3)O-), 4.30–4.35 (m, 4H,-OCH_2CH_2SS-), 2.91–2.82 (m, 4H, -S-S-CH_2CH_2O-), 1.88 (s, 3H, CH_2=C(CH_3)-), 1.52 (d, 3H, -(C=O)CH(CH_3)O-). Figure 3b is the FT-IR spectrum of the macromonomer HSEMA-PDLA. As shown in the figure, the peaks at 3631 cm^{-1} and 3026–2871 cm^{-1} are the stretching vibrations of -OH and -CH_2-CH_3 respectively. The peaks at 1758 cm^{-1} and 1187 cm^{-1} are the stretching vibration of -C=O and –C-O-C- respectively, and the peak at 1450 cm^{-1} is the bending vibration of -CH_2, -CH_3. Since the disulfide bond is in the FT-IR fingerprint region. Therefore, combining the above data, the macromonomer HSEMA-PDLA was successfully synthesized, and the relevant data of HSEMA-PLLA are basically the same as HSEMA-PDLA, and no explanation is given here.

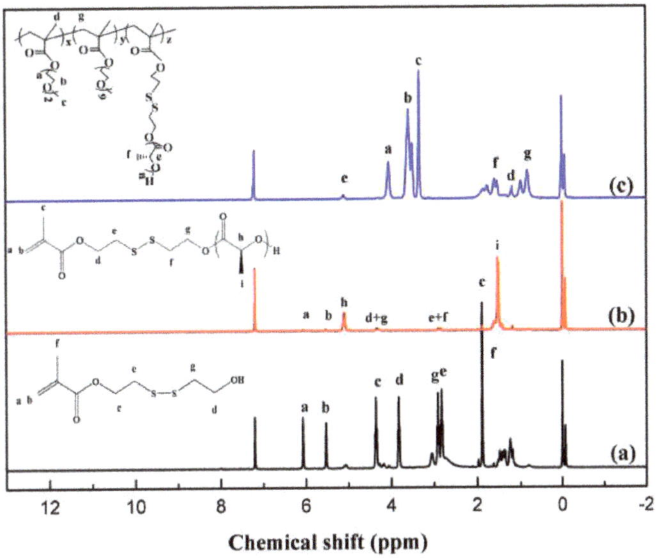

Figure 2. ^{1}H-NMR spectra of (**a**) initiator HSEMA; (**b**) macromonomer HSEMA-PDLA; (**c**) graft copolymer P(MEO$_2$MA-*co*-OEGMA)-*g*-(HSEMA-PDLA).

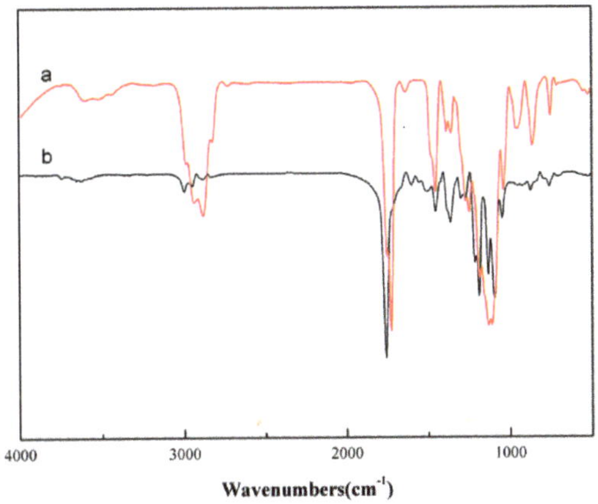

Figure 3. FTIR spectra of (**a**) gel4; (**b**) macromonomer HSEMA-PDLA.

Figure 2c is the 1H-NMR spectrum of the graft copolymers (1H-NMR (300 MHz, $CDCl_3$)), the specific chemical shift is: 5.12–5.06 (m, 1H, -(C=O)CH(CH_3)O-), 4.04 (s, 2H, -O(CH_2CH_2O)2CH_3), 3.5 (d, 2H, -O(CH_2CH_2O)$_2$$CH_3$), 3.33 (s, 3H, CH_2=C(CH_3)-),1.52–1.49 (d, 3H, -(C=O)CH(CH_3)O-), 1.19 (s,3H,-CH_2C(CH_3)CH_2-), 0.83 (s,2H, -CH_2C(CH_3)CH_2-). Figure 3a is the FT-IR spectrum of the gels. It can be seen from the figure that the C=O stretching vibration peak of the gels at 1729 cm^{-1} shifts to a lower wavelength compared to the C=O (1758 cm^{-1}) of the macromonomer. This is due to the complex hydrogen bonding between the PLA groups, which makes the C=O shift, it also shows that the stereocomplex gels were successfully synthesized.

3.2. Analysis of Complex Structure

We used a wide-angle X-ray diffraction (WAXD) to confirm the stereocomplexation of the gels in a macroscopic state. Figure 4a is the WAXD pattern of the macromonomer (red line) and gel4 (black line). As shown in the figure, the macromonomerHSEMA-PLLA$_{25}$ has the isomorphous peaks at 2θ = 17°, 19° and gel4 has a relatively obvious stereocomplex peak at 2θ = 12°. According to the relevant literature, the synthesis of stereocomplex gel is successful [38]. Figure 4b is the WXRD pattern of the gels with different PLA polymerization degrees. It can be seen from the figure that the stereocomplexed gels with different PLA polymerization degree have different degrees of peaks at 2θ = 12°, 21° and 24°. where 2θ = 12°, 21° is more obvious and the diffraction peaks are obviously enhanced with the increase of PLA polymerization degree, this is because the increase of PLA polymerization degree increases the content of stereocomplex and the intensity of complex peak increases.

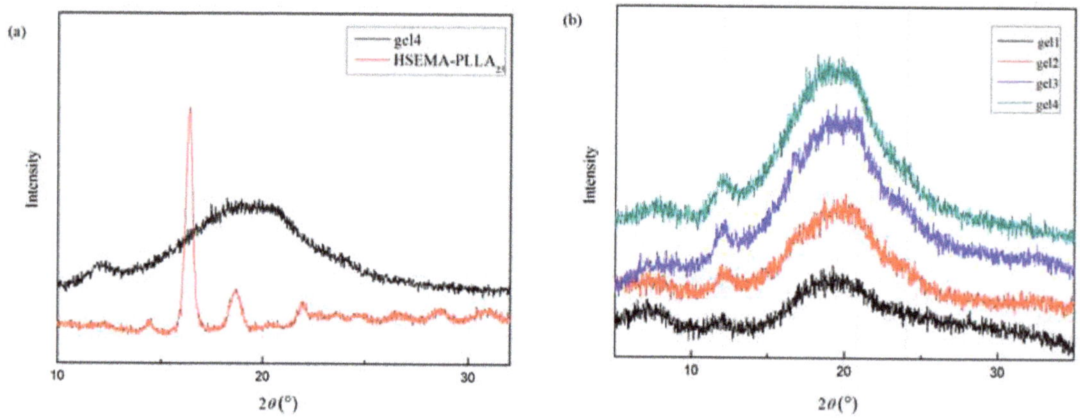

Figure 4. WXRD patterns of (**a**) macromonomer HSEMA-PLLA$_{25}$ and gel4, (**b**) gels with different polymerization degrees of PLA.

3.3. Micellar Properties of Hydrogels

The addition of temperature-sensitive monomer MEO$_2$MA and OEGMA makes the synthesized nanogels temperature sensitive. Figure 5 shows a digital camera diagram of gels 1–4 at different temperatures. As shown in photos, gels 1–4 show different clarity at room temperature due to different hydrophobic chain lengths. and become turbid with increasing hydrophobic chain segments. When the temperature rises to 35 °C, the nanogels become further turbid. As the temperature rises further, the turbidity increases and the clarity decreases further, this is because with the increase of temperature, the hydrogen bond between the hydrophilic part of the nanogel and the water molecule weakens. The enhanced hydrophobic effect makes the solution turbid, this also means that the gel is temperature sensitive.

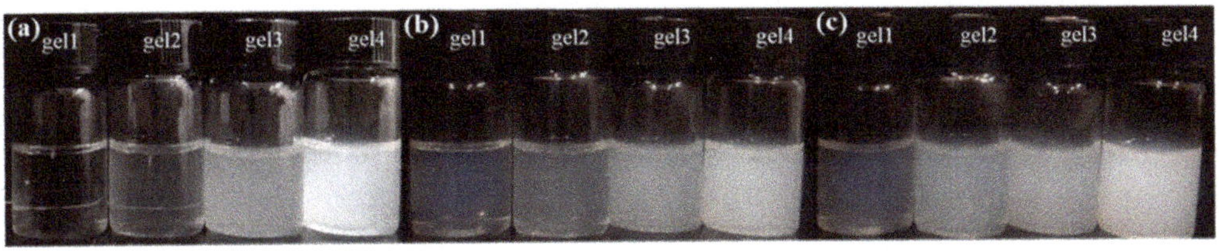

Figure 5. Digital photo of temperature sensitivity of gel 1–4 (2 mg/mL) (**a**) 25 °C, (**b**) 35 °C, (**c**) 45 °C.

By adjusting the ratio between the temperature-sensitive monomers MEO_2MA and OEGMA, the low critical solution temperature (LCST) can be adjusted to near the physiological temperature of the human body. In this study, the content of the hydrophilic part is fixed, and the variable is the degree of polymerization of the hydrophobic section of PLA. As shown in Figure 6, due to the different content of the hydrophobic part, it has different light transmittance at 25 °C, from largest to smallest, 96%, 77%, 56%, 45%, as the degree of polymerization of the hydrophobic part increases, the light transmittance gradually decreases. This is because when the content of hydrophilic chains in the polymer molecules is the same, the more the content of hydrophobic chains, the greater the degree of phase separation and the more turbid and the lower the light transmittance; Due to the same content of the hydrophilic part, the LCST values of nanohydrogels with different hydrophobic polymerization degrees are different with different hydrophobic polymerization degrees. The LCST values of gel1, gel2, gel3, and gel4 are 35 °C, 34 °C, 33 °C, 30 °C, and the LCST value decreases with the increase of the degree of polymerization of the hydrophobic part. This is because when the content of the hydrophilic part of the same, the higher the degree of polymerization of the PLA, the higher the content of the hydrophobic moiety, the lower temperature required when the hydrophobic interaction formed between the polymers is broken. In other words, the corresponding lower critical solution temperature is lower.

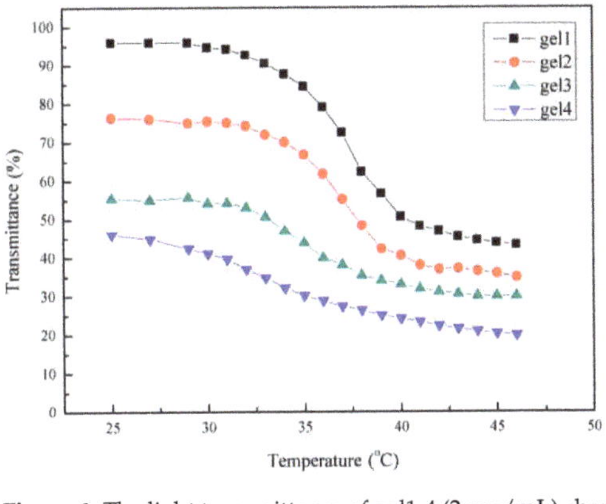

Figure 6. The light transmittance of gel1-4 (2 mg/mL) changes with temperature.

3.4. Analysis of the Particle Size and Micro Morphology of Nanogel in Solution

Nanogels are realized by self-recombination of an equal amount of enantiomeric graft copolymer solution, so nanogels still exist in the form of micelles [35]. DLS can be used to measure the particle size of nano-scale polymers in different states. Figure 7 is a graph of the particle size of nanohydrogels with different degrees of polymerization of polylactic acid. As shown in the figure, when the degree of polymerization of the hydrophobic part is different, the particle sizes of the corresponding nanohydrogels are also different. When the degree of polymerization increases, the particle size of the corresponding nanohydrogel is 150 nm, 174 nm, 255 nm, and 345 nm in sequence. As the degree of polymerization increases, the corresponding particle size increases in sequence. This is because the structure of the nanohydrogel is still a self-reorganized micellar structure of the polymer, which is composed of a hydrophobic core and a hydrophilic shell. At the same hydrophilic content, the greater the degree of polymerization of the hydrophobic part of the polymer, the larger the hydrophobic core of the corresponding micelle, so that the particle size increases with the increase of the degree of polymerization of polylactic acid.

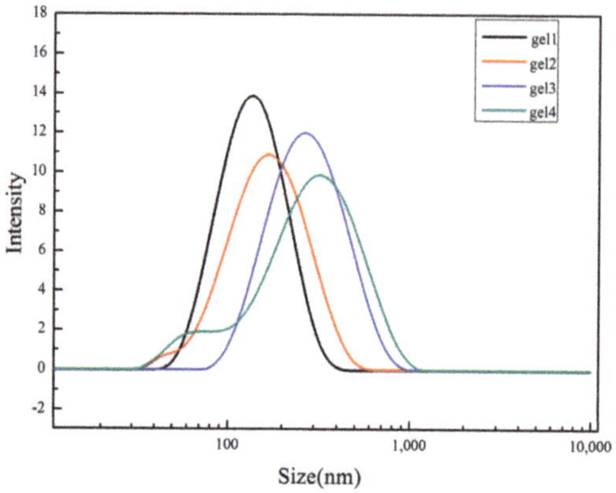

Figure 7. The particle size diagrams of nanogels at different degrees of PLA.

The morphology of the nanohydrogel can be analyzed bytransmission electron microscope. Figure 8 is the transmission electron micrographs of the nano hydrogel under different conditions. To have more intuitive data on the particle size, we use dynamic light scattering (DLS) to corroborate the corresponding particle size. As shown in the figure, the particle size of gel1 at room temperature is about 150 nm, while the particle size at 37 °C is about 100 nm. This is because in the core-shell structure of the nanohydrogel, the hydrophilic shell shrinks as the temperature increases due to its temperature sensitivity, which reduces the particle size of the nanohydrogel; The particle size after reduction with 10 mM DTT is about 270 nm, which is larger than the particle size at room temperature. This is because the addition of the reducing agent causes the disulfide bond connecting the hydrophobic segment in the nanohydrogel to be reduced to a sulfhydryl group and breaks. The number of broken free hydrophobic groups increases, and aggregation is more likely to occur, forming larger aggregates to increase the particle size of the nanohydrogel. The change of particle size under different conditions also shows that the synthesized nanohydrogel has dual stimulus responsiveness of temperature and reduction.

3.5. Determination of Reduced Sulfhydryl Group by Ultraviolet-Visible Spectroscopy

The disulfide bond in the nanohydrogel gives it the characteristic of reduction response. When it is used as a drug carrier, it can be targeted for release under the reduced environment of tumor cells. There are many detection methods for disulfide bonds, among which Ellman reagent detection is one of the detection methods widely used in many fields such as biochemistry. The main detection mechanism is to first reduce the disulfide bond to a sulfhydryl group with a strong reducing agent, and then use the color reaction of the Ellman reagent with the free sulfhydryl group and the special peak of the sulfhydryl group in the ultraviolet spectrum to measure, therefore further confirming the reduction responsiveness of the gel. Figure 9 is the UV spectrum of the hydrogel in the initial state and the reduced state, as shown in the figure, compared with the spectrum of the initial nano hydrogel without reducing agent, the reduced nanohydrogel has a characteristic peak of free sulfhydryl at 412 nm, and the reduction conversion rate is greater than 98% calculated according to the formula of absorbance and concentration. This also shows that the disulfide bond in the nanohydrogel is successfully reduced to a sulfhydryl group, i.e., the nanohydrogel has reduction responsiveness.

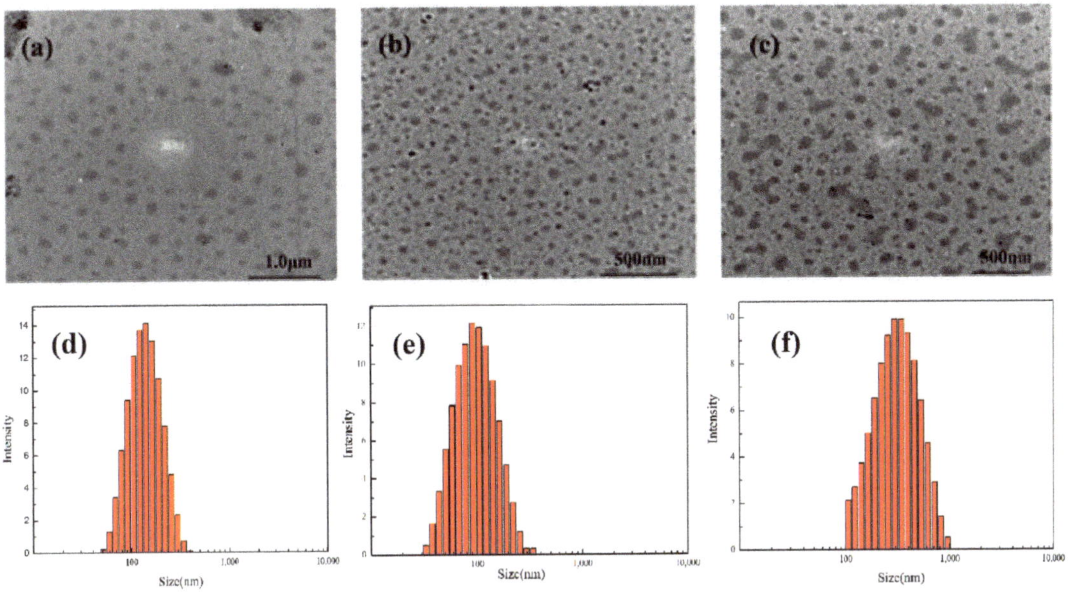

Figure 8. The transmission electron microscope image of nanohydrogel gel1 at 25 °C (**a**), 37 °C (**b**), 25 °C 10 mM DTT(**c**), the particle size diagrams of nanohydrogel gel1 at 25 °C (**d**), 37 °C (**e**), 25 °C 10 mM DTT(**f**).

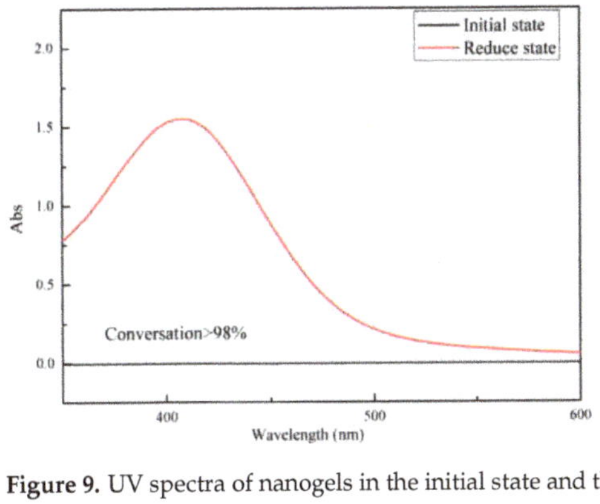

Figure 9. UV spectra of nanogels in the initial state and the reduced state.

3.6. Analysis of Swelling Kinetics of Hydrogel

Swelling is one of the methods to measure the performance of the gel. To illustrate the water absorption performance of the hydrogel, a swelling kinetics experiment was carried out using a fully dried hydrogel. Figure 10 shows the change of the swelling ratio of the gel with time at 25 °C. As shown in the figure, the gels of different polymerization degrees have different swelling ratios, and the swelling equilibrium is reached at about 210 min. Among them, the swelling ratio of gel1 is the largest, about 2.45, and the swelling ratio of gel4 is the smallest, about 0.95. This is because the hydrophilic segment P (MEO_2MA-*co*-OEGMA) in the gel can combine with water molecules to form hydrogen bonds, gel1 has the lowest degree of polymerization of PLA, the physical crosslinking point was the least and the three-dimensional complexation was the weakest, thus forming a relatively loose three-dimensional network of gels with a high swelling ratio. In gel4, PLA has the highest degree of polymerization and the strongest three-dimensional complexation, so the gel structure formed is more compact and the swelling ratio is low.

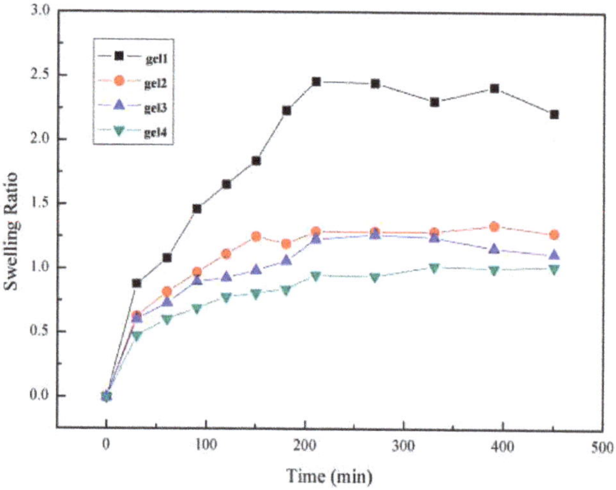

Figure 10. The swelling kinetics of gel at 25 °C.

Figure 11 is the physical image and scanning electron microscope image of gel before and after swelling. The size of each square in the background is 1 cm. It can be seen from the figure that the volume of gel increases significantly after swelling with water. Combined with the scanning electron microscope image, it can be seen that gel has an obvious three-dimensional network structure.

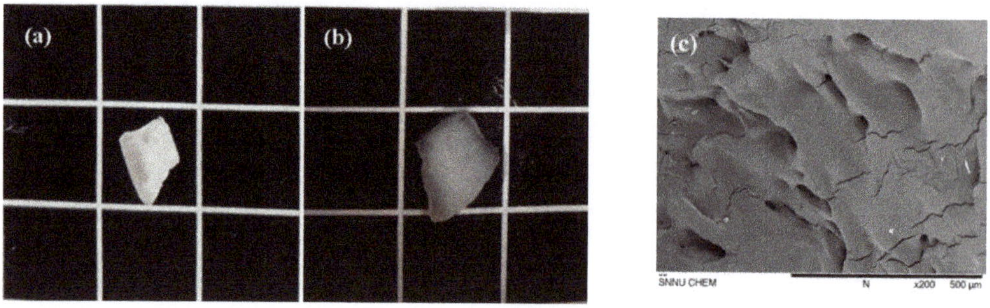

Figure 11. The physical image of gel2 hydrogel at dry state (**a**), fully swollen at 25 °C (**b**), the scanning electron microscope image of the fully swollen hydrogel at 25 °C (**c**).

3.7. Rheological Analysis

The rheological behavior is an important method to study gels. We conduct the gelation process and mechanics of complex gels through rheological experiments on equal amounts of nanohydrogels at the same concentration (10 w/v%). As we all know, G' and G'' represent the storage modulus and loss modulus of the material respectively. When $G' > G''$, the elastic deformation is dominant and the solution is in a gel state. When $G' < G''$, the viscosity deformation is dominant, and the solution is in a sol state. When $G' = G''$, the corresponding independent variable is the sol-gel transition point of the micellar solution. Among them, the storage modulus G' related to the mechanical strength of the gel. The larger the storage modulus is, the greater the mechanical strength of the corresponding gel is [39]. Figure 12 shows the analysis of the gelation process and the oscillation stress scanning diagram of the gel at 37 °C. As shown in Figure 12a, gelation time of gel2 and gel3 is different due to their different PLA block lengths, the gelation time is 101 s and 58 s respectively. This is because the greater the degree of PLA polymerization, the more PLLA or PDLA groups on the grafted chain, which increases the physical cross-linking points, reduces the gelation time, and makes the three-dimensional complexation more obvious. To further verify the impact of the degree of polymerization on the mechanical properties of the gel, we performed a stress scan on the resulting gels. Figure 12b shows the oscillation stress scan curve of the gel. As shown in the figure, different polymerization degrees gels have different yield values (the corresponding stress values when $G' = G''$ in the figure). The corresponding yield values of gel2 and gel3 are 64 Pa and 70 Pa, respectively, and the storage modulus of gel2 is always higher than gel3 before the yield value. This is because the increase in the degree of PLA polymerization increases the physical crosslinking points of the gel, and the resulting gel. The tighter the structure, the better the mechanical properties of the gel.

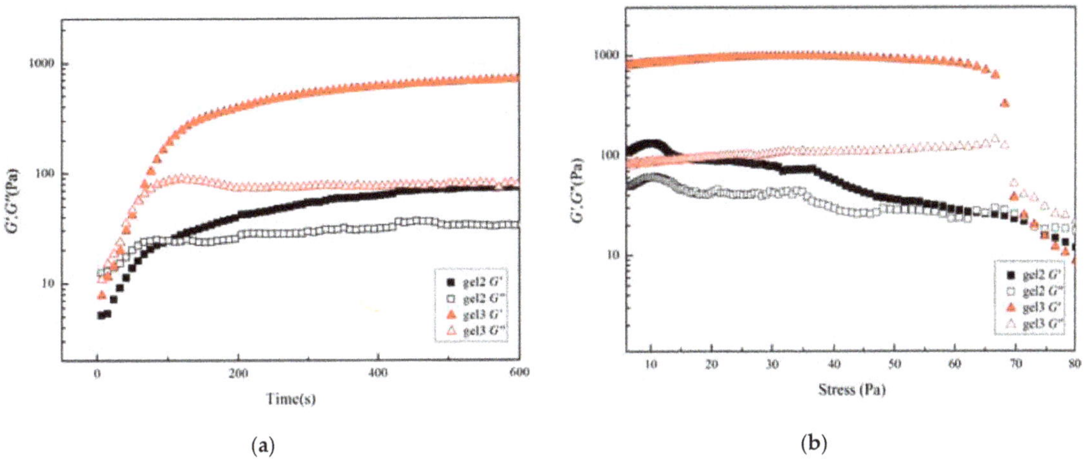

Figure 12. Rheological images of gel2 and gel3 at a constant temperature of 37 °C with (**a**) time scanning, (**b**) stress scanning.

4. Conclusions

A novel type physically crosslinked nanogel was synthesized by self-recombination of enantiomeric micellar solution under mild conditions and in situ stereocomplexation. Thermosensitive monomers MEO_2MA and OEGMA make it thermally responsive. The introduction of the disulfide bond (-S-S-) makes it reduction responsive and can be degraded within two weeks. This was proven in DLS experiments. The existence of stereocomplexation in the hydrogel was confirmed by WAXD. A rheometer was used to study the gelation process and mechanical strength, and it was confirmed that it was related to the degree of PLA polymerization. In the case of the same hydrophilic part content, the influence of the degree of polymerization of the hydrophobic segment (iepolylactic acid) on the micelle properties, swelling properties and rheological behavior of the nanohydrogel was studied. Experimental results show that the performance of nanohydrogels can be adjusted by changing the degree of polymerization of polylactic acid; at the same time, ultraviolet spectroscopy and dynamic light scattering are used to verify the reduction response and temperature-sensitive response of nanohydrogels. The preparation of physical crosslinking temperature reduction dual responsive hydrogel provides a new method.

Author Contributions: Conceptualization, S.L.; Data curation, Z.W., F.S., Y.F. and Q.W.; Formal analysis, F.S., Y.F. and Q.W.; Funding acquisition, S.L.; Investigation, W.G. and Z.W.; Methodology, W.G. and Z.W.; Writing–original draft, W.G. and Z.W.; Writing–review and editing, S.L. All authors have read and agreed to the published version of the manuscript.

Funding: The research was supported by the National Natural Science Foundation of China (No. 21773147).

Conflicts of Interest: The authors declare no conflict of interest.

References

1. Zhu, Y.; Akagi, T.; Akashi, M. Self-assembling stereocomplex nanoparticles by enantiomeric poly(g-glutamic acid)-poly(lactide) graft copolymers as a protein delivery carriera. *Macromol. Biosci.* **2014**, *14*, 576–587. [CrossRef] [PubMed]
2. Vader, P.; Mol, E.A.; Pasterkamp, G.; Schiffelers, R.M. Extracellular vesicles for drug delivery. *Adv. Drug Deliv. Rev.* **2016**, *106*, 148–156. [CrossRef]
3. Hu, X.L.; Zhang, Y.Q.; Xie, Z.G.; Jing, X.B.; Bellotti, A.; Gu, Z. Stimuli-responsive polymersomes for biomedical applications. *Biomacromolecules* **2017**, *18*, 649–673. [CrossRef] [PubMed]
4. Hu, X.L.; Tian, J.; Liu, T.; Zhang, G.Y.; Liu, S.Y. Photo-triggered release of caged camptothecin prodrugs from dually responsive shell cross-linked micelles. *Macromolecules* **2013**, *46*, 6243–6256. [CrossRef]
5. Zhang, L.S.; Liu, Y.C.; Zhang, K.; Chen, Y.W.; Luo, X.L. Redox-responsive comparison of diselenide micelles with disulfide micelles. *Colloid Polym. Sci.* **2019**, *297*, 225–238. [CrossRef]
6. Cai, M.T.; Cao, J.; Wua, Z.Z.; Cheng, F.R.; Chen, Y.W.; Luo, X.L. In vitro and in vivo anti-tumor efficiency comparison of phosphorylcholine micelles with PEG micelles. *Colloids Surf. B Biointerfaces* **2017**, *157*, 268–279. [CrossRef]
7. Jo, I.; Lee, S.; Zhu, J.T.; Shim, T.S.; Yi, G. Soft patchy micelles. *Curr. Opin. Colloid Interface Sci.* **2017**, *30*, 97–105. [CrossRef]
8. Guo, J.X.; Kaletunç, G. Dissolution kinetics of pH responsive alginate-pectin hydrogel particles. *Food Res. Int.* **2016**, *88*, 129–139. [CrossRef]
9. Sun, Y.W.; Zhang, Y.P.; Liu, J.J.; Nie, F.Q. Integrated microfluidic device for the spherical hydrogel pH sensor fabrication. *RSC Adv.* **2016**, *6*, 11204–11210. [CrossRef]
10. Cao, Q.C.; Wang, X.; Wu, D.C. Controlled cross-linking strategy for formation of hydrogels, microgels and nanogels. *Chin. J. Polym. Sci.* **2018**, *36*, 8–17. [CrossRef]
11. Wang, Y.C.; Li, Y.; Sun, T.M.; Xiong, M.H.; Wu, J.; Yang, Y.Y.; Wang, J. Core–shell–corona micelle stabilized by reversible cross-linkage for intracellular drug delivery. *Macromol. Rapid Commun.* **2010**, *31*, 1201–1206. [CrossRef]
12. Na, J.H.; Koo, H.; Lee, S.; Min, K.H.; Park, K.; Yoo, H.; Lee, S.H.; Park, J.H.; Kwon, I.C.; Jeong, S.Y.; et al. Real-time and non-invasive optical imaging of tumor-targeting glycol chitosan nanoparticles in various tumor models. *Biomaterials* **2011**, *32*, 5252–5261. [CrossRef]
13. Grimaudo, M.A.; Concheiro, A.; Alvarez-Lorenzo, C. Nanogels for regenerative medicine. *J. Control. Release* **2019**, *313*, 148–160. [CrossRef] [PubMed]
14. Hashimoto, Y.; Mukai, S.; Sasaki, Y.; Akiyoshi, K. Nanogel tectonics for tissue engineering: Protein delivery systems with nanogel chaperones. *Adv. Healthc. Mater.* **2018**, *7*, 1800729. [CrossRef]
15. Chapla, R.; Alhaj Abed, M.; West, J. Modulating Functionalized poly(ethylene glycol) diacrylate hydrogel mechanical properties through competitive crosslinking mechanics for soft tissue applications. *Polymers* **2020**, *12*, 3000. [CrossRef] [PubMed]
16. Slaughter, B.V.; Khurshid, S.S.; Fisher, O.Z.; Khademhosseini, A.; Peppas, N.A. Hydrogels in regenerative medicine. *Adv. Mater.* **2009**, *21*, 3307–3329. [CrossRef] [PubMed]
17. Konieczynska, M.D.; Grinstaff, M.W. On- demand dissolution of chemically cross-linked hydrogels. *Acc. Chem. Res.* **2017**, *50*, 151–160. [CrossRef] [PubMed]
18. Hao, Y.; Fowler, E.W.; Jia, X.Q. Chemical synthesis of biomimetic hydrogels fortissue engineering. *Polym. Int.* **2017**, *66*, 1787–1799. [CrossRef]
19. Chung, H.J.; Park, T.G. Self-assembled and nanostructured hydrogels for drug delivery and tissue engineering. *Nano Today* **2009**, *4*, 429–437. [CrossRef]
20. Sytze, J.; Buwalda, P.J.; Dijkstra, J.F. In situ forming stereocomplexed and post-photocrosslinked acrylated star poly(ethylene glycol)-poly(lactide) hydrogels. *Eur. Polym. J.* **2017**, *94*, 152–161.
21. Akhlaq, M.; Azad, A.K.; Ullah, I.; Nawaz, A.; Safdar, M.; Bhattacharya, T.; Uddin, A.B.M.H.; Abbas, S.A.; Mathews, A.; Kundu, S.K.; et al. Methotrexate-loaded gelatin and polyvinyl alcohol (Gel/PVA) hydrogel as a pH-sensitive matrix. *Polymers* **2021**, *13*, 2300. [CrossRef]
22. Guo, H.; Mussault, C.; Marcellan, A.; Hourdet, D.; Sanson, N. Hydrogels with dual thermoresponsive mechanical performance. *Macromol. Rapid Commun.* **2017**, *38*, 1700287. [CrossRef] [PubMed]
23. Wu, J.; Shi, X.Y.; Wang, Z.D.; Song, F.; Gao, W.L.; Liu, S.X. Stereocomplex poly(Lactic Acid) amphiphilic conetwork gel with Temperature and pH dual sensitivity. *Polymers* **2019**, *11*, 1940. [CrossRef] [PubMed]
24. Hu, X.Z.; Zou, C.J. Pentaerythrityl tetra-β-cyclodextrin: Synthesis, characterization and application in multiple responses hydrogel. *Colloids Surfaces A* **2017**, *529*, 571–579. [CrossRef]

25. Yang, H.; Li, C.H.; Tang, J.D.; Suo, Z.G. Strong and degradable adhesion of hydrogels. *ACS Appl. Bio Mater.* **2019**, *2*, 1781–1786. [CrossRef]
26. Kamata, H.; Kushiro, K.; Takai, M.; Chung, U.; Sakai, T. Non-osmotic hydrogels: A rational strategy for safely degradable hydrogels. *Angew. Chem. Int. Ed.* **2016**, *55*, 9282–9286. [CrossRef]
27. Patenaude, M.; Hoare, T. Injectable, degradable thermoresponsive poly(*N*-isopropylacrylamide) hydrogels. *ACS Macro Lett.* **2012**, *1*, 409–413. [CrossRef]
28. Pei, Y.; Molley, T.G.; Kilian, K.A. Enzyme responsive inverse opal hydrogels. *Macromol. Rapid Commun.* **2020**, *41*, 1900555. [CrossRef] [PubMed]
29. Peschke, T.; Bitterwolf, P.; Gallus, S.; Hu, Y.; Oelschlaeger, C.; Willenbacher, N.; Rabe, K.S.; Niemeyer, C.M. Self-assembling all-enzyme hydrogels for flow biocatalysis. *Angew. Chem. Int. Ed.* **2018**, *57*, 17028–17032. [CrossRef] [PubMed]
30. Dutta, S.; Samanta, P.; Dhara, D. Temperature, pH and redox responsive cellulose based hydrogels for protein delivery. *Int. J. Biol. Macromol.* **2016**, *87*, 92–100. [CrossRef]
31. Yildirim, T.; Traeger, A.; Preussger, E.; Stumpf, S.; Fritzsche, C.; Hoeppener, S.; Schubert, S.; Schubert, U.S. Dual responsive nanoparticles from a RAFT copolymer library for the controlled delivery of doxorubicin. *Macromolecules* **2016**, *49*, 3856–3868. [CrossRef]
32. An, X.N.; Zhu, A.J.; Luo, H.H.; Ke, H.T.; Chen, H.B.; Zhao, Y.L. Rational design of multi-stimuli-responsive nanoparticles for precise cancer therapy. *ACS Nano* **2016**, *10*, 5947–5958. [CrossRef]
33. Hu, X.L.; Hu, J.M.; Tian, J.; Ge, Z.S.; Zhang, G.Y.; Luo, K.F.; Liu, S.Y. Polyprodrug amphiphiles: Hierarchical assemblies for shape-regulated cellular internalization, trafficking, and drug delivery. *J. Am. Chem. Soc.* **2013**, *135*, 17617–17629. [CrossRef] [PubMed]
34. Zhou, Q.; Xu, L.; Liu, F.; Zhang, W.A. Construction of reduction-responsive photosensitizers based on amphiphilic block copolymers and their application for photodynamic therapy. *Polymer* **2016**, *97*, 323–334. [CrossRef]
35. Abebe, D.G.; Fujiwara, T. Controlled thermoresponsive hydrogels by stereocomplexed PLA-PEG-PLA prepared via hybrid micelles of pre-mixed copolymers with different PEG lengths. *Biomacromolecules* **2012**, *13*, 1828–1836. [CrossRef] [PubMed]
36. Pholharn, D.; Cheerarot, O.; Baimark, Y. Stereocomplexation and mechanical properties of polylactide-*b*-poly(propylene glycol)-*b*-polylactide blend films: Effects of polylactide block length and blend ratio. *Chin. J. Polym. Sci.* **2017**, *35*, 1391–1401. [CrossRef]
37. Che, Y.J.; Zschoche, S.; Obst, F.; Appelhans, D.; Voit, B. Double-crosslinked reversible redox-responsive hydrogels based on disulfide-thiol interchange. *J. Polym. Sci. Part A Polym. Chem.* **2019**, *57*, 2590–2601. [CrossRef]
38. Baimark, Y.; Pasee, S.; Rungseesantivanon, W.; Prakymoramas, N. Flexible and high heat-resistant stereocomplex PLLA-PEG-PLLA/PDLA blends prepared by melt process: Effect of chain extension. *J. Polym. Res.* **2019**, *26*, 218. [CrossRef]
39. Hiemstra, C.; Zhong, Z.Y.; Li, L.B.; Dijkstra, P.J.; Feijen, J. In-situ formation of biodegradable hydrogels by stereocomplexation of PEG-$(PLLA)_8$ and PEG-$(PDLA)_8$ star block copolymers. *Biomacromolecules* **2006**, *7*, 2790–2795. [CrossRef]

polymers

MDPI

Review

Recent Advances in Thermoresponsive OEGylated Poly(amino acid)s

Chao Geng, Shixue Wang * and Hongda Wang

State Key Laboratory of Electroanalytical Chemistry, Changchun Institute of Applied Chemistry, Chinese Academy of Sciences, Jilin 130022, China; cgeng@ciac.ac.cn (C.G.); hdwang@ciac.ac.cn (H.W.)
* Correspondence: wangsx@ciac.ac.cn; Tel.: +86-431-85262070

Abstract: Thermoresponsive polymers have been widely studied in the past decades due to their potential applications in biomedicine, nanotechnology, and so on. As is known, poly(*N*-isopropylacrylamide) (PNIPAM) and poly(oligo(ethylene glycol)methacrylates) (POEGMAs) are the most popular thermoresponsive polymers, and have been studied extensively. However, more advanced thermoresponsive polymers with excellent biocompatibility, biodegradability, and bioactivity also need to be developed for biomedical applications. OEGylated poly(amino acid)s are a kind of novel polymer which are synthesized by attaching one or multiple oligo(ethylene glycol) (OEG) chains to poly(amino acid) (PAA).These polymers combine the great solubility of OEG, and the excellent biocompatibility, biodegradability and well defined secondary structures of PAA. These advantages allow them to have great application prospects in the field of biomedicine. Therefore, the study of OEGylated poly(amino acid)s has attracted more attention recently. In this review, we summarized the development of thermoresponsive OEGylated poly(amino acid)s in recent years, including the synthesis method (such as ring-opening polymerization, post-polymerization modification, and Ugi reaction), stimuli-response behavior study, and secondary structure study. We hope that this periodical summary will be more conducive to design, synthesis and application of OEGylated poly(amino acid)s in the future.

Keywords: thermoresponsive; oligo(ethylene glycol); OEGylated; poly(amino acid); ring-opening polymerization; post-polymerization modification; Ugi reaction

Citation: Geng, C.; Wang, S.; Wang, H. Recent Advances in Thermoresponsive OEGylated Poly(amino acid)s. *Polymers* **2021**, *13*, 1813. https://doi.org/10.3390/polym13111813

Academic Editors: Mattia Sponchioni and Florian J. Stadler

Received: 15 April 2021
Accepted: 25 May 2021
Published: 31 May 2021

1. Introduction

Stimuli-responsive polymers, so called "smart polymers" that can be triggered by a variety of external environmental stimuli such as temperature, pH, light, ionic, chemical and biological stimuli etc., and consequently with the change of physical and chemical properties, have been extensively investigated because of their potential applications in the past few decades [1–12]. Among them, thermoresponsive polymers, which exhibit a reversible phase transition to temperature, have attracted much attention due to their easy to control stimulus and potential biomedical and tissue engineering applications [13–15]. Thermoresponsive behaviors of polymers can be generally classified into two categories, lower critical solution temperature (LCST) type and upper critical solution temperature (UCST) type based on the equilibrium phase separation [16]. In both types, phase separation will occur and result in a turbid mixture of the two phases at a concentration dependent cloud point temperature (T_{cp}), with $T_{cp} \geq$ LCST for separation with increasing temperature or $T_{cp} \leq$ UCST for separation with decreasing temperature, and a single phase for temperatures intermediate to these two regimes. Since the phase transition temperatures are closely dependent on the polymer structure (e.g., backbone, side-chain, topological architecture), it is very important to obtain the desired thermoresponsive temperature through reasonable structure design for specific applications. As is known, poly(*N*-isopropylacrylamide) (PNIPAM) is generally considered to be the gold standard because of its LCST around physiological temperature (≈32 °C) in water together with a low concentration and pH dependency, making it a prime candidate for in vivo biomedical applications [17,18]. Meanwhile, other types of polymers are increasingly being investigated for their thermoresponsive behavior in recent years; especially, polymers bearing an oligo(ethylene glycol) (OEG) side chain have been shown to combine the biocompatibility of OEG with a versatile and controllable LCST behavior, such as poly(oligo(ethylene glycol)methacrylate)s (POEGMA)s [19–24]. However, backbone of these thermoresponsive polymers are nondegradable, the negative effects of in vivo enrichment are unclear. As such, it is still necessary to search for novel thermoresponsive polymers with excellent biocompatibility, degradability and a tunable critical transition temperature to meet the needs of related fields.

OEGylated poly(amino acid)s are a kind of nonionic hydrophilic polymer obtained by covalently attaching one or multiple OEG chains to poly(amino acid) (PAA). These polymers combine the great solubility of OEG, and the excellent biocompatibility, biodegradability and well defined secondary structures of PAA (e.g., α-helix or β-sheet) [25–29] The biggest advantage of this combination is that it overcomes commonly water soluble poly(amino acid)s (e.g., poly(L-glutamic acid), poly(L-glutamic acid)), which suffer from pH-dependent solubility and limited circulation lifetime because of their aggregation with oppositely charged polymers [29–31]. Although OEGylated poly(amino acid) is very important, its related research started late. Deming group reported the first OEGylated poly(L-lysine) (PLL), which showed excellent water-solubility and completely α-helical in solution in 1999 [30]. Then, they also synthesized methylated mono and di(ethylene glycol)-functionalized poly(L-serine) and poly(L-cysteine) [32]. Subsequently, Zhang and coworkers synthesized OEGylated poly(L-glutamic acid) encoding pendant alkyne side groups that were amendable to further modifications [33]. Inspired by these pioneer works, the OEGylated poly(amino acid)s were prepared as thermoresponsive materials with difference amino acids, and different OEG topological structures (e.g., linear, Y-shaped) and length; the expansion of structural diversity provides more possibilities for its applications

This review summarizes the thermoresponsive OEGylated poly(amino acid)s over recent years (Table 1). Specifically, it focuses on the synthesis method, stimuli-response behavior study, and secondary structure study of OEGylated poly(amino acid)s. These will be discussed in the next two sections in detail. We hope that this periodical summary will be more conducive to the design, synthesis and application of OEGylated poly(amino acid)s in the future.

Table 1. Summary of thermoresponsive OEGylated poly(amino acid)s.

Entry	Polymer Structure	Amino Acids	Synthetic Method	T_{cp} (°C)	Secondary Structure	Ref.
1	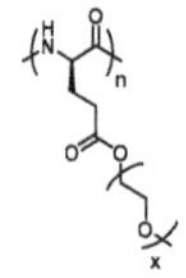	L-Glutamic acid	ROP	x = 2, T_{cp} = 32 °C x = 3, T_{cp} = 57 °C	x = 2, 100% α-helix in freshly prepared aqueous solution x = 3, 100% helix	[34]
2	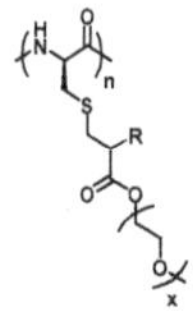	L-Cysteine	ROP	R = CH_3, x = 3, T_{cp} = 50 °C x = 4, T_{cp} = 65 °C R = H, x = 3, T_{cp} = 51 °C	α-helix, β-sheet, and random coil	[35]
3	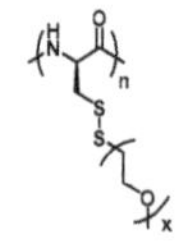	L-Cysteine	ROP	x = 3, T_{cp} = 34 °C x = 4, T_{cp} = 45 °C	α-helix, β-sheet, and random coil	[36]

Table 1. *Cont.*

Entry	Polymer Structure	Amino Acids	Synthetic Method	T_{cp} (°C)	Secondary Structure	Ref.
4	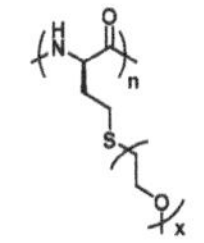	L-Homocysteine	ROP	x = 4, T_{cp} = 40 °C	>95% α-helix	[28]
5	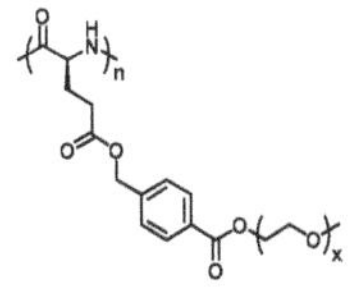	L-Glutamic acid	ROP	x = 4, T_{cp} = 25 °C x = 6, T_{cp} = 36 °C x = 8, T_{cp} = 52 °C	x = 4, 69% α-helix x = 6, 29% α-helix x = 8, 17% α-helix	[27]
6	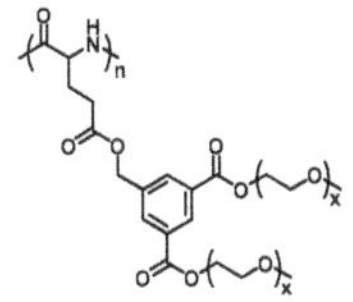	L-Glutamic acid	ROP	x = 2, T_{cp} = 17 °C x = 3, T_{cp} = 34 °C x = 4, T_{cp} = 60 °C	x = 2, 69% α-helix x = 3, 37% α-helix x = 4, 30% α-helix	[27]
7	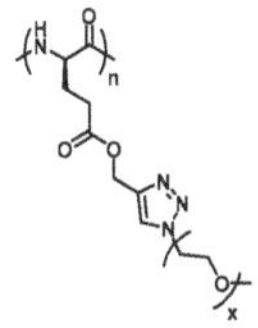	L-Glutamic acid	Post-polymerization modification	x = 2 or 3, T_{cp} = 22.3–74.1 °C by varying the molecular weight	100% α-helix	[37]
8	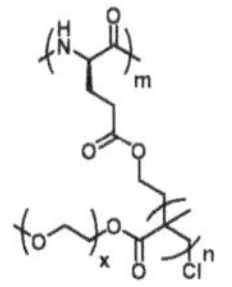	L-Glutamic acid	Post-polymerization modification	x = 3, T_{cp} = 45.7–51.3 °C by varying the molecular weight	78.4–100% α-helix	[38]
9	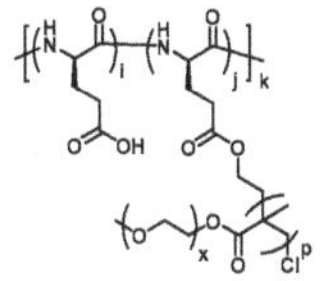	L-Glutamic acid	Post-polymerization modification	x = 3, T_{cp} = 44.1–62.1 °C by varying the molecular weight and monomer ratio	0–65.1% α-helix	[38]
10	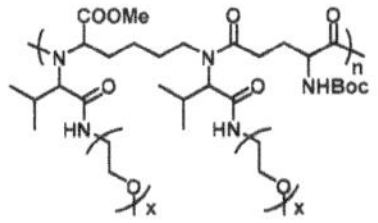	L-Lysine and L-Glutamic acid	Ugi multicomponent polymerization	x = 3, T_{cp} = 27 °C x = 4, T_{cp} = 37 °C	random coil	[39]
11	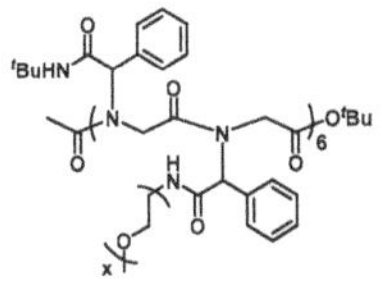	Glycine	Ugi multicomponent polymerization	x = 4, T_{cp} = 12–17.5 °C with difference sequence	random coil	[40]

2. Synthetic Strategies of OEGylated Poly(amino acid)s

So far, OEGylated poly(amino acid)s can be generally synthesized by three ways: controlled ring-opening polymerization (ROP) of OEGylated *N*-carboxyanhydride (NCA) monomers, post-polymerization modification (PPM) of poly(amino acid) precursors and the Ugi multicomponent polymerization (Figure 1). These methods have their own advantages and disadvantages. The controlled ROP of OEGylated NCAs can obtain well-defined OEGylated poly(amino acid)s with controlled molecular weights (MWs), and are easy to scale up, which is beneficial for accurately exploring structure-property relationships and widespread applications. However, it has to take a lot of time to synthesize and purify the unstable OEGylated NCA monomers. In comparison, post-polymerization modification of poly(amino acid)s has the facility to construct OEGylated poly(amino acid)s with tunable OEG side-chain and properties, yet it need to use highly efficient reactions (usually the "click chemistry") and sacrifice a large number of reactants to obtain a relatively perfect polymer structure. In recent years, multicomponent reactions (MCRs) such as the Passerini three-component reaction, Ugi reaction, Biginelli reaction, and so on, have drawn great attention and been utilized in the fields of polymer chemistry because of the mild reaction conditions, high efficiency, atom economy, and structural diversity of products [41–44]. Therefore, a series of polymers with a new backbone, side-chains, and topologies have been successfully prepared. An Ugi reaction is considered to be a flexible method to construct amide bonds using amine and carboxylic acid as the starting materials. Meier and coworkers firstly used this strategy to obtain polyamides with finely tunable structures with the diamine (AA monomer) and dicarboxylic acid (BB monomer) monomers [45]. From this background, Tao and coworkers have employed the natural amino acids as AB monomer and oligo(ethylene glycol) isocyanide and aldehyde as other two components to make sequence-specific OEGylated poly(amino acid)s, and studied the relationship of sequence structure and thermoresponsive behaviors [39,40,46]. This work provided a key example for the effect of sequence structure on thermoresponsive behaviors, which we will discuss in detail later.

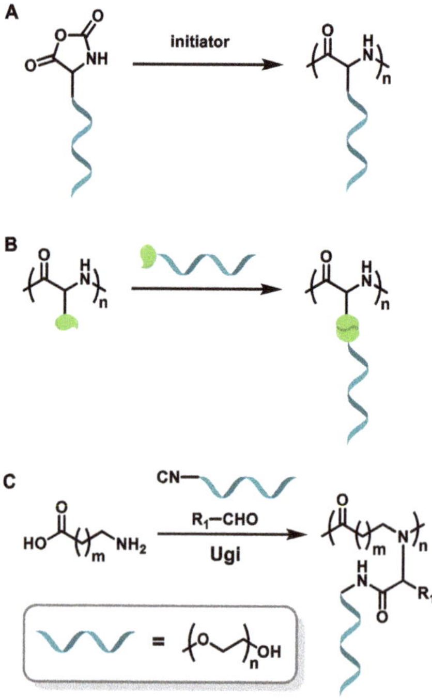

Figure 1. Synthesis strategies of OEGylated poly(amino acid)s. (**A**) ROP of OEGylated NCAs, (**B**) post-polymerization modification of poly(amino acid) precursors, and (**C**) Ugi multicomponent polymerization.

2.1. Thermoresponsive OEGylated Poly(amino acid)s from ROP of NCA Monomers

In the past decades, NCA polymerization has developed rapidly; in addition to the traditional amine initiator [47], a few novel and efficient initiators have been developed, such as transition metal complex and hexamethyldisilazane (HMDS) [48–51]. This has helped to achieve better control of polymer structures and facilitate downstream material applications. In this context, several thermoresponsive OEGylated poly(amino acid)s which were synthesized via ROP of NCAs have been reported. In 2011, Li et al. firstly reported the synthesis and characterization of new thermoresponsive OEGylated poly-L-glutamate (poly-L-EG$_x$Glu) (Figure 2) [34]. They synthesized the OEGylated NCAs via direct coupling between methylated ethyleneglycols and L-glutamate, then converted it into corresponding α-amino acid NCAs using triphosgene in THF. The obtained NCAs were viscous oils and purified by flash a column chromatography method. Then, different OEGylated homopolymers and random copolymers with narrow PDIs (<1.2) were prepared using Ni(COD)depe as initiator and DMF as solvent.

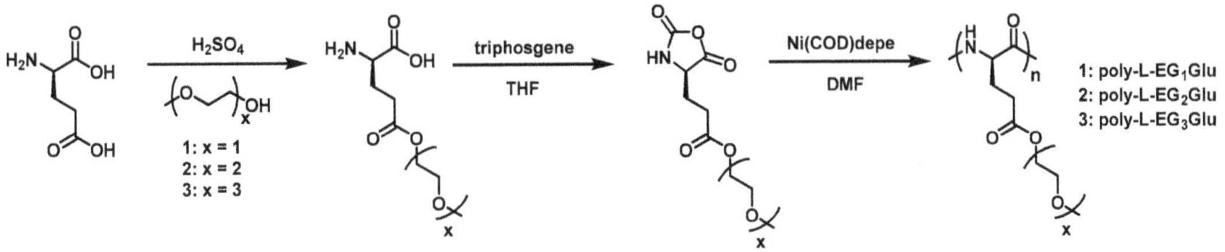

Figure 2. Synthetic route to poly-L-EGxGlu by Li et al. Reproduced with permission [34]. Copyright 2011, American Chemical Society.

OEGylated poly-L-EGxGlus obtained by Li et al. display reversible LCST transition in water, except poly-L-EG$_1$Glu (insoluble in water), the LCST of poly-L-EG$_2$Glu and poly-L-EG$_3$Glu are 32 °C and 57 °C, respectively. They also found that both poly-L-EG$_2$Glu and poly-L-EG$_3$Glu displayed hysteresis in phase transition during cooling processes, which was probably due to the redissolution of the OEG unit, requiring slight overcooling to overcome the energy barriers (Figure 3a). It is worth noting that it is easy to make random OEGylated copoly(amino acid)s with different EG$_2$Glu/EG$_3$Glu ratios in the Ni(COD)depe catalytic system with nanarrow molecular weight distribution (<1.2); meanwhile, the LCST can be varied from 36 °C with 80 mol% L-EG$_2$Glu to 54 °C with 30 mol% L-EG$_2$Glu (Figure 3b). It is noteworthy that the physiological temperature is just in this LCST range.

Subsequently, Li et al. studied the secondary structure of poly-L-EG$_x$Glus in water using circular dichroism (CD) spectra. Poly-L-EG$_2$Glu purified by dialysis did not have a well-defined secondary structure, which was composed of 16% α-helix, 32% β-strand, 20% turns, and 32% random coil, respectively. However, it almost formed 100% α-helix in freshly prepared aqueous solution. Heating the same solution above its LCST did not cause an obvious change in corresponding secondary structures (Figure 4a); it indicated that the secondary structure of poly-L-EG$_2$Glu strongly depended on sample history. In contrast, poly-L-EG$_3$Glu formed stable 100% α-helix and its secondary structure was also independent of temperature (Figure 4b). These results reveal that the secondary structure of poly-L-EG$_x$Glus is OEG chain length dependent. The longer OEG side-chain is beneficial to the stability of α-helix.

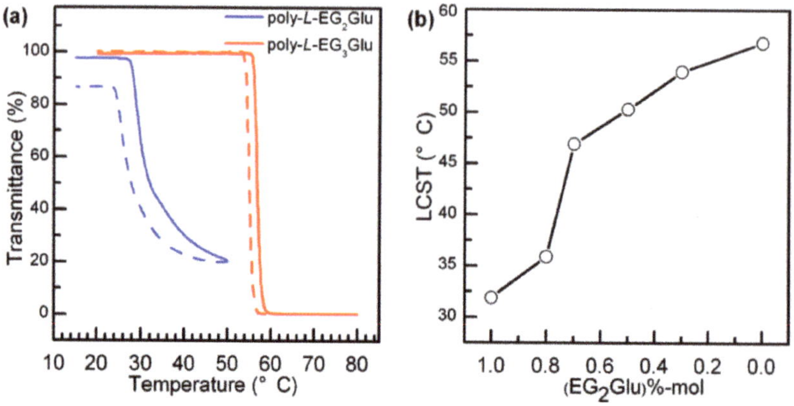

Figure 3. (**a**) Plots of transmittance as a function of temperature for aqueous solutions (2 mg/mL of poly-L-EG$_2$Glu and poly-L-EG$_3$Glu. Solid line: heating; dashed line: cooling. (**b**) LCST of poly(EG$_2$Glu-EG$_3$Glu) copolymers as a function of sample composition. Reproduced with permission [34]. Copyright 2011, American Chemical Society.

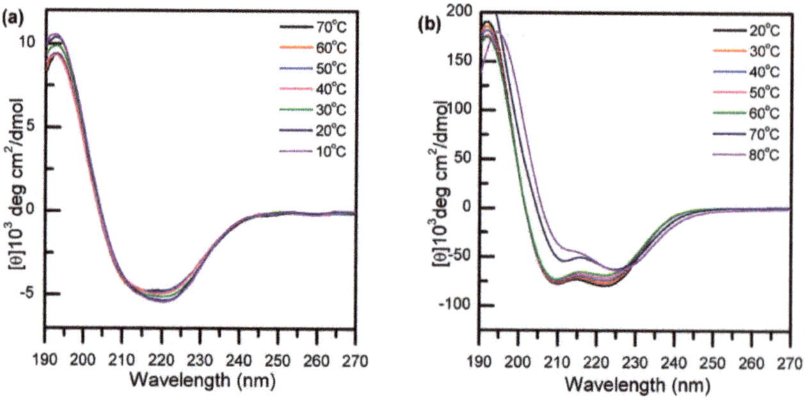

Figure 4. CD spectra of (**a**) poly-L-EG$_2$Glu and (**b**) poly-L-EG$_3$Glu as a function of temperature (heating scan). Reproduced with permission [34]. Copyright 2011, American Chemical Society.

Finally, Li et al. studied the driving force of LCST behaviors of poly-L-EGxGlu via temperature-dependent ^{1}H NMR (Figure 5). With the increase in temperature, they found that the protons of end methoxy and methylene groups of OEG units became more and more broad, accompanying a substantial decrease in signal intensity; a further increase above their corresponding LCST caused almost disappearance of their resonances. These results indicated that temperature increase induced dehydration of ethylene glycol groups and caused the phase separation.

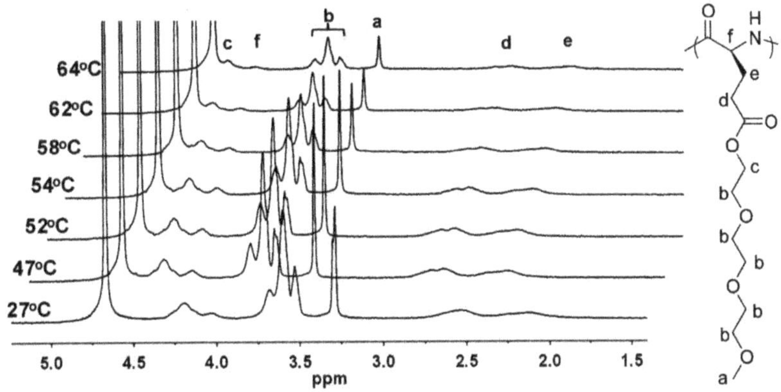

Figure 5. ^{1}H NMR spectra of poly-L-EG$_3$Glu as a function of temperature in D_2O. Reproduced with permission [34]. Copyright 2011, American Chemical Society.

The work of Li et al. developed a facile and economic strategy to prepare biodegradable thermoresponsive OEGylated poly(amino acid)s with narrow molecular weight distribution. The LCST of these materials can be tuned by changing the length of OEG units. CD characterization suggested that the secondary structures of poly-L-EGxGlus relied on the chain length of OEG side chains. These thermoresponsive poly(amino acid)s with tunable LCST will have great promise to construct new intelligent biomaterials for biomedical applications.

The above work shows a successful example for constructing OEGylated poly(amino acid)s as thermoresponsive materials. Therefore, it is of particular interest and importance to further develop novel and highly efficient methods to make thermoresponsive OEGylated poly(amino acid)s from easily available chemicals. Cysteine/homocysteine are amino acids with a thiol group; thiol is more easily derivatized by thiol-ene Michael addition or nucleophilic substitution reactions, because of their extremely good nucleophilicity. In this context, a series of novel OEGylated poly(amino acid)s were synthesized by Li's group and Deming's group using cystine and homocysteine as starting materials (Figure 6). In 2013, Li et al. reported a series of new functional amino acids which were prepared via thiol-ene Michael addition between L-cysteine and OEG functionalized methacrylates (OEG$_x$MA) or acrylate (OEG$_x$A) in a high yield [35]. These OEGylated cysteine derivatives were converted into NCA monomers using triphosgene. Subsequently, triethylamine (Et_3N) was used to catalyze the ROP of these NCA monomers to give a series of OEGylated poly-L-cysteines (poly-EG$_x$MA-C or poly-EG$_x$A-C) (Figure 6A). The resulting poly-EG$_x$MA-C and poly-EG$_x$A-C displayed OEG length dependent solubility and secondary structure in water. More importantly, when the x value is between three and five, the obtained polymers can display reversible thermoresponsive properties in water, such as poly-EG$_3$A-C, poly-EG$_3$MA-C, and poly-EG$_{4/5}$MA-C, the LCSTs are 50 °C, 65 °C, and 51 °C (Figure 7), respectively. The synthetic strategy represents a highly efficient method to prepare OEGylated poly(amino acid)s with tunable thermoresponsive properties.

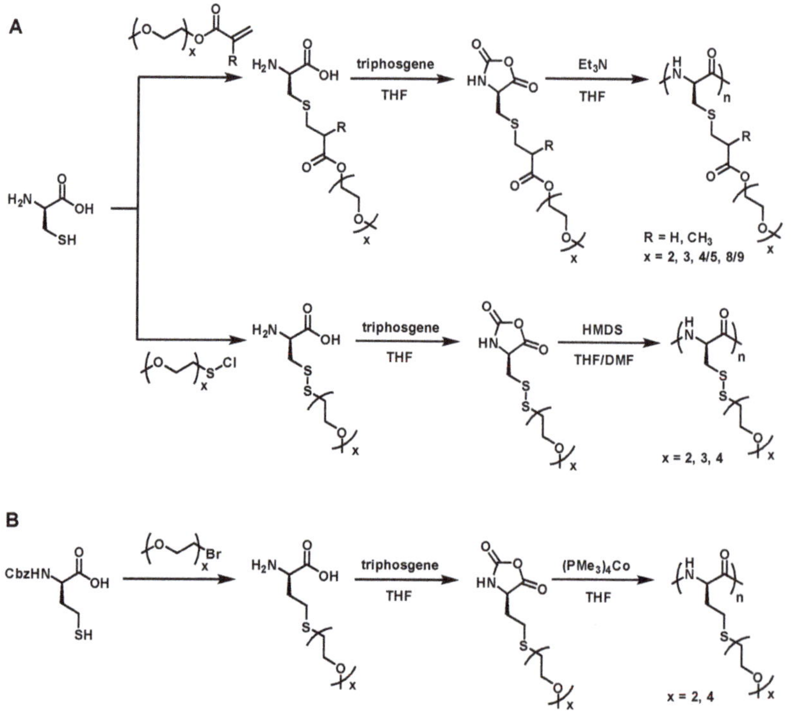

Figure 6. Synthetic routes to OEGylated poly(amino acid)s from cysteine/homocysteine by (**A**) Li's group [35,36], and (**B**) Deming's group [28].

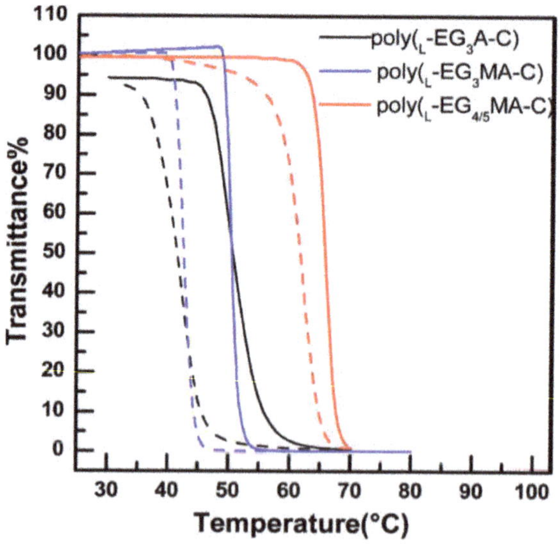

Figure 7. Plots of transmittance as a function of temperature for aqueous solutions (2 mg/mL) of (poly-L-EG3MA-C), (poly-L-EG4/5MA-C), and (poly-L-EG3A-C). Solid line: heating; dashed line: cooling. Reproduced with permission [35]. Copyright 2013, American Chemical Society.

It was known that poly-L-cysteine was a β-sheet forming polypeptide [52]; previous studies showed that conjugation of di(ethylene glycol)thioester to the poly-L-cysteine side chain did not disrupt its β-sheet conformation [32]. However, poly-L-cysteine conjugated with hydrophilic sugars adopted helical conformation [53]. In this report, Li et al. also investigated the effects of OEG side chain length on the secondary structures of poly-L-cysteine derivatives using CD spectroscopy. In Figure 8, the results revealed that both series of samples formed mixed conformation, in which a random coil was the major conformation. This result does not agree with the previous reports; the authors had analyzed that there were two possible reasons for the mixed secondary conformation. One of these was that the synthetic method of poly-L-EG_XMA-C or poly-L-EG_XA-C made them have longer side chains than poly-L-EG_XGlu, and the long side chain could destabilize the stability of secondary structure. Another possible reason was that the MWs were not high enough, which might seriously affect the content of secondary conformation. Li's report provides a new reference for secondary structures of poly-L-cysteine derivatives.

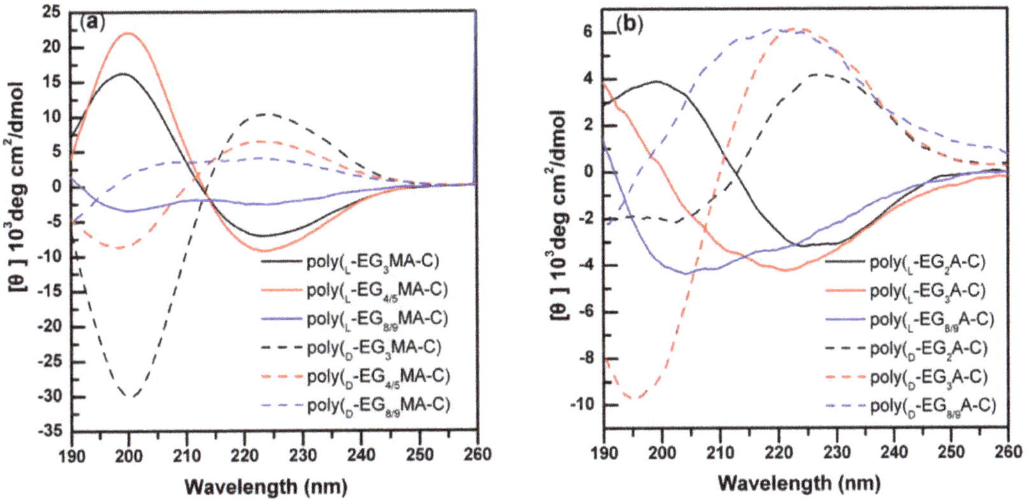

Figure 8. CD spectra of different OEGylated polycysteine homopolypeptides: (**a**) poly-L-EG_XMA-C (solid line) and poly-D-EG_XMA-C (dashed line); (**b**) poly-L-EG_XA-C (solid line) and poly-D-EG_XA-C (dashed line). Reproduced with permission [35]. Copyright 2013, American Chemical Society.

Furthermore, Li et al. also synthesized three cysteine derivatives in high yields by ligating OEG to thiol group of L-cysteine using sulfenyl chlorides [36]. These OEG groups containing di-, tri-, and tetra-OEG units were linked with L-cysteine via disulfide bonds. The three monomers were then converted into corresponding NCAs, and subsequently poly-EG_X-L-cysteines via ROP with HMDS as catalyst (Figure 6A). The obtained poly-EG_X-L-cysteine with $x = 3$ and 4 displayed thermoresponsive behaviors in water, but the temperature-induced phase transition was found to be surprisingly irreversible (Figure 9). Such irreversible thermoresponsive behaviors were attributed to cross-linking arising from disulfide bonds exchanges. Using PEG-NH_2 as macro-initiator, they also prepared two PEG-*b*-poly-EG_X-L-cysteine diblock copolymers, which could undergo irreversible thermal-induced sol-gel transition. These hydrogels displayed partially shear-thinning and rapid recovery properties, allowing new capabilities to construct stimuli-responsive injectable hydrogels in biomedical applications.

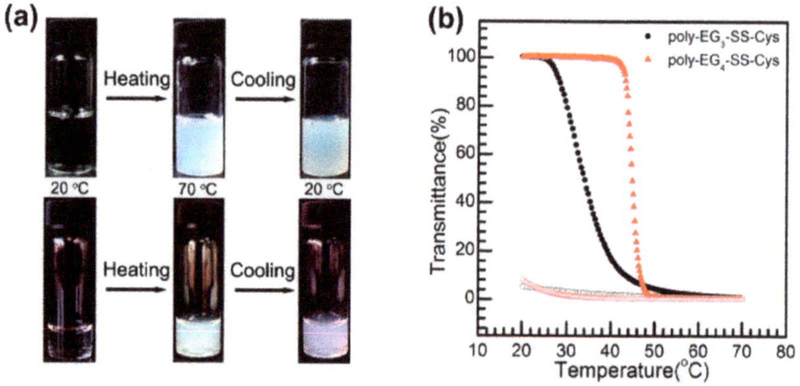

Figure 9. (**a**) Photos of temperature induced phase transition for poly-EG_3-SS-Cys (top) and poly-EG_4-SS-Cys (bottom) aqueous solutions at 2 mg/mL. (**b**) Transmittance as a function of temperature for aqueous solutions (2 mg/mL) of poly-EG_3-SS-Cys (black) and poly-EG_4-SS-Cys (red). Solid symbols: heating ramp. Open symbols: cooling ramp. Reproduced with permission [36]. Copyright 2014, American Chemical Society.

In 2014, Deming et al. reported the design and synthesis of poly(*S*-alkyl-L-homocysteine)s through ROP of homocysteine derived NCAs (Figure 6B). These are a new class of readily prepared, multi-responsive polymers that possess the unprecedented ability to respond to different stimuli, either through a change in conformation or in water solubility [28]. Among them, heating aqueous samples of poly(OEG_4-CH)$_{150}$, sharp transitions from clear solutions to opaque suspensions were observed, indicating the presence of LCST for these OEGylated poly(amino acid)s (Figure 10A). These transitions were completely reversible and could be repeated multiple times with no observable persistent precipitation or other changes to the sample (Figure 10B). In addition, authors also studied the thermoresponsive properties in the presence of different Hofmeister anions in detail, since anions are known to affect thermoresponsive properties of polymers more than cations (Figure 10C). The effects of different salt concentrations on the LCST of poly(OEG_4-CH)$_{150}$ followed trends similar to those seen with other thermoresponsive polymers, and allow tuning of the transition temperature [54,55]. The thermoresponsive properties of OEGylated poly(L-homocysteine)s, combined with their potential adjustability, makes them promising candidates for a broad range of stimuli responsive material challenges.

Despite the crowning achievements in linear OEGylated poly(amino acid)s and manipulation of the properties, exhaustive understanding of the topological architecture of OEG side-chains remains a work in progress. The distinctive topological architecture has distinct properties from their linear analogues, such as solubility, viscosity, and so forth [56]. In 2019, Tao et al. designed and synthesized a series of new linear and Y-shaped OEGylated poly(glutamic acid)s (Figure 11). They have systematically characterized and compared the thermoresponsiveness and secondary structures of several poly(glutamic acid) conjugates including linear and Y-shaped OEGs [27]. The results revealed that the LCST of OEGylated poly(glutamic acid)s could be turned by the length of OEG numbers. More importantly, the LCST of OEGylated poly(glutamic acid)s was firmly correlative to the OEG architecture (Figure 12). For example, the LCST of the Y-shaped poly(YOEG$_8$Glu) was higher than that of its linear analogue poly(LOEG$_8$Glu) (e.g., 60° for poly(YOEG$_8$Glu) and 52° for poly(LOEG$_8$Glu)). This observation is consistent with what one would forecast based on the steric repulsion influence, as the Y-shaped OEGylated polypeptides are more sterically congested than linear OEG because of the dense pendants, which would lead to a greater extent of hydration shell and thus higher LCST. Indeed, steric repulsion may result in the elevated LCST, as already demonstrated by Bitton [56]. However, it appears that this effect is OEG length-dependent and grows pronounced only when the number of the OEG units is ≥6. Notably, the Y-shaped OEGylated poly(glutamic acid)s exhibit higher α-helical conformation than linear ones (Figure 13), which is critically essential in respect to constructing nonionic water-soluble poly(amino acid)s with stable secondary structures. Collectively, this contribution not only provides an appealing route toward Y-shaped OEGylated poly(amino acid)s, but also affords us abundant knowledge to understand how the OEG architecture interferes with the performances of poly(amino acid)s.

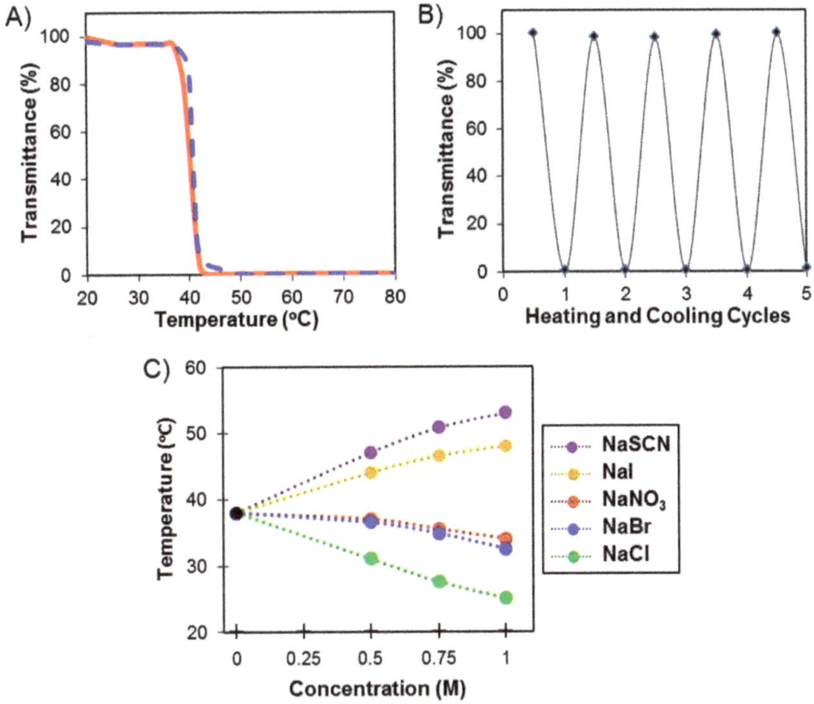

Figure 10. (**A**) Influence of temperature on light transmittance (500 nm) through a sample of aqueous poly(OEG_4-CH)$_{150}$. Solid red line = heating; dashed blue line = cooling; 1 °C/min. (**B**) Reversible change in optical transmittance of aqueous poly(OEG_4-CH)$_{150}$ when temperature was alternated between 30 °C (high transmittance) and 45 °C (low transmittance); 5 min per each heating/cooling cycle. (**C**) Cloud point temperatures of poly(OEG_4-CH)$_{150}$ measured in different Hofmeister salts (Na^+ counterion) at concentrations up to 1.0 M. All poly(amino acid)s were prepared at 3 mg/mL. Reproduced with permission [28]. Copyright 2014, American Chemical Society.

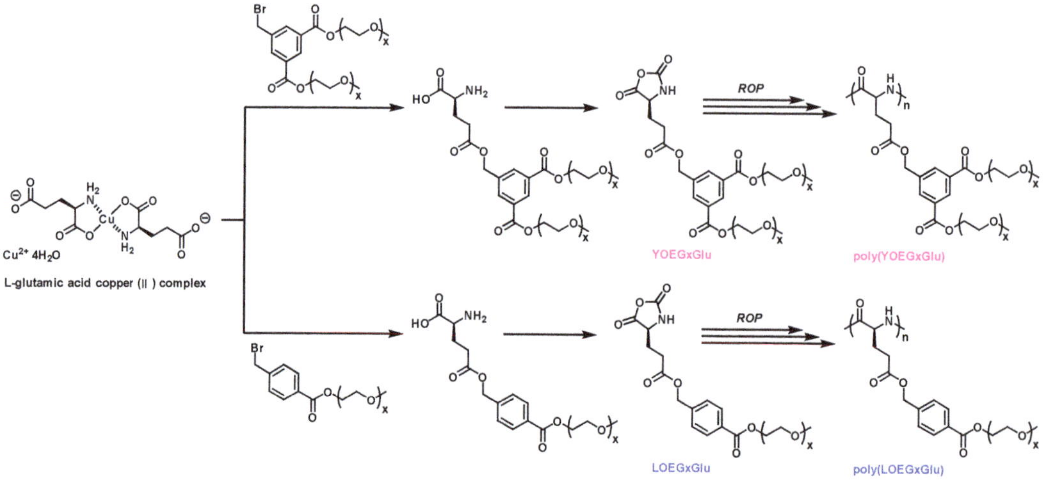

Figure 11. Synthesis of linear and Y-shaped OEGylated poly(YOEG$_x$Glu)s and poly(LOEG$_x$Glu)s via NCA polymerization. Reproduced with permission [27]. Copyright 2019, American Chemical Society.

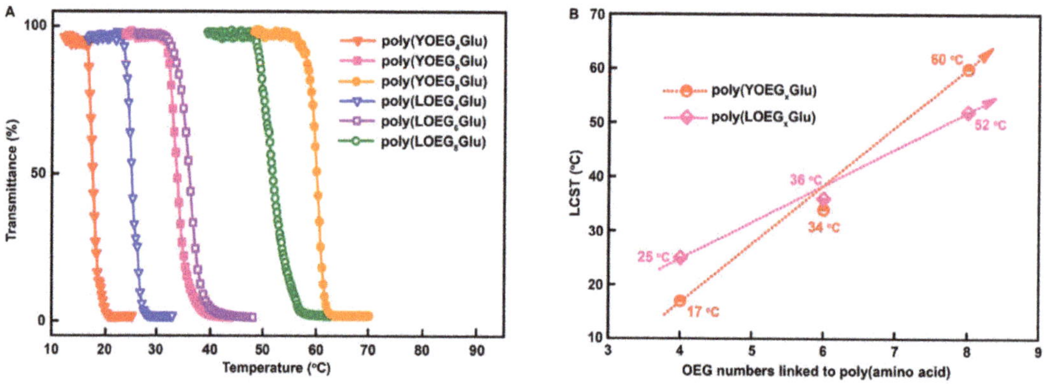

Figure 12. (**A**) Profiles of transmittance vs temperature for the aqueous solutions (2 mg/mL) of poly(LOEG$_x$Glu)s and poly(YOEG$_x$Glu)s. (**B**) LCSTs of poly(LOEGxGlu)s (ding117) and poly(YOEGxGlu)s (•); x represents the number of OEG units. Reproduced with permission [27]. Copyright 2019, American Chemical Society.

2.2. *Thermoresponsive OEGylated Poly(amino acid)s from Post-Polymerization Modification*

Post-polymerization modification of poly(amino acid)s is a facility method to construct OEGylated poly(amino acid)s with tunable OEG side-chain length and properties. "Click chemistry" is undoubtedly the most suitable method. In this respect, Chen's group firstly synthesized a series of novel alkyne functionalized poly(L-glutamic acid) via ROP of alkyne functionalized NCAs. Subsequently, the pendant alkyne groups coupled with 1-(2-methoxyethoxy)-2-azidoethane (MEO_2-N_3) or 1-(2-(2-methoxyethoxy)ethoxy)-2-azidoethane (MEO_3-N_3) by the efficient azide-alkyne "Click chemistry" to obtain OEGylated poly(amino acid)s (Figure 14) [37]. These were named PPLG$_n$-*g*-MEO$_x$. The graft copolymers exhibited sharp temperature dependent phase transitions, and the LCST could be adjusted from 22.3 °C to 74.1 °C by varying the molecular weight and the length of the OEG side chains (Figure 15). In addition, these OEG graft poly(amino acid)s were confirmed to be biocompatible and non-toxic using the methyl thiazolyl tetrazolium (MTT) method, and they were degradable in the presence of proteinase K. Drug loading and release was also conducted with these thermosensitive nanoparticles using doxorubicin as the model drug, and a temperature-dependent sustained release profile was observed. Therefore, it is believed that these novel OEG graft poly(L-glutamic acid)s with tunable temperature responsiveness should be promising for smart biomedical applications.

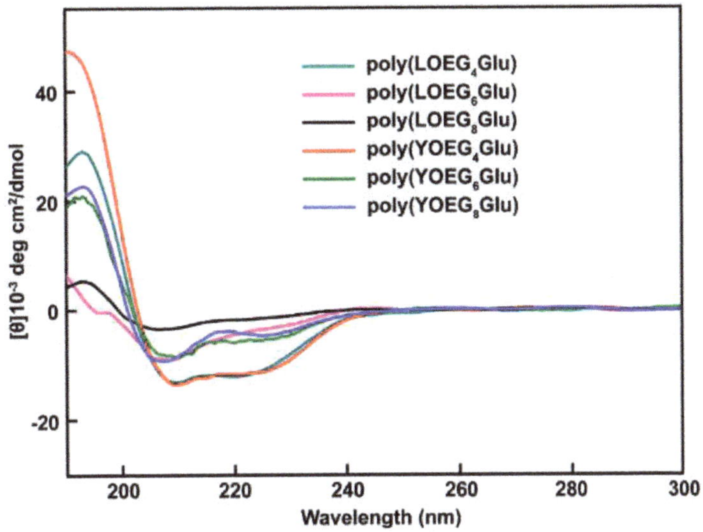

Figure 13. CD spectra of poly(LOEG$_x$Glu)s and poly(YOEG$_x$Glu)s measured in CH_3CN at 25 °C (c = 0.5 mg/mL). Reproduced with permission [27]. Copyright 2019, American Chemical Society.

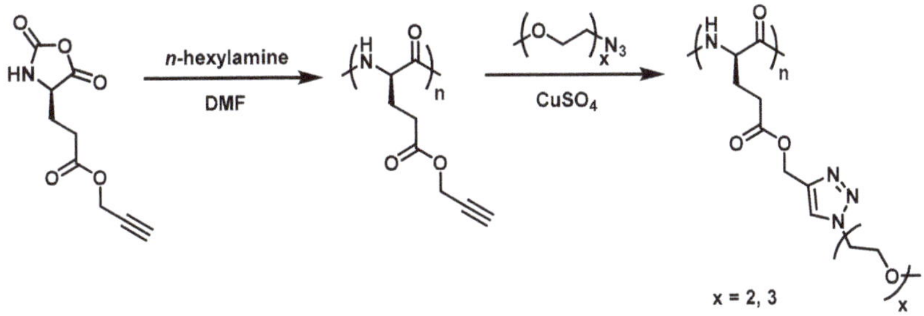

Figure 14. Synthesis route of the OEG graft poly(L-glutamate) by Chen's group. Reproduced with permission [37]. Copyright 2011, Royal Society of Chemistry.

Thermoresponsive and pH-responsive graft co-polymers, PLG-*g*-OMEO$_3$MA and P(LGA-*co*-(LG-*g*-OMEO$_3$MA)), were also synthesized by Chen and coworkers though ROP of NCA monomers and subsequent atom transfer radical polymerization of 2-(2-(2-methoxyethoxy)ethoxy)ethyl methacrylate in 2011 (Figure 16) [38]. The thermoresponsive of OEG graft co-polymers could be tuned by the MWs of OMEO$_3$MA, the composition of poly(L-glutamic acid) (PLGA) and the pH of the aqueous solution. The α-helical contents of graft copolymers could be influenced by OMEO$_3$MA length and pH of the aqueous solution. In addition, the graft copolymers exhibited tunable self-assembly behavior. The hydrodynamic radius (R_h) and critical micellization concentration values of micelles were relevant to the length of OMEO$_3$MA and the composition of the biodegradable PLGA backbone. The R_h could also be adjusted by the temperature and pH values. Lastly, in vitro MTT assay revealed that the graft copolymers were biocompatible to HeLa cells.

Therefore, these graft copolymers with good biocompatibility, well-defined secondary structure, and mono- dual-responsiveness, are promising stimuli-responsive materials for biomedical applications.

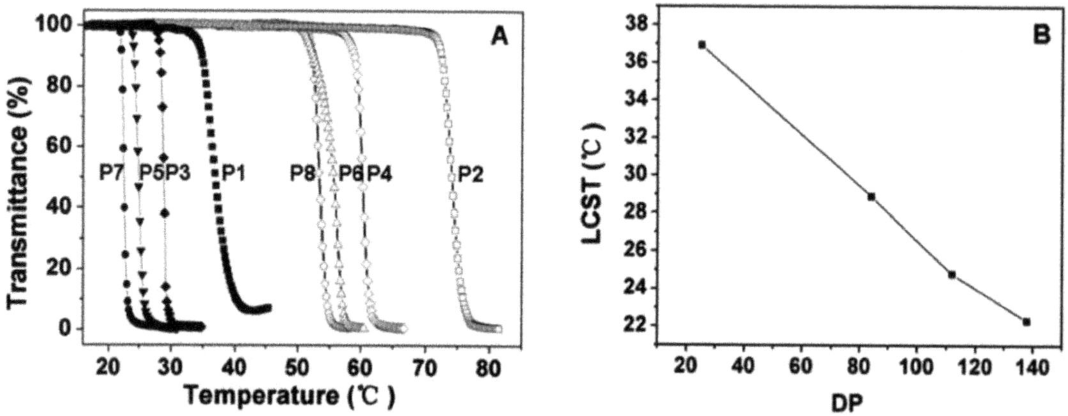

Figure 15. (**A**) Thermal phase transitions of P1 (PPLG$_{25}$-*g*-MEO$_2$), P2 (PPLG$_{25}$-*g*-MEO$_3$), P3 (PPLG$_{84}$-*g*-MEO$_2$), P4 (PPLG$_{84}$-*g*-MEO$_3$), P5 (PPLG$_{112}$-*g*-MEO$_2$), P6 (PPLG$_{112}$-*g*-MEO$_3$), P7 (PPLG$_{138}$-*g*-MEO$_2$) and P8 (PPLG$_{138}$-*g*-MEO$_3$), and the polymer concentration was 10 g/L. (**B**) The LCSTs of PPLG-*g*-MEO$_2$ as a function of the degree of polymerization. Reproduced with permission [37]. Copyright 2011, Royal Society of Chemistry.

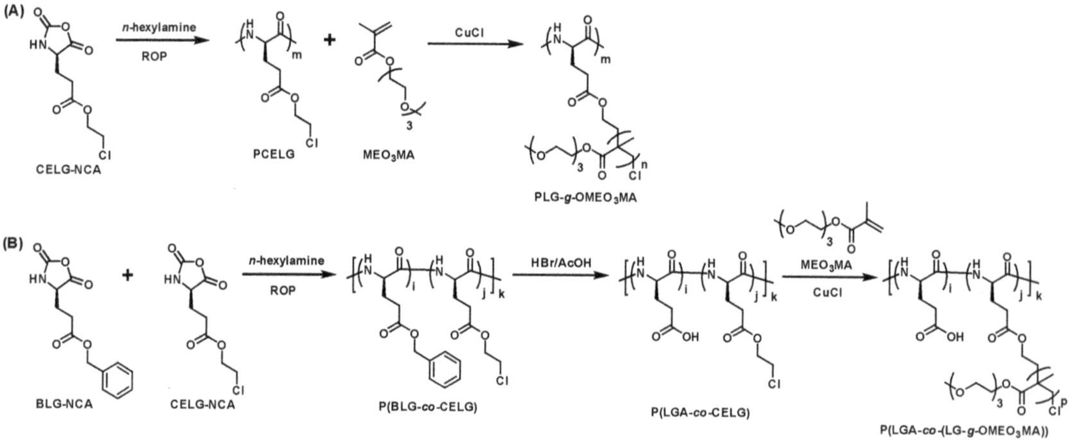

Figure 16. Synthetic routes of copolymers (**A**) PLG-*g*-OMEO$_3$MA and (**B**) P(LGA-*co*-(LG-*g*-OMEO$_3$MA)) by Chen et al. Reproduced with permission [38]. Copyright 2011, Wiley Periodicals, Inc.

2.3. *Thermoresponsive OEGylated Poly(amino acid)s from Ugi Multicomponent Polymerization*

Ugi reaction is a four-component reaction, and the reactants are commonly acid, amine, isocyanide, and aldehyde [57,58]. This reaction has gained great attention and been utilized in the fields of combinatorial chemistry, pharmaceutics, and life science due to its mild reaction conditions, high efficiency, functional group tolerance, and atom economy [59,60]. Recently, Meier demonstrated a very efficient and modular approach to synthesizing diversely substituted polyamides via the Ugi four-component reaction [45]. In contrast to conventional polyamide synthesis, this approach proceeds under very mild reaction conditions and without the use of a catalyst in a one-pot reaction. Subsequently, Tao and coworkers reported the synthesis of structurally diverse poly(amino acid)s (also called polypeptoids) by Ugi polymerization of natural amino acids under mild conditions, and this strategy offered a general methodology toward facile preparation of functionalized poly(amino acid)s [46]. Based on this work, Tao et al. also designed and synthesized a series of new alternating poly(amino acid)s via the Ugi reaction of readily available natural amino acids. Among them, the thermoresponsive OEGylated poly(amino acid)s have been prepared using oligo(ethylene glycol) isocyanide (Figure 17), and exhibited cloud points (T_{cp}) between 27 °C and 37 °C (Figure 18). The alternating structure and diverse polymer properties described here offer a new direction for the synthesis of novel OEGylated poly(amino acid)s materials [39].

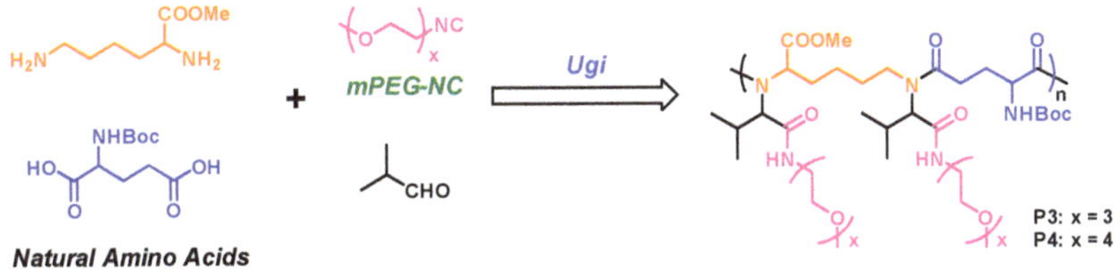

Figure 17. Synthesis of alternating OEGylated poly(amino acid)s via Ugi reaction of natural amino acids. Reproduced with permission [39]. Copyright 2018, American Chemical Society.

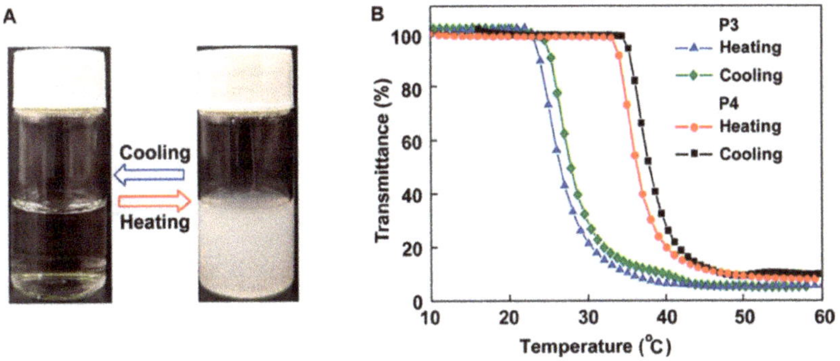

Figure 18. (**A**) Visual turbidity change of P4 upon heating the aqueous solution. (**B**) Temperature dependence of transmittance for the aqueous solutions (2 mg/mL) of P3 and P4 (500 nm, heating or cooling at a rate of 1 °C/min). Reproduced with permission [39]. Copyright 2018, American Chemical Society.

Recently, sequence-controlled synthetic polymers have been drawing great interest because the specific sequences can endow more advanced functions to the polymers, such as DNA and proteins [61–65]; however, the control remains a great challenge in polymer science, and this drives people to find more concise and efficient methods to construct polymers with accurate sequence structure. Additionally, the molecular functions and properties determined by the sequence structure are less studied. In this context, Tao et al. firstly reported the development of amino acid building blocks coupled with iterative Ugi reactions for the efficient and multigram-scale assembly of sequence-defined poly(amino acid)s (Figure 19) [40]. This efficient chemistry provides much feasibility for structural diversity, synthetic varying and sequencing both the side chains and the backbones. Using this advanced method, they coupled the OEG units in the sequence-defined polymers, and further demonstrated that the alteration in the overall hydrophobicity and LCST behaviors of these precisely defined OEGylated poly(amino acid)s could be accordingly changed by variation of the sequence (Figure 20). Regulation of sequence-specific hydrophobic aggregation within a polymer is a significant result. This versatile strategy may afford new materials for application in therapeutics, and as supramolecular foldamers or simple protein mimics for the investigation of advanced self-assembly driven by hydrophobic or other supramolecular interactions.

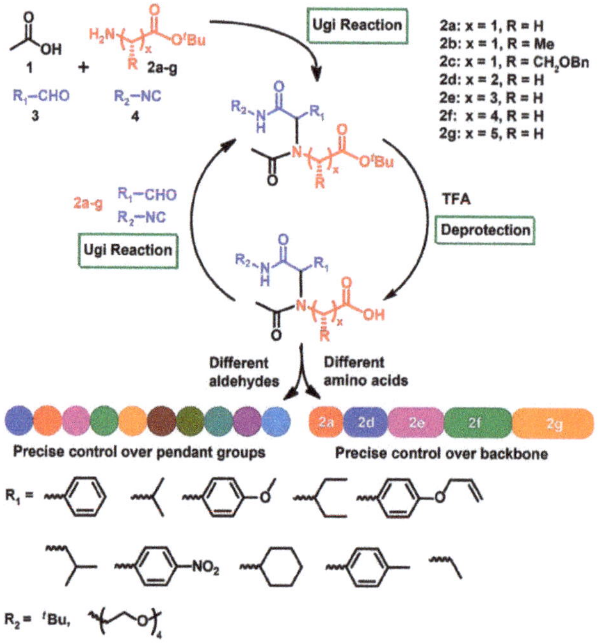

Figure 19. Synthesis strategy towards sequence-defined poly(amino acid)s via amino acid building blocks and iterative Ugi reactions. Reproduced with permission [40]. Copyright 2019, Royal Society of Chemistry.

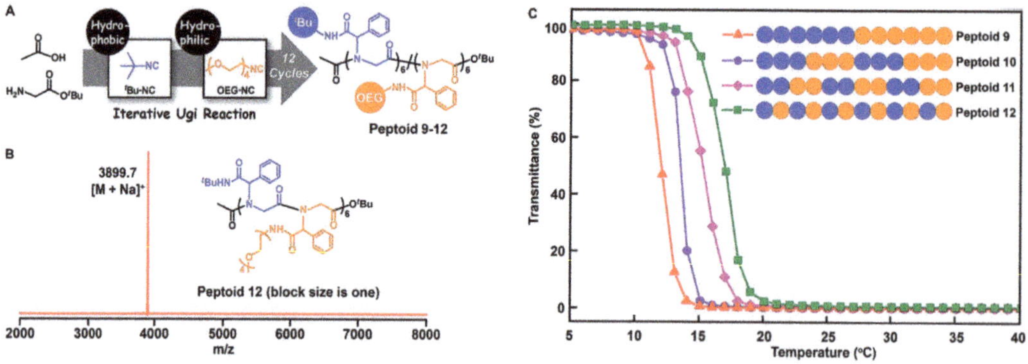

Figure 20. (**A**) Illustration of the side-chain sequence-regulated OEGylated poly(amino acid)s 9–12 synthesized by iterative Ugi reactions. (**B**) MALDI-TOF-MS spectrum of 12. (**C**) Temperature dependence of transmittance for the aqueous solutions (2 mg/mL) of sequence-regulated OEGylated poly(amino acid)s 9–12 (500 nm, heating at a rate of 1 °C/min). Reproduced with permission [40]. Copyright 2019, Royal Society of Chemistry.

3. Applications

Applications of thermoresponsive polymers have been studied in many fields, such as biomedicine, tissue engineering, and in sensors [15,66,67]. For example, PNIPAM, as the most widely studied thermoresponsive polymer, has been extensively used in many biomedical applications. Andrew and coworkers loaded the insulin on PNIPAM based thermoresponsive microgels, and studied the release using variable temperature ^{1}H NMR [68]. This type of direct release investigation could prove to be a useful method in the future design of controlled macromolecule drug delivery devices. Additionally, thermoresponsive polypeptides and polypeptoids also show a good application prospect. Lu et al. synthesized a series of thermoresponsive polypeptides incorporated with various functional side groups [69]. They found that these polypeptides showed perfect hemostatic properties and healing effects which are expected to be potential candidates for medical applications, such as tissue adhesives. Zhang and coworkers demonstrated that water-soluble horseradish peroxidases can be easily encapsulated in a thermoresponsive triblock copolypeptoid hydrogels for extended period of time with the retained enzymatic activity, and the hydrogels show low cytotoxicity towards human adipose-derived stem cells [70]. These results indicate the potential utilization of the polypeptoid hydrogel as tissue engineering material.

For thermoresponsive OEGylated Poly(amino acid)s, little application research has been conducted. Chen et al. evaluated the cytotoxicity of graft copolymers $PPLG_{112}$-g-MEO_2 in vitro by MTT assay [37]. It was observed that the HeLa cells treated with $PPLG_{112}$-g-MEO_2 remained almost 100% viable at all test concentrations up to 1 mg/mL, indicating non-cytotoxicity and good biocompatibility of the graft copolymers. Then, they investigated the temperature-dependent drug release behavior of the drug-loaded co-polymer nanoparticles, and doxorubicin (Dox) was used as a model drug. The results suggested that the drug release from the temperature-sensitive amphiphilic nanoparticles could be accelerated by increasing the temperature above their LCSTs. In addition, the release of Dox displayed a constant rate in the first 24 h at 37 °C. This demonstrated that an ideal constant Dox release could be obtained in the nanoparticle system. The result suggested that the graft co-polymers could be promising candidates as drug carriers for controlled drug delivery.

4. Conclusions and Outlook

Thermoresponsive OEGylated poly(amino acid)s combined the advantages of OEG and poly(amino acid)s with great solubility, excellent biocompatibility and well-defined secondary structures. These advantages allow it to have great application prospects in the field of biomedicine, tissue engineering, and sensors. However, it is still in the initial research stage. In this review, we summarized the research progress of thermoresponsive OEGylated poly(amino acid)s in recent years, including the synthesis methods, stimuli-response behavior study, and secondary structure study of these OEGylated poly(amino acid)s. We hope that this periodical summary will be more conducive to the design, synthesis and application of OEGylated poly(amino acid)s in the future.

In addition, the design of the structure and function for practical applications is the future development direction of OEGylated poly(amino acid)s; it is suggested to carry out targeted research in the following aspects. In respect to polymer design, the topological structure of side-chains (such as cyclic, star-shape) need to be expanded to bring about the diversity of structures and functions. Moreover, it is necessary to select the appropriate amino acids according to the secondary structure requirements of OEGylated poly(amino acid) materials. In respect to the synthesis method, although ROP of NCAs and the post-polymerization modification are the typical methods to construct OEGylated poly(amino acid)s, some novel strategies need to be developed for efficient preparation of the sequence-defined polymers, such as the Ugi multi-component polymerization. Finally, its application in the field of medicine should be further strengthened due to its great application prospect.

Author Contributions: The manuscript was written through contributions by all authors. All authors have read and agreed to the published version of the manuscript.

Funding: This work is supported by the National Natural Science Foundation of China (Grant No. 52073274).

Institutional Review Board Statement: Not applicable.

Informed Consent Statement: Not applicable.

Data Availability Statement: Not applicable.

Conflicts of Interest: The authors declare no conflict of interest.

References

1. Mura, S.; Nicolas, J.; Couvreur, P. Stimuli-responsive nanocarriers for drug delivery. *Nat. Mater.* **2013**, *12*, 991–1003. [CrossRef] [PubMed]
2. Stuart, M.A.C.; Huck, W.T.S.; Genzer, J.; Müller, M.; Ober, C.; Stamm, M.; Sukhorukov, G.B.; Szleifer, I.; Tsukruk, V.V.; Urban, M.; et al. Emerging applications of stimuli-responsive polymer materials. *Nat. Mater.* **2010**, *9*, 101–113. [CrossRef] [PubMed]
3. Hoffman, A.S. Stimuli-responsive polymers: Biomedical applications and challenges for clinical translation. *Adv. Drug Deliv. Rev.* **2013**, *65*, 10–16. [CrossRef] [PubMed]
4. Nath, N.; Chilkoti, A. Creating "Smart" Surfaces Using Stimuli Responsive Polymers. *Adv. Mater.* **2002**, *14*, 1243–1247. [CrossRef]
5. Zhao, C.; Nie, S.; Tang, M.; Sun, S. Polymeric pH-sensitive membranes—A review. *Prog. Polym. Sci.* **2011**, *36*, 1499–1520. [CrossRef]
6. Dong, J.; Wang, Y.; Zhang, J.; Zhan, X.; Zhu, S.; Yang, H.; Wang, G. Multiple stimuli-responsive polymeric micelles for controlled release. *Soft Matter* **2013**, *9*, 370–373. [CrossRef]
7. Schmaljohann, D. Thermo-and pH-responsive polymers in drug delivery. *Adv. Drug Deliv. Rev.* **2006**, *58*, 1655–1670. [CrossRef]
8. Bawa, P.; Pillay, V.; Choonara, Y.E.; Du Toit, L.C. Stimuli-responsive polymers and their applications in drug delivery. *Biomed. Mater.* **2009**, *4*, 022001. [CrossRef]
9. Gil, E.S.; Hudson, S. Stimuli-reponsive polymers and their bioconjugates. *Prog. Polym. Sci.* **2004**, *29*, 1173–1222. [CrossRef]
10. Wang, S.; Liu, Q.; Li, L.; Urban, M.W. Recent Advances in Stimuli-Responsive Commodity Polymers. *Macromol. Rapid Commun.* **2021**, 2100054. [CrossRef]
11. Wang, Y.; Weng, J.; Wen, X.; Hu, Y.; Ye, D. Recent advances in stimuli-responsive in situ self-assembly of small molecule probes for in vivo imaging of enzymatic activity. *Biomater. Sci.* **2021**, *9*, 406–421. [CrossRef]
12. Li, L.; Yang, Z.; Chen, X. Recent Advances in Stimuli-Responsive Platforms for Cancer Immunotherapy. *Accounts Chem. Res.* **2020**, *53*, 2044–2054. [CrossRef]
13. Pasparakis, G.; Tsitsilianis, C. LCST polymers: Thermoresponsive nanostructured assemblies towards bioapplications. *Polymer* **2020**, *211*, 123146. [CrossRef]
14. Vanparijs, N.; Nuhn, L.; De Geest, B.G. Transiently thermoresponsive polymers and their applications in biomedicine. *Chem. Soc. Rev.* **2017**, *46*, 1193–1239. [CrossRef]
15. Hogan, K.J.; Mikos, A.G. Biodegradable thermoresponsive polymers: Applications in drug delivery and tissue engineering. *Polymer* **2020**, *211*, 123063. [CrossRef]
16. Zhu, Y.; Batchelor, R.; Lowe, A.B.; Roth, P.J. Design of Thermoresponsive Polymers with Aqueous LCST, UCST, or Both: Modification of a Reactive Poly(2-vinyl-4,4-dimethylazlactone) Scaffold. *Macromolecules* **2016**, *49*, 672–680. [CrossRef]
17. Schild, H.G. Poly(N-isopropylacrylamide): Experiment, theory and application. *Prog. Polym. Sci.* **1992**, *17*, 163–249. [CrossRef]
18. Ko, C.-H.; Claude, K.-L.; Niebuur, B.-J.; Jung, F.A.; Kang, J.-J.; Schanzenbach, D.; Frielinghaus, H.; Barnsley, L.C.; Wu, B.; Pipich, V.; et al. Temperature-Dependent Phase Behavior of the Thermoresponsive Polymer Poly(N-isopropylmethacrylamide) in an Aqueous Solution. *Macromolecules* **2020**, *53*, 6816–6827. [CrossRef]
19. Lutz, J.-F.; Akdemir, Ö.; Hoth, A. Point by Point Comparison of Two Thermosensitive Polymers Exhibiting a Similar LCST: Is the Age of Poly(NIPAM) Over? *J. Am. Chem. Soc.* **2006**, *128*, 13046–13047. [CrossRef]
20. Lutz, J.-F.; Andrieu, J.; Üzgün, S.; Rudolph, C.; Agarwal, S. Biocompatible, Thermoresponsive, and Biodegradable: Simple Preparation of "All-in-One" Biorelevant Polymers. *Macromolecules* **2007**, *40*, 8540–8543. [CrossRef]
21. Qiao, Z.-Y.; Du, F.-S.; Zhang, R.; Liang, D.-H.; Li, Z.-C. Biocompatible Thermoresponsive Polymers with Pendent Oligo(ethylene glycol) Chains and Cyclic Ortho Ester Groups. *Macromolecules* **2010**, *43*, 6485–6494. [CrossRef]
22. Roth, P.J.; Jochum, F.D.; Theato, P. UCST-type behavior of poly[oligo(ethylene glycol) methyl ether methacrylate] (POEGMA) in aliphatic alcohols: Solvent, co-solvent, molecular weight, and end group dependences. *Soft Matter* **2011**, *7*, 2484–2492. [CrossRef]
23. Chua, G.B.H.; Roth, P.J.; Duong, H.T.T.; Davis, T.P.; Lowe, A.B. Synthesis and Thermoresponsive Solution Properties of Poly[oligo(ethylene glycol) (meth) acrylamide]s: Biocompatible PEG Analogues. *Macromolecules* **2012**, *45*, 1362–1374. [CrossRef]
24. Vancoillie, G.; Frank, D.; Hoogenboom, R. Thermoresponsive poly (oligo ethylene glycol acrylates). *Prog. Polym. Sci.* **2014**, *39*, 1074–1095. [CrossRef]

25. Jia, F.; Lu, X.; Tan, X.; Wang, D.; Cao, X.; Zhang, K. Effect of PEG Architecture on the Hybridization Thermodynamics and Protein Accessibility of PEGylated Oligonucleotides. *Angew. Chem. Int. Ed.* **2017**, *56*, 1239–1243. [CrossRef]
26. Harris, J.M.; Chess, R.B. Effect of pegylation on pharmaceuticals. *Nat. Rev. Drug Discov.* **2003**, *2*, 214–221. [CrossRef]
27. Wang, S.; He, W.; Xiao, C.; Tao, Y.; Wang, X. Synthesis of Y-Shaped OEGylated Poly(amino acid)s: The Impact of OEG Architecture. *Biomacromolecules* **2019**, *20*, 1655–1666. [CrossRef]
28. Kramer, J.R.; Deming, T.J. Multimodal Switching of Conformation and Solubility in Homocysteine Derived Polypeptides. *J. Am. Chem. Soc.* **2014**, *136*, 5547–5550. [CrossRef]
29. Lu, H.; Wang, J.; Bai, Y.; Lang, J.W.; Liu, S.; Lin, Y.; Cheng, J. Ionic polypeptides with unusual helical stability. *Nat. Commun.* **2011**, *2*, 206. [CrossRef]
30. Yu, M.; Nowak, A.P.; Deming, T.J.; Pochan, D.J. Methylated Mono-and Diethyleneglycol Functionalized Polylysines: Nonionic, α-Helical, Water-Soluble Polypeptides. *J. Am. Chem. Soc.* **1999**, *121*, 12210–12211. [CrossRef]
31. Sternhagen, G.L.; Gupta, S.; Zhang, Y.; John, V.; Schneider, G.J.; Zhang, D. Solution Self-Assemblies of Sequence-Defined Ionic Peptoid Block Copolymers. *J. Am. Chem. Soc.* **2018**, *140*, 4100–4109. [CrossRef]
32. Hwang, J.; Deming, T.J. Methylated Mono- and Di(ethylene glycol)-Functionalized β-Sheet Forming Polypeptides. *Biomacromolecules* **2000**, *2*, 17–21. [CrossRef]
33. Cao, J.; Hu, P.; Lu, L.; Chan, B.A.; Luo, B.-H.; Zhang, D. Non-ionic water-soluble "clickable" [small alpha]-helical polypeptides: Synthesis, characterization and side chain modification. *Polym. Chem.* **2015**, *6*, 1226–1229. [CrossRef]
34. Chen, C.; Wang, Z.; Li, Z. Thermoresponsive Polypeptides from Pegylated Poly-l-glutamates. *Biomacromolecules* **2011**, *12*, 2859–2863. [CrossRef]
35. Fu, X.; Shen, Y.; Fu, W.; Li, Z. Thermoresponsive Oligo(ethylene glycol) Functionalized Poly-l-cysteine. *Macromolecules* **2013**, *46*, 3753–3760. [CrossRef]
36. Ma, Y.; Fu, X.; Shen, Y.; Fu, W.; Li, Z. Irreversible Low Critical Solution Temperature Behaviors of Thermal-responsive OEGylated Poly(l-cysteine) Containing Disulfide Bonds. *Macromolecules* **2014**, *47*, 4684–4689. [CrossRef]
37. Cheng, Y.; He, C.; Xiao, C.; Ding, J.; Zhuang, X.; Chen, X. Versatile synthesis of temperature-sensitive polypeptides by click grafting of oligo(ethylene glycol). *Polym. Chem.* **2011**, *2*, 2627–2634. [CrossRef]
38. Ding, J.; Xiao, C.; Zhao, L.; Cheng, Y.; Ma, L.; Tang, Z.; Zhuang, X.; Chen, X. Poly(L-glutamic acid) grafted with oligo(2-(2-(2-methoxyethoxy)ethoxy)ethyl methacrylate): Thermal phase transition, secondary structure, and self-assembly. *J. Polym. Sci. Part A Polym. Chem.* **2011**, *49*, 2665–2676. [CrossRef]
39. Tao, Y.; Wang, S.; Zhang, X.; Wang, Z.; Tao, Y.; Wang, X. Synthesis and Properties of Alternating Polypeptoids and Polyampholytes as Protein-Resistant Polymers. *Biomacromolecules* **2018**, *19*, 936–942. [CrossRef]
40. Wang, S.; Tao, Y.; Wang, J.; Tao, Y.; Wang, X. A versatile strategy for the synthesis of sequence-defined peptoids with side-chain and backbone diversity via amino acid building blocks. *Chem. Sci.* **2018**, *10*, 1531–1538. [CrossRef]
41. Brauch, S.; van Berkel, S.S.; Westermann, B. Higher-order multicomponent reactions: Beyond four reactants. *Chem. Soc. Rev.* **2013**, *42*, 4948–4962. [CrossRef]
42. Ruijter, E.; Scheffelaar, R.; Orru, R.V.A. Multicomponent Reaction Design in the Quest for Molecular Complexity and Diversity. *Angew. Chem. Int. Ed.* **2011**, *50*, 6234–6246. [CrossRef]
43. Slobbe, P.; Ruijter, E.; Orru, R.V.A. Recent applications of multicomponent reactions in medicinal chemistry. *MedChemComm* **2012**, *3*, 1189–1218. [CrossRef]
44. Dömling, A.; Wang, W.; Wang, K. Chemistry and Biology of Multicomponent Reactions. *Chem. Rev.* **2012**, *112*, 3083–3135. [CrossRef]
45. Sehlinger, A.; Dannecker, P.-K.; Kreye, O.; Meier, M.A.R. Diversely Substituted Polyamides: Macromolecular Design Using the Ugi Four-Component Reaction. *Macromolecules* **2014**, *47*, 2774–2783. [CrossRef]
46. Zhang, X.; Wang, S.; Liu, J.; Xie, Z.; Luan, S.; Xiao, C.; Tao, Y.; Wang, X. Ugi Reaction of Natural Amino Acids: A General Route toward Facile Synthesis of Polypeptoids for Bioapplications. *ACS Macro Lett.* **2016**, *5*, 1049–1054. [CrossRef]
47. Aliferis, T.; Iatrou, H.; Hadjichristidis, N. Living Polypeptides. *Biomacromolecules* **2004**, *5*, 1653–1656. [CrossRef]
48. Lu, H.; Cheng, J. Hexamethyldisilazane-Mediated Controlled Polymerization of α-Amino Acid N-Carboxyanhydrides. *J. Am. Chem. Soc.* **2007**, *129*, 14114–14115. [CrossRef]
49. Deming, T.J. Transition Metal−Amine Initiators for Preparation of Well-Defined Poly(γ-benzyl L-glutamate). *J. Am. Chem. Soc.* **1997**, *119*, 2759–2760. [CrossRef]
50. Deming, T.J. Facile synthesis of block copolypeptides of defined architecture. *Nature* **1997**, *390*, 386–389. [CrossRef]
51. Deming, T.J. Amino Acid Derived Nickelacycles: Intermediates in Nickel-Mediated Polypeptide Synthesis. *J. Am. Chem. Soc.* **1998**, *120*, 4240–4241. [CrossRef]
52. Berger, A.; Noguchi, J.; Katchalski, E. Poly-L-cysteine. *J. Am. Chem. Soc.* **1956**, *78*, 4483–4488. [CrossRef]
53. Kramer, J.R.; Deming, T.J. Glycopolypeptides via Living Polymerization of Glycosylated-L-lysine *N*-Carboxyanhydrides. *J. Am. Chem. Soc.* **2010**, *132*, 15068–15071. [CrossRef] [PubMed]
54. Zhang, Y.; Furyk, S.; Bergbreiter, D.E.; Cremer, P.S. Specific Ion Effects on the Water Solubility of Macromolecules: PNIPAM and the Hofmeister Series. *J. Am. Chem. Soc.* **2005**, *127*, 14505–14510. [CrossRef]
55. Deyerle, B.A.; Zhang, Y. Effects of Hofmeister Anions on the Aggregation Behavior of PEO–PPO–PEO Triblock Copolymers. *Langmuir* **2011**, *27*, 9203–9210. [CrossRef]

56. Navon, Y.; Zhou, M.; Matson, J.B.; Bitton, R. Dendritic Elastin-like Peptides: The Effect of Branching on Thermoresponsiveness. *Biomacromolecules* **2016**, *17*, 262–270. [CrossRef]
57. Ugi, I.; Steinbrückner, C. Über ein neues Kondensations-Prinzip. *Angew. Chem.* **1960**, *72*, 267–268. [CrossRef]
58. Ugi, I. From Isocyanides via Four-Component Condensations to Antibiotic Syntheses. *Angew. Chem. Int. Ed. Engl.* **1982**, *21*, 810–819. [CrossRef]
59. Tao, Y.; Tao, Y. Ugi Reaction of Amino Acids: From Facile Synthesis of Polypeptoids to Sequence-Defined Macromolecules. *Macromol. Rapid Commun.* **2020**, *42*, 2000515.
60. Tao, Y.; Wang, Z.; Tao, Y. Polypeptoids synthesis based on Ugi reaction: Advances and perspectives. *Biopolymers* **2019**, *110*, e23288. [CrossRef]
61. Lutz, J.-F.; Lehn, J.-M.; Meijer, E.W.; Matyjaszewski, K. From precision polymers to complex materials and systems. *Nat. Rev. Mater.* **2016**, *1*, 16024. [CrossRef]
62. Stross, A.E.; Iadevaia, G.; Núñez-Villanueva, D.; Hunter, C.A. Sequence-Selective Formation of Synthetic H-Bonded Duplexes. *J. Am. Chem. Soc.* **2017**, *139*, 12655–12663. [CrossRef]
63. Badi, N.; Lutz, J.-F. Sequence control in polymer synthesis. *Chem. Soc. Rev.* **2009**, *38*, 3383–3390. [CrossRef]
64. Ouchi, M.; Badi, N.; Lutz, J.-F.; Sawamoto, M. Single-chain technology using discrete synthetic macromolecules. *Nat. Chem.* **2011**, *3*, 917–924. [CrossRef]
65. Lutz, J.-F.; Ouchi, M.; Liu, D.R.; Sawamoto, M. Sequence-Controlled Polymers. *Science* **2013**, *341*, 628–637. [CrossRef]
66. Lau, K.H.A. Peptoids for biomaterials science. *Biomater. Sci.* **2014**, *2*, 627–633. [CrossRef]
67. Hu, J.; Liu, S. Responsive Polymers for Detection and Sensing Applications: Current Status and Future Developments. *Macromolecules* **2010**, *43*, 8315–8330. [CrossRef]
68. Nolan, C.M.; Gelbaum, L.T.; Lyon, L.A. H NMR Investigation of Thermally Triggered Insulin Release from Poly(N-isopropylacrylamide) Microgels. *Biomacromolecules* **2006**, *7*, 2918–2922. [CrossRef]
69. Lu, D.; Wang, H.; Li, T.E.; Li, Y.; Wang, X.; Niu, P.; Guo, H.; Sun, S.; Wang, X.; Guan, X.; et al. Versatile Surgical Adhesive and Hemostatic Materials: Synthesis, Properties, and Application of Thermoresponsive Polypeptides. *Chem. Mater.* **2017**, *29*, 5493–5503. [CrossRef]
70. Xuan, S.; Lee, C.-U.; Chen, C.; Doyle, A.B.; Zhang, Y.; Guo, L.; John, V.T.; Hayes, D.; Zhang, D. Thermoreversible and Injectable ABC Polypeptoid Hydrogels: Controlling the Hydrogel Properties through Molecular Design. *Chem. Mater.* **2015**, *28*, 727–737. [CrossRef]

polymers MDPI

rticle

[T]hermocontrolled Reversible Enzyme Complexation-[I]nactivation-Protection by Poly(*N*-acryloyl glycinamide)

[P]avel I. Semenyuk [1,*], Lidia P. Kurochkina [1], Lauri Mäkinen [2], Vladimir I. Muronetz [1] and Sami Hietala [2]

[1] Belozersky Institute of Physico-Chemical Biology, Lomonosov Moscow State University, 119234 Moscow, Russia; lpk@belozersky.msu.ru (L.P.K.); vimuronets@belozersky.msu.ru (V.I.M.)
[2] Department of Chemistry, University of Helsinki, FIN-00014 Helsinki, Finland; Lauri.Makinen@helsinki.fi (L.M.); sami.hietala@helsinki.fi (S.H.)
* Correspondence: psemenyuk@belozersky.msu.ru

Abstract: A prospective technology for reversible enzyme complexation accompanied with its inactivation and protection followed by reactivation after a fast thermocontrolled release has been demonstrated. A thermoresponsive polymer with upper critical solution temperature, poly(*N*-acryloyl glycinamide) (PNAGA), which is soluble in water at elevated temperatures but phase separates at low temperatures, has been shown to bind lysozyme, chosen as a model enzyme, at a low temperature (10 °C and lower) but not at room temperature (around 25 °C). The cooling of the mixture of PNAGA and lysozyme solutions from room temperature resulted in the capturing of the protein and the formation of stable complexes; heating it back up was accompanied by dissolving the complexes and the release of the bound lysozyme. Captured by the polymer, lysozyme was inactive, but a temperature-mediated release from the complexes was accompanied by its reactivation. Complexation also partially protected lysozyme from proteolytic degradation by proteinase K, which is useful for biotechnological applications. The obtained results are relevant for important medicinal tasks associated with drug delivery such as the delivery and controlled release of enzyme-based drugs.

Keywords: thermosensitive polymers; enzyme complexation; controlled release; reversible inactivation; UCST polymers; stimuli-responsive polymers

[C]itation: Semenyuk, P.I.; [K]urochkina, L.P.; Mäkinen, L.; [M]uronetz, V.I.; Hietala, S. [T]hermocontrolled Reversible Enzyme [C]omplexation-Inactivation-Protection [b]y Poly(*N*-acryloyl glycinamide). [Po]lymers **2021**, *13*, 3601. https://[d]oi.org/10.3390/polym13203601

[A]cademic Editor: Florian J. Stadler

[R]eceived: 21 September 2021
[A]ccepted: 14 October 2021
[P]ublished: 19 October 2021

1. Introduction

With a growing number of peptide-based and enzyme-based drugs accepted for clinical trials and medicinal use, the development of the approaches for targeted delivery is of special importance. Plenty of approaches such as polymeric nanoparticles or nanogels [1], liposome-based delivery systems [2], protein conjugates, and other nanocarriers [3] have been suggested.

Stimuli-responsive polymers are frequently used as a platform for the construction of new drug delivery systems with an aim at the controlled release of various drugs [4–6]. Among such stimuli relevant for biological use, one can mention pH or concentration of specific molecules, light [7,8], and temperature. Thermosensitive polymers provide an opportunity to control the interaction with other macromolecules, especially proteins, by temperature. Thus, the temperature-dependent interaction of polymers with lower critical solution temperature (LCST) with proteins has allowed the construction of artificial chaperones, which are capable of recognizing the unfolded state of the enzymes [9,10]. In addition to actual chaperones, encapsulation or conjugation approaches have been used to immobilize and stabilize various enzymes for catalytic applications [11–16]. However, with LCST type of systems, the thermal denaturation of biocomponents at elevated temperature remains an issue. Examples concerning polymers with upper critical solution temperature (UCST) are less numerous and include some techniques with crosslinking stages required for hydrogel or nanoparticle production [17–19]. A simple noncrosslinking cooling-induced protein capturing by UCST-type polymers was suggested as an approach for protein extraction with some specificity to protein charge [20]. Noteworthy, many of such polymers are nontoxic and are prospective for biological use [21,22].

In the present study, we tested the interaction of UCST-type polymer, poly(*N*-acryloyl glycinamide) (PNAGA) [23–25], which was already suggested for medicinal use [21,26] with lysozyme as a model enzyme at different temperatures. Lysozyme is an enzyme of the hydrolase class that cleaves to the peptidoglycan component of bacterial cell walls which leads to cell death; therefore, it is widely used as an antimicrobial agent. We have demonstrated the reversible binding of PNAGA and lysozyme at low temperatures followed by a dissociation after heating that can be used as a platform for the creation of a protein delivery system with controlled release. A key advantage of this strategy is that complexation is accompanied by reversible enzyme inactivation and protection from proteolytic digestion; the heating-induced release of the enzyme is accompanied by its fast reactivation.

2. Methods

2.1. Materials

For NAGA, monomer synthesis glycinamide hydrochloride (Bachem, Bubendorf Switzerland) and acryloyl chloride (Sigma-Aldrich, Saint Louis, MO, USA) were used as received. The initiator 2,2′-azobis[2-(2-imidazolin-2-yl)propane] dihydrochloride (VA-044 Wako Specialty Chemicals) was recrystallized from methanol. The monomer and polymer syntheses have been reported earlier by [25]. In brief, polymerizations of NAGA were carried out using a thermal radical initiator VA-044 in DMSO at 60 °C. The structure and main characteristics of the polymer are shown in Scheme 1.

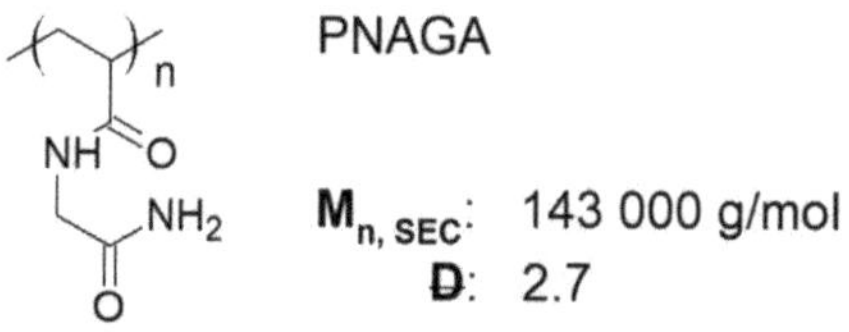

Scheme 1. Structure and characteristics of the polymer.

Chicken egg lysozyme was purchased from Sigma-Aldrich. Protein concentration was measured spectrophotometrically using $A_{280}^{0.1\%}$ value of 2.6 (for sample preparation) and by measuring SDS-PAGE bands intensity (for analysis of the complex composition).

All experiments were performed in 10 mM potassium phosphate-buffered saline, pH 7.4.

2.2. Dynamic Light Scattering

The phase-transition behavior of the polymer and its complexes with lysozyme was studied using dynamic light scattering with the ZetaSizer NanoZS instrument (Malvern, UK). The PNAGA solution with a concentration of 10 mg/mL in the absence as well as in the presence of lysozyme with a concentration of 5 mg/mL was incubated overnight on ice before the measurements. The samples were heated up in the instrument with an average heating rate of 0.7°/min to 45 °C and then cooled down with the same rate. Each point was determined as an average over three runs. Temperature of the cloud point was estimated as a temperature of inflection point from a sigmoidal fitting of the curves.

2.3. Isothermal Titration Calorimetry

ITC experiments were performed using a VP-ITC calorimeter (MicroCal, Northampton, MA, USA) at 10 and 25 °C. A solution of the polymer (0.8 or 2 mg/mL) was titrated by successive 20 µL injections of lysozyme solution (3 or 2 mg/mL), with a time interval between the injections of 5 min. To compare the heat effect with a heat effect of the dilution of the polymers, the same polymer solutions were titrated with the buffer without lysozyme. All samples were degassed before the experiment. The binding isotherms were fitted with the "one set of sites" model using MicroCal Origin 7.0 software. For the fitting, the concentration of PNAGA was expressed in terms of the molar concentration of NAGA groups.

2.4. Preparation of the Complexes

The stable PNAGA*Lysozyme complexes were prepared using the following simple procedure (Figure 1A). The enzyme and the polymer solutions were mixed at room temperature in 10 mM phosphate buffer, pH 7.4, and cooled down to +4 or 0 °C (i.e., on ice). After overnight incubation, the formed complexes were separated from unbound lysozyme by centrifugation and washed with pure phosphate buffer. Amount of the bound and unbound protein was measured using Bradford protein assay. For testing complex stability, the washed complexes were incubated for 20 h in pure 10 mM phosphate buffer, pH 7.4 and separated from released protein in the same manner. All experiments were performed at least three times to obtain statistical data.

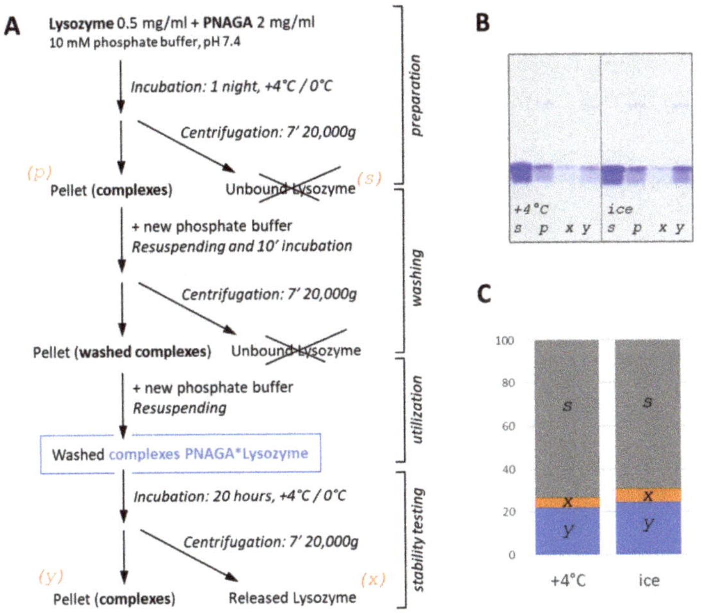

Figure 1. (**A**) The scheme of a simple procedure of mixing and cooling down followed by washing used to prepare stable PNAGA*Lysozyme complexes, as well as an additional step to test the complex stability. (**B**) SDS-PAGE of supernatant (s) and pellet (p) of the first centrifugation step, and supernatant (x) and pellet (y) of the last centrifugation after a stability test for complexes of PNAGA with Lysozyme obtained at +4 and at 0 °C (on ice). (**C**) Amount of lysozyme in the above samples determined using Bradford protein assay and expressed as a percentage of total amount of lysozyme.

2.5. Lysozyme Activity Assay

Enzymatic activity of lysozyme was determined from a decrease in absorbance of cell suspension due to addition of the enzyme. The *E. coli* SupF cells treated by freeze were used as a substrate. Sample aliquots containing 0.2–1 μg of lysozyme were mixed with 150 μL of cell suspension, and optical density was measured at 400 nm for 2 min using a VersaMax microplate reader (Molecular Devices, San Jose, CA, USA). Negative control (buffer without enzyme) was subtracted from sample measurements. The activity values were determined as a slope of linear part of the time dependence and then divided by actual lysozyme concentration determined from SDS-PAGE bands intensity. The values were averaged among at least three measurements and expressed as a percentage from the specific activity of free lysozyme at 25 °C.

2.6. Proteinase K Proteolysis Assay

Proteolysis was initiated by the addition of 2.5 μL proteinase K (Eurogene, Moscow, Russia) to a concentration of 67, 42, 26, 16, and 10 μg/mL into 20 μL aliquots of sample (PNAGA*Lysozyme complexes). Lysozyme solution with a concentration of 0.1 mg/mL was used as a control. Proteolysis was performed at 4 °C and quenched after 4 h incubation by addition of 1 mM phenylmethylsulfonyl fluoride in isopropanol. The samples were separated on 16% SDS-PAGE. The amount of intact lysozyme was determined from the SDS-PAGE bands intensity using ImageJ software and expressed as a percentage from an initial value.

As an additional control for a possible effect of the polymer on proteinase K activity, the same experiment was performed in 50 mM Tris-HCl buffer, pH 7.4.

3. Results

3.1. Polymer-Enzyme Complexes Formed by the Mixture Cooling Are Stable in Cold but Dissolute When Heated

The thermosensitive polymer with upper critical solution temperature, namely, poly(*N*-acryloyl glycinamide) homopolymer (PNAGA), was tested for interaction with lysozyme, selected as a model enzyme. The synthesis of the PNAGA polymer used in this study has been reported earlier [25], and its relevant characteristics are reported in Scheme 1. The phase-transition behavior of the 10 mg/mL polymer solution is shown in Figure 2A: a soluble form with the particles diameter of 43 nm at room temperature but larger particles (~160 nm) in cold were detected. The temperature of phase transition for the heating of the precooled sample was 15 °C. As for a cooling experiment, an increase in the particle diameter was not observed, indicating that aggregation is slow. The hysteresis of phase transition is generally observed for PNAGA and is considered to be related to the kinetics of formation of the hydrogen bond network [23,24]. In the presence of 5 mg/mL lysozyme, the aggregation behavior and phase-transition temperature of the polymer were significantly altered: the phase-transition temperature increased up to 22.5 and 12.5° for heating and cooling, respectively; the size of the particles in the cold became much higher (Figure 2A). Since a solution of free lysozyme does not show any phase transition in this temperature interval, the observed effect of lysozyme on the behavior of the polymer clearly indicates interaction between the polymer and the enzyme. However, after a heating up, the preincubated in the cold mixture of PNAGA and lysozyme demonstrated disaggregation and became transparent again.

The data of light scattering agreed well with the visual observation of the systems with lower concentrations of lysozyme and PNAGA, which are more suitable for handling enzymes (Figure 2B). The mixtures of the polymer and lysozyme as well as a solution of free polymer were transparent at 25 °C. However, after a 2 h incubation at +4 °C, the mixture of PNAGA (1 mg/mL) with lysozyme (0.5 mg/mL) became turbid, in contrast to the solution of free polymer and the mixture of PNAGA (1 mg/mL) with lysozyme (0.2 mg/mL). Overnight incubation caused both mixtures of the polymer and lysozyme to become turbid (the system with a higher concentration of the lysozyme became more turbid), whereas free polymer solution was just slightly turbid. Since the solution of free lysozyme is completely transparent, the difference between free polymer and its mixtures with lysozyme indicates the binding of the polymer and lysozyme and formation of complexes, which are larger than particles formed by free PNAGA, which collapsed due to phase transition. When heated back up, all systems became transparent again, suggesting the dissolution of the complexes. Such a cooling–heating cycle can be repeated with the same result.

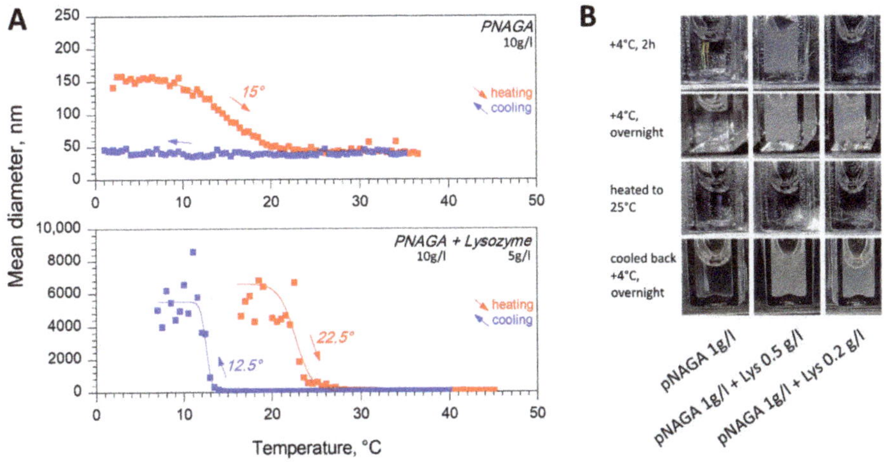

Figure 2. Phase-separation behavior of PNAGA in a solution is altered in the presence of lysozyme. (**A**) Mean diameter of particles determined using DLS for heating (red curves) and cooling (blue curves) of free polymers solutions (top) and their mixtures with lysozyme (bottom). Temperature values of cloud point are presented near the curves. Here, 10 mM phosphate buffer, pH 7.4. (**B**) Cooling down of PNAGA and lysozyme mixtures results in reversible formation of large complexes and decrease in the system transmittance; cooled solutions of free polymers are almost transparent.

3.2. PNAGA Binds Lysozyme Only at LOW Temperature

The binding of lysozyme with PNAGA polymer was tested directly at different temperature using isothermal titration calorimetry. The polymer efficiently binds lysozyme at 10 °C but does not bind it at 25 °C (Figure 3; compare curves with filled and empty circles, which represent the titration of the polymer solution with a protein solution and buffer solution, respectively). The binding is exothermic process (binding enthalpy < 0) with the binding constant of $3.1 \pm 0.6 \times 10^5$ M^{-1}; the stoichiometry is 3100 ± 700 NAGA monomers per one protein molecule. Such a high value indicates that few polymer chains (in average, 2–3 chains) bind to one protein molecule.

3.3. Lysozyme in the Complexes Is Inactive

Based on the presented results, a simple procedure was used to prepare and separate stable PNAGA*Lysozyme complexes (Figure 1B). In brief, solutions of the enzyme and the polymer were mixed at room temperature, cooled down to +4 or 0 °C (i.e., on ice), and incubated overnight. Then, the formed complexes were separated from unbound lysozyme by centrifugation and washed with pure phosphate buffer. Although most of the protein remained unbound, some amount of the lysozyme was captured by the polymer (Figure 1B,C). The complexes obtained at 0 °C (on ice) contain a larger amount of the protein compared to those obtained at +4 °C. The prepared complexes are stable and therefore are appropriate for further usage. Although a 20 h incubation in pure phosphate buffer resulted in the release of a small amount of lysozyme, most of it remained bound (Figure 1B,C).

The effect of complexation on enzymatic activity of lysozyme (i.e., lysis of bacterial cells) was analyzed (Figure 4A). In the cold, where the prepared complexes PNAGA*Lysozyme are stable, the specific enzymatic activity was about 35% of specific activity of free lysozyme, whilst heating to 25 °C followed by release of the enzyme from the complexes resulted in its almost complete reactivation.

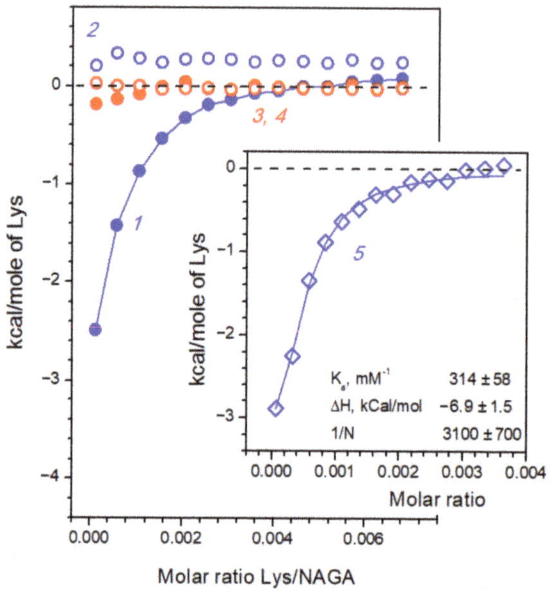

Figure 3. PNAGA binds lysozyme at 10 °C (blue circles) but does not bind it at 25 °C (red circles). ITC data for titration of polymer solutions with lysozyme solutions (curves 1 and 3, filled circles) and buffer solutions (curves 2 and 4, empty circles). The inset represents titration with lower molar ratio and the values of binding constant (K_a), enthalpy (ΔH), and stoichiometry (1/N, in terms of bound NAGA units per a protein molecule) of the binding. Polymer concentration is expressed in terms of molar concentration of NAGA repeated units. 10 mM phosphate buffer, pH 7.4.

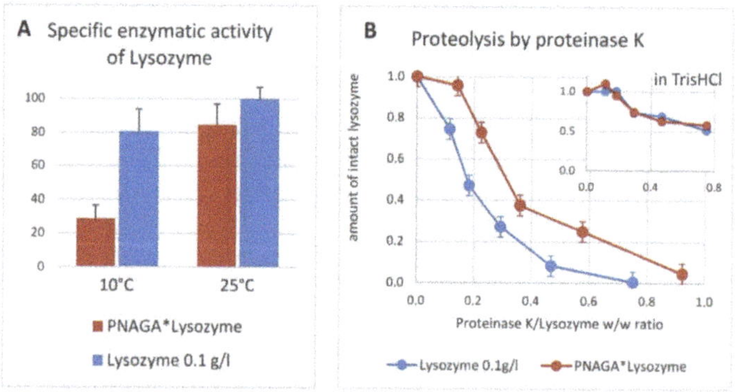

Figure 4. (**A**) Specific enzymatic activity of lysozyme in a free form and complexed with PNAGA. (**B**) Proteolytic digestion of lysozyme by proteinase K. Amount of intact lysozyme determined from SDS-PAGE bands intensity versus protease/lysozyme *w*/*w* ratio; red and blue line for complexes and free lysozyme, respectively. Here, 10 mM phosphate buffer, pH 7.4, +4 °C. Inset represents control experiments in 50 mM TrisHCl buffer, pH 7.4.

3.4. Encapsulation Protects Lysozyme from Proteolytic Degradation

Encapsulated into the complexes with PNAGA, lysozyme was shown to be partially protected from proteolytic cleavage by proteinase K (Figure 4B). The prepared complexes PNAGA*Lysozyme incubated for 4 h at +4 °C in the presence of different concentrations of proteinase K were digested by a significantly lower extent compared to free lysozyme at a similar concentration. To check if the polymer can affect the activity of proteinase K, a similar control experiment was performed in the Tris-HCl buffer, where large complexes of PNAGA and lysozyme are not formed. No effect of the polymer on the proteolysis level was observed (Figure 4B, inset). Thus, the data clearly indicate that the decrease in a proteolysis level is a direct protection of the lysozyme inside the complexes but not an inhibition of the protease by the polymer.

4. Discussion

To summarize, a prospective technology for reversible enzyme complexation accompanied with its inactivation and protection followed by the reactivation after a thermocontrolled release was demonstrated (Figure 5). A thermosensitive polymer with upper critical solution temperature, poly(*N*-acryloyl glycinamide), was shown to bind lysozyme at cold and to do not bind it at room temperature. Since the binding is reversible, the cooling–heating cycle allows for forming the complexes and performing a controlled release of the enzyme. Noteworthy, the complexes are easy to prepare by a simple mixing of two solutions at room temperature followed by cooling down to the binding temperature and washing of the complexes at cold. No crosslinking steps that might (directly or indirectly due to further purification procedures) affect enzyme structure and activity are required. Thus prepared, the complexes are stable in cold and can be easily handled.

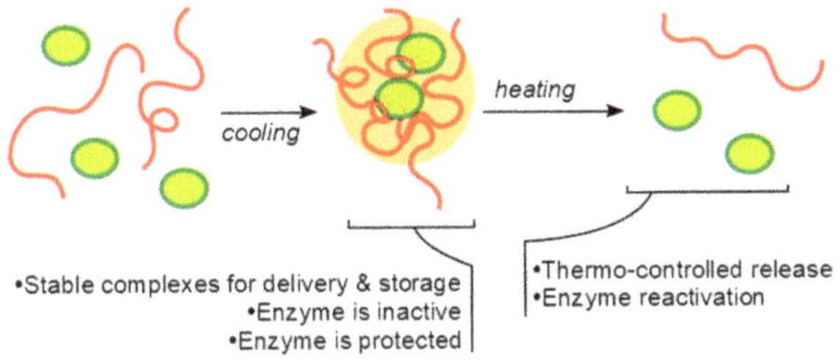

Figure 5. The scheme represents two key advantages of the suggested approach: simple preparation of the complexes with inactivated and protected enzyme and temperature-controlled release of the enzyme accompanied by its reactivation.

Inside the complexes, lysozyme is inactive, but its enzymatic activity is restored after release from the complexes. It provides an opportunity for reversible and controlled switching off/on the activity of the enzyme. The reversible switching off of the enzyme activity in complex with synthetic polymers was earlier shown in a few papers [27,28], but the release of the enzyme from the complexes was managed by additional polymers, which might complicate the use of such an approach in multicomponent systems. In such systems, the heating-induced enzyme release seems to be preferable since it is gentler. One more advantage of the suggested strategy is a combination of controlled reversible protein inactivation and its partial protection from proteolytic degradation that should facilitate storage of the protein. It can be relevant for the biotechnological use of enzymes at elevated temperature. Thus, a specific enzyme can be stored at a low temperature for a long time in a complexed form (inactive and stabilized) but released from the complexes and performed catalysis when transferred to the system at an elevated temperature (addition of the complexes into bioreactor system or direct heating of the system, which initially contains the complexed enzyme).

Temperature-controlled release is relevant for some important medicinal tasks associated with drug delivery. Indeed, the enzyme inside the complexes is inactive and partially protected from proteolytic degradation but can be easily reactivated due to controlled release. Focused on biological approaches, we used physiological buffer system, namely potassium phosphate buffer, pH 7.4. Of course, our model system with phase-transition temperature lower than 25 °C looks to be not optimal for medicinal usage, although some tasks, including skin delivery, allow for system formulation in a very wide temperature range due to a high ability to cooling without the occurrence of undesired side effects [29,30]. Regardless, the temperature of phase transition of PNAGA-based polymers can be tuned in a wide range by copolymerization or by adjusting the molar mass or end-groups [31–35]. For example, using a hydrophobic dodecyl end-group in RAFT polymerization of NAGA and changing the molar mass, the phase transition upon heating could be tuned in the range from 24 to 43 °C [32], which covers physiologic temperature range. Copolymeric PNAGA-poly(*N*-phenylacrylamide) brushes, which sorb and release cells at 30 and 37 °C, respectively, can be one more example [35]. A fine-tuning of the temperature of the transition allows one to perform a selective release of a complexed enzyme only in particular organs or tissues, the temperature of which increased due to pathologies such as inflammation, and to preserve the complexes stable (and inactive) in healthy tissues. Unfortunately, it might be difficult to obtain such a fine selectivity since the difference between the temperatures of normal and inflamed tissues is small, namely fewer than a few degrees. However, the selective heating of specific regions and tissues (in particular, cancer tumors) can be achieved using lasers as well as a radio frequency radiation combined with gold or silicon nanoparticle-based sensitizers [36]. We suppose that a combination of selectively induced hyperthermia and the introducing of thermoresponsive polymer–enzyme complexes, which release the destructive enzyme (for example, proteinases) in a heating-driven manner, should be very promising.

One more problem important for biological and medicinal use of a polymer-based complex as carrier vectors is its biocompatibility and biodegradability. Though it is not a trivial question how PNAGA can be degraded after enzyme delivery and release, polymers with a mass less than 40 kDa (which is true for short PNAGA-based copolymers with phase transition at physiological temperature) can be expected to be filtered by the kidneys [37]. Some clearance from PNAGA-based hydrogel was shown in [38]. In addition, PNAGA does not exhibit significant toxicity according to previous works [21,39]. However, the biocompatibility and biodegradability of such polymers require additional direct studies.

5. Conclusions and Perspectives

A thermoresponsive polymer, poly(*N*-acryloyl glycinamide), which is soluble in water at elevated temperatures but phase-separates at low temperatures, is shown to capture lysozyme at temperature lower than 10 °C and form stable polymer–enzyme complexes. Heating to room temperature (around 25 °C) resulted in the complex dissociation and release of the enzyme. Being almost inactive in a complexed form, lysozyme restored its enzymatic activity after a thermocontrolled release. In addition, capturing by the polymer partially protected lysozyme against proteolytic degradation, which is useful for biotechnological application. The reversible capturing-inactivation with a thermocontrolled release is promising for the medicinal use of poly(*N*-acryloyl glycinamide)-based polymers as drug vehicles to deliver enzyme-based therapeutics. The development of particular carriers with optimal phase-transition behavior should be an issue of future research, and our results taken together with the data on tuning phase transition temperature of the polymer by other groups suggest an outstanding potential for such carriers. In addition, such polymers can be used as a protein-capturing part of complex carrier combined with tags for targeted drug delivery. The biodegradability of the developed polymeric carriers also should be investigated directly.

Author Contributions: Conceptualization, P.I.S., V.I.M. and S.H.; investigation, P.I.S., L.P.K. and L.M.; writing—original draft preparation, P.I.S.; writing—review and editing, P.I.S., L.P.K., V.I.M. and S.H.; supervision, V.I.M. and S.H.; project administration, P.I.S.; funding acquisition, P.I.S. All authors have read and agreed to the published version of the manuscript.

Funding: This research was funded by Russian Foundation for Basic Research, grant number 20-34-70012, and also supported by the exchange agreement between Lomonosov Moscow State University and the University of Helsinki.

Data Availability Statement: Data available on request.

Conflicts of Interest: The authors declare no conflict of interest.

References

1. Kabanov, A.V.; Vinogradov, S.V. Nanogels as Pharmaceutical Carriers: Finite Networks of Infinite Capabilities. *Angew. Chem. Int. Ed Engl.* **2009**, *48*, 5418–5429. [CrossRef]
2. Torchilin, V.P. Recent Advances with Liposomes as Pharmaceutical Carriers. *Nat. Rev. Drug Discov.* **2005**, *4*, 145–160. [CrossRef]
3. Duncan, R. The Dawning Era of Polymer Therapeutics. *Nat. Rev. Drug Discov.* **2003**, *2*, 347–360. [CrossRef] [PubMed]
4. Gil, E.S.; Hudson, S.M. Stimuli-Reponsive Polymers and Their Bioconjugates. *Prog. Polym. Sci.* **2004**, *29*, 1173–1222. [CrossRef]
5. Cheng, R.; Meng, F.; Deng, C.; Klok, H.-A.; Zhong, Z. Dual and Multi-Stimuli Responsive Polymeric Nanoparticles for Programmed Site-Specific Drug Delivery. *Biomaterials* **2013**, *34*, 3647–3657. [CrossRef] [PubMed]
6. Mane, S.R.; Sathyan, A.; Shunmugam, R. Biomedical Applications of PH-Responsive Amphiphilic Polymer Nanoassemblies. *ACS Appl. Nano Mater.* **2020**, *3*, 2104–2117. [CrossRef]
7. Rösler, A.; Vandermeulen, G.W.M.; Klok, H.-A. Advanced Drug Delivery Devices via Self-Assembly of Amphiphilic Block Copolymers. *Adv. Drug Deliv. Rev.* **2012**, *64*, 270–279. [CrossRef]
8. Zhang, S.; Wang, C.; Chang, H.; Zhang, Q.; Cheng, Y. Off-on Switching of Enzyme Activity by near-Infrared Light-Induced Photothermal Phase Transition of Nanohybrids. *Sci. Adv.* **2019**, *5*, eaaw4252. [CrossRef]
9. Lu, D.; Liu, Z.; Zhang, M.; Wang, X.; Liu, Z. Dextran-Grafted-PNIPAAm as an Artificial Chaperone for Protein Refolding. *Biochem. Eng. J.* **2006**, *27*, 336–343. [CrossRef]
10. Semenyuk, P.; Tiainen, T.; Hietala, S.; Tenhu, H.; Aseyev, V.; Muronetz, V. Artificial Chaperones Based on Thermoresponsive Polymers Recognize the Unfolded State of the Protein. *Int. J. Biol. Macromol.* **2019**, *121*, 536–545. [CrossRef]
11. Welsch, N.; Wittemann, A.; Ballauff, M. Enhanced Activity of Enzymes Immobilized in Thermoresponsive Core−Shell Microgels. *J. Phys. Chem. B* **2009**, *113*, 16039–16045. [CrossRef]
12. Schachschal, S.; Adler, H.-J.; Pich, A.; Wetzel, S.; Matura, A.; van Pee, K.-H. Encapsulation of Enzymes in Microgels by Polymerization/Cross-Linking in Aqueous Droplets. *Colloid Polym. Sci.* **2011**, *289*, 693–698. [CrossRef]
13. Gawlitza, K.; Wu, C.; Georgieva, R.; Wang, D.; Ansorge-Schumacher, M.B.; von Klitzing, R. Immobilization of Lipase B within Micron-Sized Poly-N-Isopropylacrylamide Hydrogel Particles by Solvent Exchange. *Phys. Chem. Chem. Phys.* **2012**, *14*, 9594–9600. [CrossRef]
14. Gawlitza, K.; Georgieva, R.; Tavraz, N.; Keller, J.; von Klitzing, R. Immobilization of Water-Soluble HRP within Poly-N-Isopropylacrylamide Microgel Particles for Use in Organic Media. *Langmuir* **2013**, *29*, 16002–16009. [CrossRef] [PubMed]
15. Li, F.; Wang, C.; Guo, W. Multifunctional Poly-N-Isopropylacrylamide/DNAzyme Microgels as Highly Efficient and Recyclable Catalysts for Biosensing. *Adv. Funct. Mater.* **2018**, *28*, 1705876. [CrossRef]
16. Reinicke, S.; Fischer, T.; Bramski, J.; Pietruszka, J.; Böker, A. Biocatalytically Active Microgels by Precipitation Polymerization of N-Isopropyl Acrylamide in the Presence of an Enzyme. *RSC Adv.* **2019**, *9*, 28377–28386. [CrossRef]
17. Cummings, C.; Murata, H.; Koepsel, R.; Russell, A.J. Dramatically Increased PH and Temperature Stability of Chymotrypsin Using Dual Block Polymer-Based Protein Engineering. *Biomacromolecules* **2014**, *15*, 763–771. [CrossRef] [PubMed]
18. Yang, D.; Tenhu, H.; Hietala, S. Bicatalytic Poly(N-Acryloyl Glycinamide) Microgels. *Eur. Polym. J.* **2020**, *133*, 109760. [CrossRef]
19. Kotsuchibashi, Y. Recent Advances in Multi-Temperature-Responsive Polymeric Materials. *Polym. J.* **2020**, *52*, 681–689. [CrossRef]
20. Shimada, N.; Nakayama, M.; Kano, A.; Maruyama, A. Design of UCST Polymers for Chilling Capture of Proteins. *Biomacromolecules* **2013**, *14*, 1452–1457. [CrossRef]
21. Xu, Z.; Liu, W. Poly(N-Acryloyl Glycinamide): A Fascinating Polymer That Exhibits a Range of Properties from UCST to High-Strength Hydrogels. *Chem. Commun.* **2018**, *54*, 10540–10553. [CrossRef] [PubMed]
22. Zhang, Z.; Li, H.; Kasmi, S.; Van Herck, S.; Deswarte, K.; Lambrecht, B.N.; Hoogenboom, R.; Nuhn, L.; De Geest, B.G. A Synthetic, Transiently Thermoresponsive Homopolymer with UCST Behaviour within a Physiologically Relevant Window. *Angew. Chem. Int. Ed.* **2019**, *58*, 7866–7872. [CrossRef] [PubMed]
23. Seuring, J.; Bayer, F.M.; Huber, K.; Agarwal, S. Upper Critical Solution Temperature of Poly(N-Acryloyl Glycinamide) in Water: A Concealed Property. *Macromolecules* **2012**, *45*, 374–384. [CrossRef]
24. Seuring, J.; Agarwal, S. Polymers with Upper Critical Solution Temperature in Aqueous Solution. *Macromol. Rapid Commun.* **2012**, *33*, 1898–1920. [CrossRef]
25. Mäkinen, L.; Varadharajan, D.; Tenhu, H.; Hietala, S. Triple Hydrophilic UCST–LCST Block Copolymers. *Macromolecules* **2016**, *49*, 986–993. [CrossRef]
26. Deng, Y.; Käfer, F.; Chen, T.; Jin, Q.; Ji, J.; Agarwal, S. Let There Be Light: Polymeric Micelles with Upper Critical Solution Temperature as Light-Triggered Heat Nanogenerators for Combating Drug-Resistant Cancer. *Small* **2018**, *14*, 1802420. [CrossRef]
27. Tomita, S.; Ito, L.; Yamaguchi, H.; Konishi, G.; Nagasaki, Y.; Shiraki, K. Enzyme Switch by Complementary Polymer Pair System (CPPS). *Soft Matter* **2010**, *6*, 5320–5326. [CrossRef]
28. Evstafyeva, D.B.; Izumrudov, V.A.; Muronetz, V.I.; Semenyuk, P.I. Tightly Bound Polyelectrolytes Enhance Enzyme Proteolysis and Destroy Amyloid Aggregates. *Soft Matter* **2018**, *14*, 3768–3773. [CrossRef]
29. Vikulina, A.S.; Feoktistova, N.A.; Balabushevich, N.G.; von Klitzing, R.; Volodkin, D. Cooling-Triggered Release from Mesoporous Poly(N-Isopropylacrylamide) Microgels at Physiological Conditions. *ACS Appl. Mater. Interfaces* **2020**, *12*, 57401–57409. [CrossRef]

30. Matsumoto, K.; Kimura, S.; Itai, S.; Kondo, H.; Iwao, Y. In Vivo Temperature-Sensitive Drug Release System Trigged by Coolin Using Low-Melting-Point Microcrystalline Wax. *J. Controlled Release* **2019**, *303*, 281–288. [CrossRef]
31. Käfer, F.; Lerch, A.; Agarwal, S. Tunable, Concentration-Independent, Sharp, Hysteresis-Free UCST Phase Transition fron Poly(N-Acryloyl Glycinamide-Acrylonitrile) System. *J. Polym. Sci. Part Polym. Chem.* **2017**, *55*, 274–279. [CrossRef]
32. Liu, F.; Seuring, J.; Agarwal, S. Controlled Radical Polymerization of N-Acryloylglycinamide and UCST-Type Phase Transition o the Polymers. *J. Polym. Sci. Part Polym. Chem.* **2012**, *50*, 4920–4928. [CrossRef]
33. Sponchioni, M.; Bassam, P.R.; Moscatelli, D.; Arosio, P.; Palmiero, U.C. Biodegradable Zwitterionic Nanoparticles with Tunabl UCST-Type Phase Separation under Physiological Conditions. *Nanoscale* **2019**, *11*, 16582–16591. [CrossRef] [PubMed]
34. Li, Z.; Hao, B.; Tang, Y.; Li, H.; Lee, T.-C.; Feng, A.; Zhang, L.; Thang, S.H. Effect of End-Groups on Sulfobetaine Homopolymers with the Tunable Upper Critical Solution Temperature (UCST). *Eur. Polym. J.* **2020**, *132*, 109704. [CrossRef]
35. Xue, X.; Thiagarajan, L.; Braim, S.; Saunders, B.R.; Shakesheff, K.M.; Alexander, C. Upper Critical Solution Temperature Thermo Responsive Polymer Brushes and a Mechanism for Controlled Cell Attachment. *J. Mater. Chem. B* **2017**, *5*, 4926–4933. [CrossRef] [PubMed]
36. Tamarov, K.P.; Osminkina, L.A.; Zinovyev, S.V.; Maximova, K.A.; Kargina, J.V.; Gongalsky, M.B.; Ryabchikov, Y.; Al-Kattan, A. Sviridov, A.P.; Sentis, M.; et al. Radio Frequency Radiation-Induced Hyperthermia Using Si Nanoparticle-Based Sensitizers fo Mild Cancer Therapy. *Sci. Rep.* **2014**, *4*, 7034. [CrossRef]
37. Fox, M.E.; Szoka, F.C.; Fréchet, J.M.J. Soluble Polymer Carriers for the Treatment of Cancer: The Importance of Molecula Architecture. *Acc. Chem. Res.* **2009**, *42*, 1141–1151. [CrossRef]
38. Boustta, M.; Colombo, P.-E.; Lenglet, S.; Poujol, S.; Vert, M. Versatile UCST-Based Thermoresponsive Hydrogels for Loco-Regiona Sustained Drug Delivery. *J. Controlled Release* **2014**, *174*, 1–6. [CrossRef]
39. Wu, Q.; Wei, J.; Xu, B.; Liu, X.; Wang, H.; Wang, W.; Wang, Q.; Liu, W. A Robust, Highly Stretchable Supramolecular Polyme Conductive Hydrogel with Self-Healability and Thermo-Processability. *Sci. Rep.* **2017**, *7*, 41566. [CrossRef]

polymers MDPI

Article

Thermoresponsive Poly(*N*,*N*-diethylacrylamide-*co*-glycidyl methacrylate) Copolymers and Its Catalytically Active α-Chymotrypsin Bioconjugate with Enhanced Enzyme Stability

György Kasza [1,*,†], Tímea Stumphauser [1,†], Márk Bisztrán [1], Györgyi Szarka [1], Imre Hegedüs [2,3], Endre Nagy [2] and Béla Iván [1,*]

1 Polymer Chemistry Research Group, Institute of Materials and Environment Chemistry, Research Centre for Natural Sciences, Magyar tudósok körútja 2., H-1117 Budapest, Hungary; stumphauser.timea@ttk.hu (T.S.); bisztmark@gmail.com (M.B.); szarka.gyorgyi@ttk.hu (G.S.)
2 Chemical and Biochemical Procedures Laboratory, Institute of Biomolecular and Chemical Engineering, Faculty of Engineering, University of Pannonia, Egyetem u. 10, H-8200 Veszprém, Hungary; hegedus@mukki.richem.hu (I.H.); nagy@mukki.richem.hu (E.N.)
3 Department of Biophysics and Radiation Biology, Semmelweis University, Tűzoltó u. 37–47, H-1094 Budapest, Hungary
* Correspondence: kasza.gyorgy@ttk.hu (G.K.); ivan.bela@ttk.hu (B.I.)
† These authors contributed equally to this work.

Citation: Kasza, G.; Stumphauser, T.; Bisztrán, M.; Szarka, G.; Hegedüs, I.; Nagy, E.; Iván, B. Thermoresponsive Poly(*N*,*N*-diethylacrylamide-*co*-glycidyl methacrylate) Copolymers and Its Catalytically Active α-Chymotrypsin Bioconjugate with Enhanced Enzyme Stability. *Polymers* **2021**, *13*, 987. https://doi.org/10.3390/polym13060987

Academic Editor: Florian J. Stadler

Received: 1 March 2021
Accepted: 20 March 2021
Published: 23 March 2021

Abstract: Responsive (smart, intelligent, adaptive) polymers have been widely explored for a variety of advanced applications in recent years. The thermoresponsive poly(*N*,*N*-diethylacrylamide) (PDEAAm), which has a better biocompatibility than the widely investigated poly(*N*,*N*-isopropylacrylamide), has gained increased interest in recent years. In this paper, the successful synthesis, characterization, and bioconjugation of a novel thermoresponsive copolymer, poly(*N*,*N*-diethylacrylamide-*co*-glycidyl methacrylate) (P(DEAAm-*co*-GMA)), obtained by free radical copolymerization with various comonomer contents and monomer/initiator ratios are reported. It was found that all the investigated copolymers possess LCST-type thermoresponsive behavior with small extent of hysteresis, and the critical solution temperatures (CST), i.e., the cloud and clearing points, decrease linearly with increasing GMA content of these copolymers. The P(DEAAm-*co*-GMA) copolymer with pendant epoxy groups was found to conjugate efficiently with α-chymotrypsin in a direct, one-step reaction, leading to enzyme–polymer nanoparticle (EPNP) with average size of 56.9 nm. This EPNP also shows reversible thermoresponsive behavior with somewhat higher critical solution temperature than that of the unreacted P(DEAAm-*co*-GMA). Although the catalytic activity of the enzyme–polymer nanoconjugate is lower than that of the native enzyme, the results of the enzyme activity investigations prove that the pH and thermal stability of the enzyme is significantly enhanced by conjugation the with P(DEAAm-*co*-GMA) copolymer.

Keywords: poly(*N*,*N*-diethylacrylamide); glycidyl methacrylate; thermoresponsive copolymer; α-chymotrypsin; polymer-enzyme conjugate nanoparticle

1. Introduction

Today, polymers with special, advanced properties and targeted functionalities, such as responsive (smart, intelligent, adaptive) polymers and macromolecules with well-defined array of functional groups belong to the most intensively investigated fields of polymer science and technology. Reactive functionalities in polymer chains can be introduced either along the chains (pendant functionalities) and/or at the chain ends (terminal functionalities). These functional polymers can be applied in various fields of polymer material science, technology, and industry, such as crosslinkers, chain extenders, and as building blocks of complex macromolecular assemblies [1–4], and life sciences and biotechnology as well, such as targeting delivery [5,6], biological sensors [7], receptors [8], and surfaces to control cell behavior [9]. Polymers with epoxide or glycidyl functional groups are among the most versatile functional polymeric materials for such purposes, because they can react with numerous nucleophiles, such as amines, thiols, phenols, carboxylic acids, or anhydrides via ring-opening reactions [10–14]. Epoxide functional polymers can be obtained by postmodification of double bond containing side or end group(s) of polymers [15–21], but such macromolecules can also be synthesized by copolymerizations with epoxy group containing monomers. Undoubtedly, glycidyl methacrylate (GMA) is the most investigated and used monomer to obtain functional macromolecules with epoxy side groups, but other epoxide-containing monomers were also studied, such as 4-vinylphenyl glycidyl ether [22]. Previously, the copolymerization of GMA with numerous monomers, e.g., 3-methylthienyl methacrylate [23], trimethylolpropane trimethacrylate [24], sulfobetaine methacrylate [25], ethylene–methyl acrylate [26], styrene [27,28], 2-hydroxyethyl methacrylate [29], by various polymerization techniques, such as free radical polymerization, ATRP, NMP, RAFT, etc., was widely investigated. These functional materials were successfully applied for several purposes, for instance, medical devices [25,29], compatibilizing agent [26], and metal ion absorbers [30–32].

Among responsive materials, LCST-type (LCST = lower critical solution temperature) and UCST-type (UCST = upper critical solution temperature) thermally responsive polymers belong to a unique class of smart materials with broad application possibilities ranging from nanotechnologies, oil recovery [33–35], to biomaterials, tissue engineering scaffolds [36,37], intelligent drug release assemblies [36,38,39], sensors [40,41], self-healing structures [42,43], responsive hybrid materials [44,45], etc. As to the use of the LCST and UCST terminology, it has to be noted that most of the authors still report incorrectly the result of a single-point measurement as LCST or UCST, i.e., the result of only one given condition with one single polymer concentration, one single heating/cooling rate and wavelength for cloud point and clearing point determination is claimed misleadingly as LCST or UCST. In contrast, the LCST or UCST are defined as the minimum or maximum, respectively, of the polymer concentration versus critical solution temperature (CST) curves, and not the single CST of a certain selected condition in terms of polymer concentration, heating/cooling rates, and wavelength for cloud point and cooling point determination. Hence, for LCST and UCST determination, the full CST versus polymer concentration (mass fraction or volume fraction) relationship should be measured, and the resulting minimum or maximum of such curves should be reported as LCST or UCST, respectively. Therefore, recently a standardization of the conditions for the measurements of CST in order to obtain comparable results of the laboratories worldwide was proposed on the basis of systematic investigations on the effect of the experimental conditions on the CST of poly(*N*-isopropylacrylamide) (PNIPAAm) solutions [46,47].

Undisputedly, poly(*N*-isopropylacrylamide) has been the most investigated temperature responsive polymer since the first report of its LCST-type behavior (see, e.g., Refs. [46–63] and references therein). Recently, intensive research has been focused on how to control the critical solution temperature (CST) of thermoresponsive polymers by using other monomers than NIPAAm (e.g., other acrylamides and *N*-vinyl lactams), by copolymerization with common monomers, especially with functional monomers, which can further increase the range of potential applications. Although NIPAAm-GMA copolymers were already synthesized and investigated for a variety of purposes [56–63], much less attention was paid to other GMA containing thermoresponsive polymers in the past. Recently, various *N*-vinyl lactam monomers were copolymerized with GMA, and the resulting copolymers were successfully applied as robust building blocks for protein conjugation, and the biohybrid nanogels of these copolymers exhibited significantly enhanced resistance against harsh storing conditions, chaotropic agents, and organic solvents [64].

It is interesting to note that due to the better biocompatibility of the LCST-type thermoresponsive poly(*N*,*N*-diethylacrylamide) (PDEAAm) than that of PNIPAAm [65], investigations in relation to the responsive and biocompatible behavior of PDEAAm, its derivatives, and gels have gained increased attention only in recent years (see, e.g., Refs. [66–92] and references therein). It should also be mentioned that aqueous PDEAAm solutions possess similar critical solution temperatures (CST) [72,93–96] in the range of ~25–40 °C than that of PNIPAAm [46,55]. On the other hand, although some block copolymers with PDEAAm and poly(glycidyl methacrylate) (PGMA) segments have been prepared and studied [91,92], random copolymers of DEAAm with GMA, which provides reactive pendant epoxy functionalities for the thermoresponsive PDEAAm, and its utilization for polymer-based protein engineering have not been reported so far according to the best of our knowledge.

Polymer-based protein engineering mainly focuses on the synthesis, characterization, and applications of conjugates of proteins, especially enzymes, with polymers for various purposes, such as stability improvement, better biodistribution, biocatalytic syntheses, purification, recovery, etc. (see, e.g., Refs. [97–113] and references therein). Among enzymes, α-chymotrypsin (CT), a peptide bond cleaving serine protease enzyme, is one of the most widely investigated proteins in terms of its bioconjugation with a variety of polymers and applications in bioengineering [106,114–120]. In general, attachment of polymer chains by covalent bonds to CT and other proteins as well can be carried out by either grafting from and grafting onto, and rarely by grafting through as well. Grafting from involves the functionalization of the protein with functional groups suitable for initiating the desired polymerization of selected monomers, usually by a living polymerization process [97–107]. This two-step or multi-step laborious and time-consuming process requires various reagents, in many cases toxic compounds (e.g., copper salts and complex forming amines for quasiliving atom transfer radical polymerization), relatively high temperatures that may deactivate the enzymes, and vigorous purification steps [99–107]. Grafting onto takes place by reacting the protein with pre-synthesized functional, mainly endfunctional, polymers, including the widely applied PEGylation with terminally functional PEGs, usually in two or more steps (see, e.g., [97–102,113] and references therein). Conjugation of proteins, especially enzymes, with thermoresponsive polymers offers additional unique possibilities for switching enzyme activity, efficient purification, enzyme recovery, etc., based on the precipitation of such polymer assemblies above their critical solution temperature [97,106–108,112]. As to grafting onto proteins with epoxy group containing polymers, only few examples can be found in the literature [111,114,115]. Recently, a grafting through approach of an enzyme macromonomer, functionalized with glycidyl methacrylate, was also reported [121]. For α-chymotrypsin, the widely applied method for polymer conjugation with epoxy-functional polymers involves amination of the epoxy group with a diamine followed by coupling the resulting amine-functionalized polymer to the enzyme by glutaraldehyde [114,115]. Although poly(*N*-isopropylacrylamide) bioconjugates have been investigated in numerous cases, its relatively large extent of hysteresis due to hydrogen bonding between PNIPAAm chains [46,47] and even full activity loss of the conjugated enzyme [122] may limit its application possibilities. In contrast, the lack of hydrogen bonding between PDEAAm chains may provide unique advantages for bioconjugations with functional PDEAAm. Considering the high reactivity of primary amines with epoxy, especially glycidyl groups, the question arises whether the biocompatible poly(*N*,*N*-diethylacrylamide) with epoxy functionalities can be conjugated to CT, containing 15 primary amine sites, directly in a simple one-step process, and if this were possible, what are the characteristics of such bioconjugates in terms of their size, thermoresponsive behavior, enzymatic activity, and stability.

Based on the above aspects and unique potentials of the pendant epoxy containing poly(*N*,*N*-diethylacrylamide-*co*-glycidyl methacrylate) (P(DEAAm-*co*-GMA)) copolymers, we aimed at exploring its thermoresponsive property, the one-step conjugation possibility with α-chymotrypsin, and the catalytic activity and stability of such bioconjugates. Herein, we present the results of our investigations on the synthesis of P(DEAAm-*co*-GMA) and on the thermoresponsive behavior, i.e., on the effect of composition of the resulting copolymers on the critical solution temperature, of the resulting copolymers. In addition, we also report on the utilization of the epoxy functionalities of P(DEAAm-*co*-GMA) for the one-step preparation of enzyme–polymer nanoparticles (EPNP) by conjugation with α-chymotrypsin (CT), its thermoresponsive characteristics, thermal and pH stability, and the enzymatic catalytic behavior of these new bioconjugates.

2. Materials and Methods

2.1. Materials

Glycidyl methacrylate and *N,N*-diethylacrylamide (both from Sigma-Aldrich, St. Louis MO, USA) were freshly distilled under reduced pressure prior to use. 2,2′-Azoisobutyronitrile (AIBN, 98%, Sigma-Aldrich, St. Louis, MO, USA) was recrystallized from hexane and methanol twice, respectively. Tetrahydrofuran (THF, >99%, Molar Chemicals Halásztelek, Hungary) was refluxed over $LiAlH_4$, distilled, and was kept under nitrogen until its use. Diethyl ether and methanol (>99%, Molar Chemicals, Halásztelek, Hungary) PBS (pH = 7.4) and phosphate buffers (pH = 6; 7; 7.8; 8; 9 both from Sigma-Aldrich, St Louis, MO, USA) were used without further purification. α-Chymotrypsin and *N*-benzoyl L-tyrosine ethyl ester (BTEE, 99%) were purchased from Sigma-Aldrich and were used as received.

2.2. Synthesis Methods

2.2.1. Synthesis of PDEAAm Homopolymer and P(DEAAm-*co*-GMA) Copolymers by Free Radical Polymerization

Poly(*N,N*-diethylacrylamide-*co*-glycidyl methacrylate) (P(DEAAm-*co*-GMA)) copolymers were prepared by free radical copolymerization initiated by AIBN with various initiator/monomer molar ratios (1:100 and 1:200) and comonomer contents (5 and 10 mol% GMA). A typical copolymer synthesis is described below. In the case, when the molar ratio of the components, i.e., AIBN:DEAAm:GMA was 1:95:5, first 1.02 mL of DEAAm (7.43 mmol) and 0.052 mL of GMA (0.39 mmol) were charged into sealed round bottom flask, and the monomers were dissolved in 9.5 mL of THF. This solution was deoxygenized by bubbling with argon for 20 min. Then the reaction mixture was warmed to 60 °C, and 0.5 mL AIBN stock solution (25.75 mg/mL, 0.078 mmol) was added. After stirring for 18 h the resulting polymer was precipitated twice from THF solution in hexane and filtered Finally, the product was dried in vacuum at 60 °C until constant weight. The poly(*N,N*-diethylacrylamide) (PDEAAm) homopolymer was synthesized by the same method using 100:1 monomer/initiator ratio.

2.2.2. Synthesis of Enzyme–Copolymer Nanoparticle (EPNP)

One selected P(DEAAm-*co*-GMA) copolymer (Sample C) was measured (15.85 mg) into a glass vial and dissolved in 2 mL water. Then, 3 mL of α-chymotrypsin stock solution (10.14 mg/mL in water) was added dropwise to the stirred polymer solution by a syringe pump (dosing rate was 1.5 mL/h). The reaction mixture was stirred overnight at room temperature. Subsequently, the reaction mixture was dialyzed (MWCO = 25 kDa) for three days against water, which was refreshed twice daily. Then, the purified dry product was obtained by lyophilization.

2.3. Characterization

2.3.1. Gel Permeation Chromatography (GPC)

Average molecular weights and molecular weight distributions (dispersity index Đ) of the produced polymers were determined by GPC. The GPC was equipped with differential refractive index detector (Agilent 390, Agilent Technologies, Santa Clara, CA USA), three 5 μm particle size Waters Styragel (columns (HR1, HR2 and HR4) and with a Waters Styragel guard column (both form Waters, Milford, MA, USA) thermostated at 35 °C THF was used as eluent with a flow rate of 0.3 mL/min. The average molecular weights and Đ were determined by using conventional calibration based on linear polystyrene standards (from PSS Polymer Standards Services GmbH, Mainz, Germany).

2.3.2. ^{1}H NMR Spectroscopy

The ratio of the incorporated comonomers was determined by ^{1}H NMR measurements. The analysis was performed on Bruker Advance 500 (Bruker, Billerica, MA, USA) equipment operating at 500 MHz ^{1}H frequency in $CDCl_3$ at 30 °C.

2.3.3. Thermoresponsive Behavior

The transmittance versus temperature curves for obtaining the critical solution temperatures (T_C), i.e., the cloud point (T_{CP}) and the clearing point (T_{CL}) were measured by a UV–Vis spectrophotometer (Jasco V-650, JASCO Corporation, Tokyo, Japan) equipped with Jasco MCB-100 (JASCO Corporation, Tokyo, Japan) mini circulation bath and Peltier thermostat. Standard 1 cm × 1 cm cuvettes were used for these measurements. Deionised water was used as reference and solvent. The polymer and the enzyme–polymer nanoparticle solutions (1 mg/mL) were heated and then cooled in the temperature range of 15 to 50 °C with 0.2 °C/min heating/cooling rate and the transmittance was recorded at 488 nm according to recent studies on the standardization of measurements for the determination of the critical solution temperatures [46,47]. The inflection points of the transmittance–temperature curves were taken as both the T_{CP} and T_{CL} values.

2.3.4. Dynamic Light Scattering (DLS)

The average hydrodynamic diameter and dispersity of the obtained enzyme–polymer nanoparticle and the applied copolymer as well as the α-chymotrypsin were determined by a dynamic light scattering (DLS) system (Malvern Zetasizer Nano ZS, Malvern, UK). The measurements were carried out at 25 °C, and the concentrations of the samples were 1 mg/mL in PBS (pH = 7.4).

2.3.5. Quantification of the Enzyme Content in the Nanoparticles

The enzyme content of the nanoparticles was determined by UV-Vis spectroscopy measurements. The absorbance of the nanoparticle aqueous solution (1 mg/mL) was recorded by UV-Vis spectrophotometer (Jasco V-650, JASCO Corporation, Tokyo, Japan) equipped with Jasco MCB-100 mini circulation bath and Peltier thermostat at 25 °C in the 200–355 nm range. The enzyme content of the nanoparticles was evaluated on basis of an α-chymotrypsin calibration curve at 283 nm, where the polymer has no absorbance.

2.3.6. Catalytic Activity Assay

The catalytic activity of α-chymotrypsin and the produced enzyme–polymer nanoparticle (EPNP) was investigated by UV-Vis spectroscopy assay [123]. In this assay, the transformation of the substrate *N*-benzoyl-L-tyrosine ethyl ester (BTEE) to *N*-benzoyl-L-tyrosine via enzymatic hydrolysis was followed by spectroscopy measurements at 256 nm. The measurements were carried out in 3 mL quartz cuvettes, where 1.5 mL buffer (pH = 6; 7; 7.4 (PBS); 7.8; 8; 9), 1.4 mL BTEE stock solution (prepared by dissolving 74.3 mg BTEE in 126.8 mL methanol and adjusted by water to 200 mL in a volumetric flask) and 0.1 mL enzyme or enzyme–polymer nanoparticle solution were mixed (the enzyme concentration of the enzyme stock solution was 0.1 mg/mL). The increment in the absorbance at 256 nm was measured for five minutes with 10 s delays. Three independent measurements were carried out with every sample with varying pH at 25 °C, and good reproducibility was observed in each case. To eliminate the error due to the autohydrolysis of BTEE, measurements were performed by using a blank at every pH with the replacement of the enzyme solutions with distilled water. The reference was distilled water. For investigations of the thermal stability of the enzyme–polymer nanoparticles, the α-chymotrypsin and the nanoparticle solutions with enzyme concentrations of 0.1 mg/mL were thermostated at 45 °C. After predetermined time (5, 15, 30, 60, 120 min) of such thermal treatment, 0.1 mL samples were withdrawn and allowed to cool to room temperature for 5 min. Then the catalytic activity was measured by the method described above using PBS buffer (pH = 7.4) at 25 °C. The enzymatic activity was calculated with the following equation:

$$\text{Activity} = ((\Delta A_T - \Delta A_B)\ V_T\ d_f)/(0.964\ \text{vs.}\ c_e), \quad (1)$$

where ΔA_T and ΔA_B are the maximum rate of increase in the absorbance in one minute for the test sample and blank, respectively, V_T is the total volume (3 mL), d_f is the dilution factor (30), 0.964 is the millimolar extinction coefficient of BTEE at 256 nm, vs. is the sample volume (0.1 mL), and c_e is the enzyme concentration (0.1 mg/mL).

3. Results and Discussion

As displayed in Scheme 1, we aimed at synthesizing poly(*N,N*-diethylacrylamide-*co*-glycidyl methacrylate) (P(DEAAm-*co*-GMA)), preparing α-chymotrypsin-P(DEAAm-*co*-GMA) bioconjugate by the utilization of the reactive pendant epoxy functional groups of this copolymer, and characterization of the resulting copolymers and enzyme–polymer nanoparticles (EPNP) in terms of their thermoresponsive behavior, enzyme activity, and stability. The P(DEAAm-*co*-GMA) copolymers were synthesized by free radical copolymerization of PDEAAm and GMA with AIBN as radical initiator by using two different monomer/initiator ratios (100 and 200) with two different comonomer contents (5 and 10 mol%). PDEAAm homopolymer was also prepared with 100:1 monomer/initiator ratio under identical conditions to that of the copolymer syntheses. As shown in Table 1, polymers with relatively high yields in the range of ~60–87% were obtained. The molecular mass distributions (MMD) of the resulting polymers, displayed in Figure 1, were determined by GPC analysis (the GPC chromatograms are shown in Figure S1 in the Supplementary Materials). As expected, these MMD curves in Figure 1, and the number average molecular weight (M_n) and the peak molecular weight values (M_p) in Table 1 clearly indicate that P(DEAAm-*co*-GMA) copolymers with higher molecular masses are formed with higher monomer/initiator ratios.

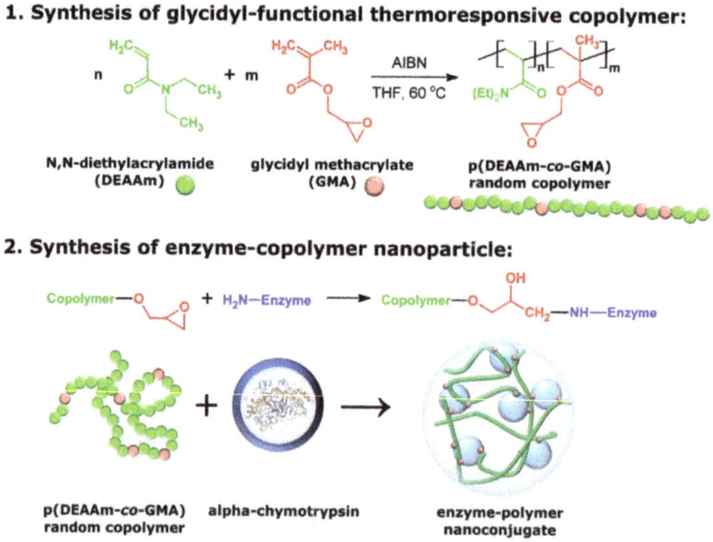

Scheme 1. Synthesis of glycidyl-functional poly(*N,N*-diethylacrylamide-*co*-glycidyl methacrylate) copolymers via free radical copolymerization (**1**) and the design of enzyme–polymer nanoconjugate with α-chymotrypsin (**2**).

Table 1. Yields and GPC results of the PDEAAm homopolymer and P(DEAAm-*co*-GMA) copolymers.

Sample	Molar Feed Ratio AIBN:DEAAm:GMA	Yield %	M_n (g/mol)	M_p (g/mol)	Đ
A	1:95:5	62.7	6650	7530	1.75
B	1:90:10	69.6	6525	7620	1.88
C	1:190:10	75.6	8620	20,170	2.31
D	1:180:20	86.8	7520	21,470	2.80
PDEAAm	1:100:0	81.5	9820	18,840	1.90

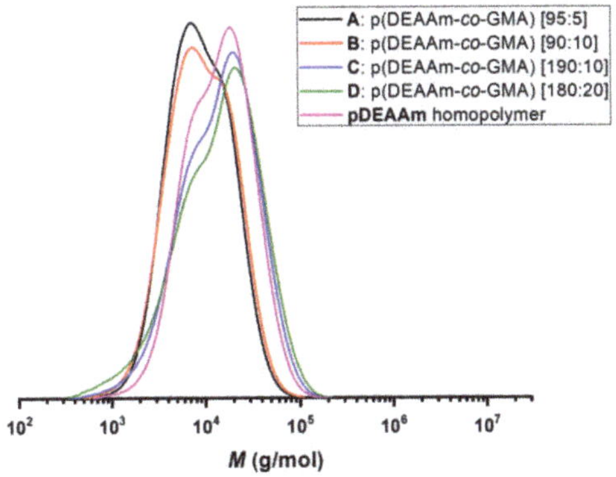

Figure 1. Molar mass distribution curves of the P(DEAAm-*co*-GMA) copolymers and PDEAAm homopolymer.

The compositions of the P(DEAAm-*co*-GMA) copolymers were determined by ^1H NMR spectroscopy. Comparing the ^1H NMR spectra of the PDEAAm homopolymer (Figure 2A) with that of the P(DEAAm-*co*-GMA) copolymers (Figure 2B and Figures S2–S4), it can be seen that with the exception of the chemical shifts of the methylene group next to the epoxy group in the GMA monomer units (dCH_2 3.7–4.1 and 4.1–4.6 ppm), the rest of the signals overlap with that of the PDEAAm homopolymer. This allows the determination of the composition of the P(DEAAm-*co*-GMA) copolymers by the integral values of the ^1H NMR signals. As shown in Table 2, the DEAAm/GMA ratios are smaller in the copolymers than in the feed. This means that the GMA contents in the P(DEAAm-*co*-GMA) copolymers are higher than that in the feed, which means that the reactivity of GMA is higher than that of DEAAm in this copolymerization reaction. This is in line with the reactivity ratios reported for the copolymerization of another alkyl acrylamide, *N*-isopropylacrylamide (NIPAAm), and GMA, according to which $r_1 = 0.39$ and $r_2 = 2.69$ [56]. Taking into account the similar structure of NIPAAm and DEAAm, higher reactivity of GMA is expected in the DEAAm-GMA copolymerization process as well, on the one hand. Considering that the product of the r_1 and r_2 values of the alkyl acrylamide copolymerization with GMA is in the range of one, random copolymerization occurs in such cases, on the other hand. Thus, it can be concluded that random copolymers of DEAAm and GMA with 5.5–11.4 mol% GMA contents were obtained in the applied copolymerization reactions as shown in Table 2.

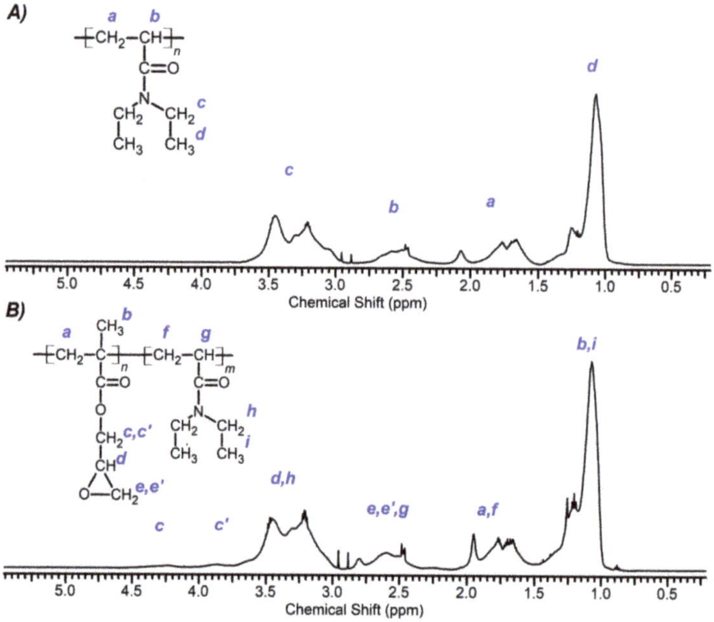

Figure 2. ^{1}H NMR spectra of the PDEAAm homopolymer (**A**) and the P(DEAAm-*co*-GMA) copolymer (Sample A, molar feed ratio AIBN:DEAAm:GMA = 1:95:5) (**B**).

Table 2. The DEAAm/GMA molar ratios in the feed and in the P(DEAAm-*co*-GMA) copolymers the molar percent of the GMA (X_{GMA}) of the polymers and the cloud point (T_{CP}) and clearing point (T_{CL}) measured by turbidimetry.

Sample	DEAAm/GMA Comonomerfeed Ratio	DEAAm/GMA Ratio in the Copolymers [a]	X_{GMA} (%)	T_{CP} (°C)	T_{CL} (°C)
A	19:1	17.06:1	5.5	31.2	30.8
B	9:1	8.91:1	10.1	27.2	26.2
C	19:1	15.25:1	6.2	30.6	30.1
D	9:1	7.80:1	11.4	24.8	24.6
PDEAAm	-	-	0	37.4	36.9

[a] Determined by ^{1}H NMR analysis.

The thermoresponsive behavior of the P(DEAAm-*co*-GMA) copolymers was investigated by turbidity measurements under the conditions proposed for standardization of the determination of critical solution temperatures of thermoresponsive LCST-type and UCST-type polymers [46,47]. As shown in Figure 3A, the transmittance–temperature curves of the heating and cooling cycles indicate reversible thermoresponsive precipitation–dissolution transitions for both the PDEAAm homopolymer and all the investigated copolymers with relatively small extent of heating–cooling hysteresis due to the lack of hydrogen bond formation between the PDEAAm chains in accordance with previous results [96]. The critical solution temperature (T_C) is defined as the temperature at the inflection point of the transmittance–temperature curves, i.e., the so-called cloud point temperature (T_{CP}) for heating and the clearing point temperature (T_{CL}) for cooling. As presented in Figure 3B and Table 2, the critical solution temperatures decrease linearly with decreasing DEAAm, i.e., with increasing GMA content, independent of the molar mass of the copolymers. It should also be noted that incorporating relatively low amounts of GMA in the P(DEAAm-*co*-GMA) copolymers results in significant decrease of the critical solution temperature (T_C) values (from 37.4 °C for the homopolymer to 24.8 °C at 11.4 mol% GMA in the copolymer), and this can be well tuned on the basis of the found linear relationship between T_C and the composition of the P(DEAAm-*co*-GMA) copolymers. Similar tendency was found for the critical solution temperature versus composition of poly(*N*,*N*-dimethylacrylamide-*co*-glycidyl methacrylate) copolymers but at much higher GMA contents (32–50 mol%) [124].

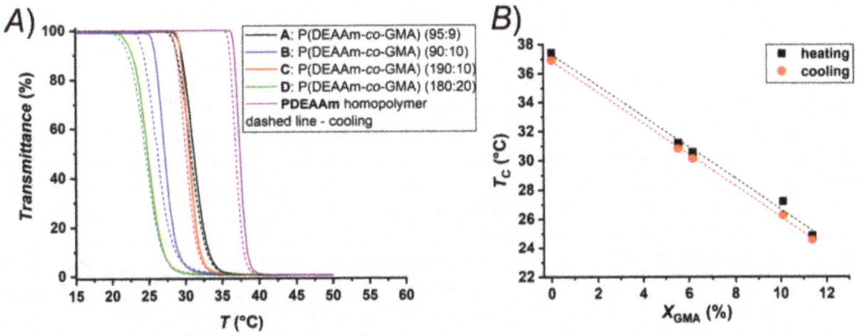

Figure 3. Transmittance vs. temperature curves of the P(DEAAm-*co*-GMA) copolymers and the PDEAAm homopolymer during heating and cooling (**A**) and the cloud points and clearing points as a function of the DEAAm content (**B**).

One of the intensively investigated application of epoxy(glycidyl)-functionalized polymers is their conjugation with various biomaterials, such as proteins and enzymes. In our work, the applicability of the produced epoxy-functional thermoresponsive P(DEAAm-*co*-GMA) copolymers for bioconjugation was investigated via a direct reaction between α-chymotrypsin, a widely used enzyme, and one selected copolymer (Sample C), as depicted in Scheme 1. Under the conditions described in the Experimental, 22.5 mg dried conjugate was obtained, which means that the yield of the conjugation was 48%. The resulting conjugate was investigated by DLS measurement and the results are compared to that of the starting copolymer sample and the enzyme. The recorded size distribution curves are presented in Figure 4. The size of the α-chymotrypsin is 3.34 nm with low dispersity, which corresponds well to the literature value [125]. The size of the P(DEAAm-*co*-GMA) copolymer is somewhat larger and has broader size distribution (d = 6.24 nm, PDI = 0.11). As observed, the size of the resulting enzyme–copolymer conjugate is in the range of 30–100 nm with average size of 56.9 nm. In addition, peaks do not appear in the range of the size of the reactants, which means that all unreacted copolymer and enzyme was removed by the applied dialysis purification method. These findings provide clear evidence that the designed one-step reaction took place successfully, and enzyme–polymer nanoparticles (EPNPs) are formed in the direct conjugation reaction between the P(DEAAm-*co*-GMA) copolymer and α-chymotrypsin. It has to be noted that usually epoxy containing carriers, i.e., polymers or inorganic particles, are first converted to amine by either treating with ammonia or a diamine, and then the conjugation (coupling) to the enzyme is carried out by glutaraldehyde [114,115]. In contrast to this widely applied two-step conjugation process, the P(DEAAm-*co*-GMA) copolymers enable an efficient one-step conjugation reaction with amine containing proteins and enzymes as proved by our results. This finding may open new routes for a variety of novel protein-polymer conjugations, especially by applying the biocompatible thermoresponsive epoxy-functionalized PDEAAm.

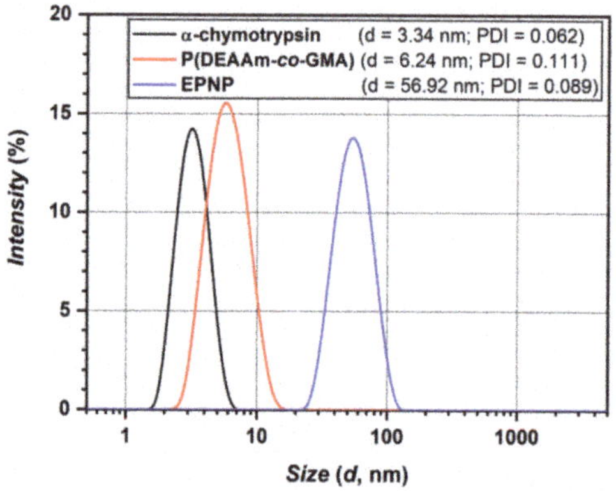

Figure 4. DLS size distribution curves of the α-chymotrypsin (black), the P(DEAAm-*co*-GMA) copolymer (Sample C) (red) and the produced enzyme–polymer conjugate (blue).

The enzyme content of the produced enzyme–polymer nanoparticle (EPNP) was determined by UV-Vis spectroscopy. The recorded UV spectra of the EPNP and the unreacted copolymer are presented in Figure S5. As can be seen, the copolymer has no absorbance above 250 nm, but a broad peak appears in the spectrum of the EPNP in the 260–300 nm range due the aromatic side groups of the enzyme component. This also confirms that the enzyme is incorporated into the EPNP. In addition, it allows the determination of the enzyme content as well, because the composition of the EPNP can be determined on the basis of calibration with the enzyme at a selected wavelength, 283 nm in this case (see Figure S6 in the Supplementary Materials). The determined enzyme content is 0.687 mg/mg EPNP, which means that the produced nanoconjugate consists of 68.7% of α-chymotrypsin and 31.3% of P(DEAAm-*co*-GMA) copolymer. Considering the composition of the EPNP gives that the molar ratio of GMA to the enzyme in the conjugate is 5.4, i.e., sufficiently high for coupling of the copolymer even to more than one CT molecule. A rough estimate can also be provided on the average number of the copolymer chains and enzyme molecules in the bioconjugate if it is assumed that the diameter (volume) of the components does not change by conjugation. On the basis of this approximation, the average numbers of the P(DEAAm-*co*-GMA) copolymer and the enzyme in their conjugate are around 6.5 and 5, respectively.

The effect of the conjugation on the thermoresponsive behavior of the EPNP was investigated by turbidimetry. The transmittance versus temperature curves (Figure 5A) and its first derivative (Figure 5B) of the EPNP are plotted and compared to that of the unmodified copolymer. As shown in this Figure, the thermal transition is slightly shifted to higher temperature by the conjugation, but the shape of the curve is similar to that of the unreacted copolymer. In the cooling cycle, the temperature range of the dissolution process is significantly broadened for the EPNP, but it has to be emphasized that the transmittance of the EPNP is returned to its maximum value (100% transmittance) by cooling, indicating that the EPNP preserved the reversible thermoresponsive behavior.

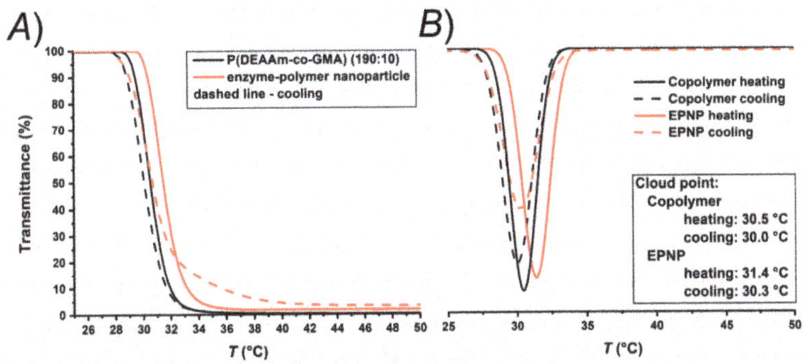

Figure 5. Transmittance vs. temperature curves (**A**) and its first derivatives (**B**) of the copolymer (black) and enzyme–polymer nanoparticle (red) in heating (full line) and cooling cycle (dashed line).

The applicability of the produced EPNP and the effect of the conjugation on the catalytic activity were investigated by enzymatic activity assay, where the enzymatic hydrolysis of BTEE was followed by UV-Vis spectroscopy at 25 °C in solutions of various pH and after thermal treatment at 45 °C. The enzymatic activity was calculated based on the rate of the conversion of the BTEE substrate. The determined enzymatic activity of the α-chymotrypsin-P(DEAAm-*co*-GMA) EPNP in the 6–9 pH range is presented and compared to the unmodified enzyme in Figure 6. As can be seen in this Figure, the enzymatic activity of the EPNP is significantly lower than that of the native enzyme, but this is a general phenomenon in the case of enzyme conjugates [50–52]. This can be explained by decreased accessibility of the substrate to the active pocket of the enzyme in the enzyme–polymer conjugates. The highest enzymatic activity was observed in the pH 7–7.5 range. The pH optimum was determined by the inflection point of the first derivative of the Gauss function fitted on the activity data. As shown in Figure 6A, on the one hand, the pH optimum of the EPNP is slightly lower (pH = 7.3) than that of the native enzyme (pH = 7.4). On the other hand, the activity is greatly decreased even with a slight change in pH in the case of the native α-chymotrypsin, but only a lower extent of change of activity was observed for the EPNP.

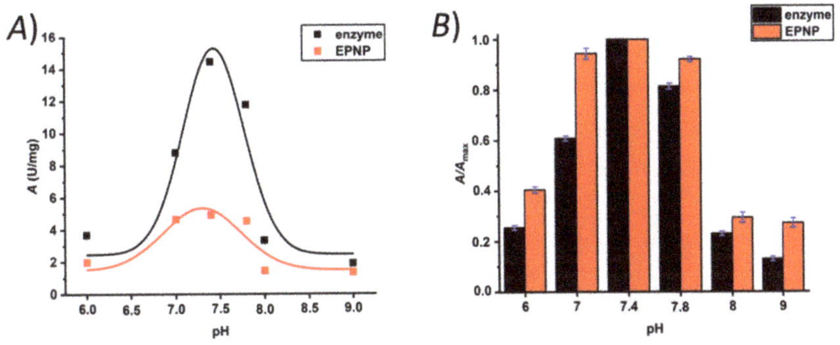

Figure 6. Enzymatic activity (**A**) and the relative activity (**B**) of α-chymotrypsin (black) and the enzyme–polymer nanoparticle (red) as a function of pH.

For better understanding, the effect of the pH change on the enzymatic activity, the relative (also called as residual) activity was expressed by the ratio of the measured activity and the maximal activity (Figure 6B). In the case of the native enzyme, the activity is decreased drastically by 0.4 pH change, namely the activity is only 80% and 60% of the maximum at pH 7.8 and 7.0, respectively. In addition, by getting away from the optimum pH value to pH 6 and pH 9, the relative activity is further decreased to ~20%. In contrast, less than 10% activity loss was observed in the pH range of 7–7.8 in the case of the EPNP and the relative activity was also significantly higher than that of the native enzyme at more extreme pH values. Thus, it can be concluded that the polymer conjugation with P(DEAAm-*co*-GMA) advantageously enhances the pH stability of α-chymotrypsin.

The thermal stability of the enzyme–polymer nanoconjugate was also studied. The solution of the EPNP and the native enzyme as well was thermostated at 45 °C, and samples were taken at predetermined treatment times. Because the results of the thermoresponsive investigation of the EPNP shows that the hydrophobic–hydrophilic transition occurs in a wider temperature range during cooling, the withdrawn samples were allowed to cool to room temperature for 5 min before the activity assay measurements at 25 °C. The obtained relative activity plotted as a function of the thermal treatment time is displayed in Figure 7. As can be seen in this Figure, the initial activity of the free α-chymotrypsin decreases to 20% after only 10 min and to 10% after 60 min, and the enzyme becomes completely inactive after 120 min thermal treatment. This finding is in good agreement with results of others [118], according to which native CT loses its activity after thermal treatment at 50 °C for 90 min. It is widely accepted that this caused by the unfolding and inactivation during thermal denaturation of the enzymes. In contrast, the residual activity of the polymer conjugated enzyme in the EPNP is much higher, namely the relative activity is around 85% after five minutes and it is still over 40% after 30 min. Furthermore, the enzymatic activity of EPNP does not fall below 20% even after two hours thermal treatment. These results clearly indicate that the conjugated polymer can reduce the thermal unfolding of the enzyme in the produced nanoconjugate. Hence, it can be concluded that the enzyme-polymer nanoparticle preserves the enzymatic activity after heating, that is, the conjugation with the thermoresponsive P(DEAAm-*co*-GMA) copolymer increases advantageously the thermal stability of α-chymotrypsin.

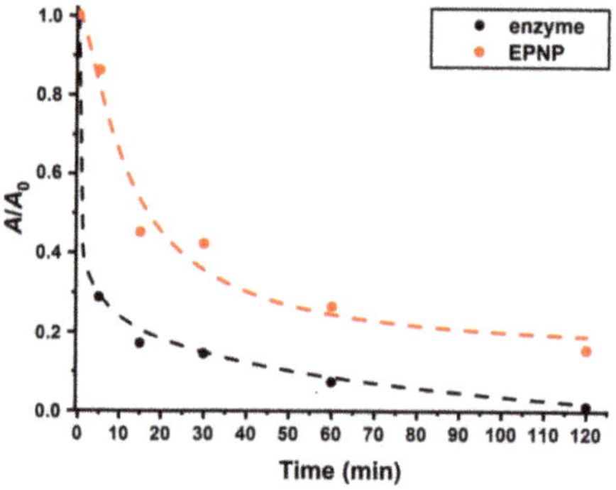

Figure 7. The relative activity of α-chymotrypsin (black) and the CT-P(DEAAm-*co*-GMA) enzyme-polymer nanoparticle (EPNP) (red) at 25 °C as a function of the time of thermal treatment at 45 °C.

4. Conclusions

Poly(*N*,*N*-diethylacrylamide-*co*-glycidyl methacrylate) (P(DEAAm-*co*-GMA)) copolymers, unreported so far, were successfully synthesized by free radical copolymerization. On the one hand, turbidity measurements revealed that the P(DEAAm-*co*-GMA) copolymers still possess reversible thermoresponsive behavior. It was found that the cloud point and clearing point temperatures of the copolymers are lower than that of the PDEAAm homopolymer and decrease linearly with increasing GMA content in the investigated composition range up to 11 mol% GMA. The reactivity of the epoxy (glycidyl) pendant groups and the applicability of such copolymers were demonstrated by a conjugation reaction with α-chymotrypsin. The formation of nanosized enzyme–polymer conjugates was confirmed by DLS with average diameter of 56.9 nm. In addition, it was also confirmed that not only the P(DEAAm-*co*-GMA) copolymers, but its enzyme–polymer nanoparticles (EPNP) also possess the reversible thermoresponsive behavior. The enzymatic activity of the produced

EPNP was investigated at various pH and after 45 °C thermal treatment, and compared to that of the native enzyme. It was found that the activity of the EPNP was lower than that of the free α-chymotrypsin, but the relative activity results proved that the activity of the EPNP is less sensitive to the changes of the pH and the temperature. On the basis of these findings, it can be concluded that the enzyme stability can be significantly enhanced by the polymer conjugation with P(DEAAm-*co*-GMA) copolymers. These findings can be utilized in a variety of applications, e.g., preparation of novel thermoresponsive protein-P(DEAAm-*co*-GMA) bioconjugates with enhanced stability in a one-step process, separation of the products from the thermoresponsive enzyme–polymer conjugates precipitating above its critical solution temperature etc.

Supplementary Materials: The following are available online at https://www.mdpi.com/2073-4360/13/6/987/s1, Figure S1: GPC chromatograms of the P(DEAAm-*co*-GMA) copolymers and PDEAAm homopolymer, Figure S2: ^{1}H NMR spectrum of *Sample B* P(DEAAm-*co*-GMA) copolymer (molar feed ratio AIBN:DEAAm:GMA = 1:90:10), Figure S3: ^{1}H NMR spectrum of *Sample C* P(DEAAm-*co*-GMA) copolymer (molar feed ratio AIBN:DEAAm:GMA = 1:190:10), Figure S4: ^{1}H NMR spectrum of *Sample D* P(DEAAm-*co*-GMA) copolymer (molar feed ratio AIBN:DEAAm:GMA = 1:180:20), Figure S5: UV spectra of the P(DEAAm-*co*-GMA) (Sample C, blue) and the produced enzyme–polymer nanoparticle (red), Figure S6: UV spectra of the α-chymotrypsin in the concentration range of 0.033–1 mg/mL (a) and the calibration curve fitted on the absorbance at 283 nm as a function of the enzyme concentration (b), Figure S7: Representative enzymatic activity investigation curves of the absorbance measurement of the enzyme (black) and EPNP (red) in time in different pH solvents (pH = 6 (A); 7 (B); 7.4 (C); 7.8 (D); 8 (E); 9 (F)), Figure S8: Representative curves of the activity measurements of the enzyme (black) and EPNP (red) in PBS buffer after thermostated at 45 °C for 0 min (A), 5 min (B), 15 min (C), 30 min (D), 60 min (E) and 120 min (F).

Author Contributions: Conceptualization: G.K., T.S., M.B., G.S., I.H., E.N., and B.I.; methodology: G.K., T.S., M.B., G.S., I.H., E.N., and B.I.; analysis: G.K., T.S., M.B., G.S., I.H., E.N., and B.I.; data evaluation: G.K., T.S., M.B., G.S., I.H., E.N., and B.I.; writing—original draft preparation: G.K., T.S., M.B., G.S., I.H., E.N., and B.I.; writing—review and editing: G.K., T.S., M.B., G.S., I.H., E.N., and B.I.; visualization: G.K., T.S., M.B., G.S., I.H., E.N., and B.I.; supervision: G.K. and B.I.; funding acquisition, B.I. All authors have read and agreed to the published version of the manuscript.

Funding: This research was supported by European Regional Development Fund, grant number GINOP-2.3.2-15-2016-00017 (BIONANO project), and the National Research, Development, and Innovation Office, Hungary.

Institutional Review Board Statement: Not applicable.

Informed Consent Statement: Not applicable.

Data Availability Statement: The data presented in this study are available on request from the corresponding authors.

Acknowledgments: The authors acknowledge the support by the European Regional Development Fund, grant number GINOP-2.3.2-15-2016-00017 (BIONANO project), and the National Research, Development and Innovation Office, Hungary (K135946).

Conflicts of Interest: The authors declare no conflict of interest.

References

1. Tasdelen, M.A.; Kahveci, M.U.; Yagci, Y. Telechelic polymers by living and controlled/living polymerization methods. *Prog. Polym. Sci.* **2011**, *36*, 455–567. [CrossRef]
2. Bokern, S.; Gries, K.; Görtz, H.H.; Warzelhan, V.; Agarwal, S.; Greiner, A. Precisely designed gold nanoparticles by surface polymerization-artificial molecules as building blocks for novel materials. *Adv. Funct. Mater.* **2011**, *21*, 3753–3759. [CrossRef]
3. Gao, H.; Matyjaszewski, K. Synthesis of functional polymers with controlled architecture by CRP of monomers in the presence of cross-linker: From stars to gels. *Prog. Polym. Sci.* **2009**, *34*, 317–350. [CrossRef]
4. Kennedy, J.P.; Iván, B. *Designed Polymers by Carbocationic Macromolecular Engineering: Theory and Practice*; Hanser Publisher: New York, NY, USA, 1992.
5. He, X.-Y.; Liu, B.-Y.; Ai, S.-L.; Xu, L.; Zhuo, R.-X.; Cheng, S.-X. Functional polymer/inorganic hybrid nanoparticles for macrophage targeting delivery of oligodeoxynucleotides in cancer immuniotherapy. *Mater. Today Chem.* **2017**, *4*, 106–116. [CrossRef]

6. Fu, H.-L.; Cheng, S.-X.; Zhang, X.-Z.; Zhuo, R.-X. Dendrimer/DNA complexes encapsulated functional biodegradable polymer for substrate-mediated gene delivery. *J. Gene Med.* **2008**, *10*, 1334–1342. [CrossRef] [PubMed]
7. Liu, M.-Q.; Wang, C.; Kim, N.-Y. High-sensitivity and low-hysteresis porous MIM-type capacitive humidity sensor using functional polymer mixed with TiO_2 microparticles. *Sensors* **2017**, *17*, 284. [CrossRef]
8. Takeuchi, T.; Hayashi, T.; Ichikawa, S.; Kaji, A.; Masui, M.; Matsumoto, H.; Sasao, R. Molecularly imprinted tailor-made functional polymer receptors for highly sensitive and selective separation and detection of target molecules. *Chromatography* **2016**, *37*, 43–64. [CrossRef]
9. Chen, L.; Yan, C.; Zheng, Z. Functional polymer surfaces for controlling cell behaviors. *Mater. Today* **2018**, *21*, 38–59. [CrossRef]
10. Zhao, Y.; Weix, D.J. Nickel-catalyzed regiodivergent opening of epoxides with aryl halides: Co-catalysis controls regioselectivity. *J. Am. Chem. Soc.* **2014**, *136*, 48–51. [CrossRef] [PubMed]
11. Ko, S.; Jang, J. Protein immobilization on aminated poly(glycidyl methacrylate) nanofibers as polymeric carriers. *Biomacromolecules* **2007**, *8*, 1400–1403. [CrossRef]
12. Gadwal, I.I.; Stuparu, M.C.; Khan, A. Homopolymer biofunctionalization through sequential thiol-epoxy and esterification reactions: An optimization, quantification, and structural elucidation study. *Polym. Chem.* **2015**, *6*, 1393–1404. [CrossRef]
13. Undin, J.; Finne-Wistrand, A.; Albertsson, A.C. Copolymerization of 2-methylene-1,3-dioxepane and glycidyl methacrylate, a well-defined and efficient process for achieving functionalized polyesters for covalent binding of bioactive molecules. *Biomacromolecules* **2013**, *14*, 2095–2102. [CrossRef] [PubMed]
14. Schmitt, S.K.; Murphy, E.L.; Gopalan, P. Crosslinked PEG mats for peptide immobilization and stem cell adhesion. *J. Mater. Chem. B* **2013**, *1*, 1349–1360. [CrossRef] [PubMed]
15. Kasza, G.Y.; Szarka, G.Y.; Bodor, A.; Kali, G.; Iván, B. In situ terminal functionalization of polystyrene obtained by quasiliving ATRP and subsequent derivatization. *ACS Symp. Ser.* **2018**, *1285*, 281–295.
16. Iván, B.; Fónagy, T. Quantitative Derivatizations of 1-Chloro-1-phenylethyl Chain End of Polystyrene Obtained by Quasiliving Atom Transfer Radical Polymerization. *ACS Symp. Ser.* **2000**, *768*, 372–383.
17. Sane, P.S.; Palaskar, D.V.; Wadgaonkar, P.P. Synthesis of bis-allyloxy functionalized polystyrene and poly(methyl methacrylate) macromonomers using a new ATRP initiator. *Eur. Polym. J.* **2011**, *47*, 1621–1629. [CrossRef]
18. Hirao, A.; Shimohara, N.; Ryu, S.W.; Sugiyama, K. Synthesis of highly branched comblike polymers having one branch in each repeating unit by linking reaction of polystyryllithium with well-defined new epoxy-functionalized polystyrene. *Macromol. Smyp.* **2004**, *214*, 17–28. [CrossRef]
19. Muzammil, E.M.; Khan, A.; Stuparu, M.C. Post-polymerization modification reactions of poly(glycidyl methacrylate)s. *RSC Adv.* **2017**, *7*, 55874–55884. [CrossRef]
20. Stuparu, M.C.; Khan, A. Thiol-epoxy "click" chemistry: Application in preparation and postpolymerization modification of polymers. *J. Polym. Sci. Part A Polym. Chem.* **2016**, *54*, 3057–3070. [CrossRef]
21. Kuroishi, P.K.; Bennison, M.J.; Dove, A.P. Synthesis and post-polymerisation modification of an epoxy-functional polycarbonate. *Polym. Chem.* **2016**, *46*, 7108–7115. [CrossRef]
22. McLeod, D.C.; Tsarevsky, N.V. 4-Vinylphenyl glycidyl ether: Synthesis, RAFT polymerization and postpolymerization modifications with alcohols. *Macromolecules* **2016**, *49*, 1135–1142. [CrossRef]
23. Gunaydin, O.; Yilmaz, F. Copolymers of glycidyl methacrylate with 3-methylthienyl methacrylate: Synthesis, characterization and reactivity ratios. *Polym. J.* **2007**, *39*, 579–588. [CrossRef]
24. Maciejewska, M. Thermal properties of TRIM-GMA copolymers with pendant amine groups. *J. Therm. Anal. Calorim.* **2016**, *126*, 1777–1785. [CrossRef]
25. Chou, Y.-N.; Wen, T.-C.; Chang, Y. Zwitterionic surface grafting of epoxylated sulfobetaine copolymers for the development of stealth biomaterial interfaces. *Acta Biomater.* **2016**, *40*, 78–91. [CrossRef] [PubMed]
26. Brito, G.F.; Agrawal, P.; Mélo, T.J.A. Mechanical and morphological properties of PLA/BioPE blen compatibilized with E-GMA and EMA-GMA copolymers. *Macromol. Symp.* **2016**, *367*, 176–182. [CrossRef]
27. Yang, X.; Wang, H.T.; Chen, J.L.; Fu, Z.A.; Zhao, X.W.; Li, Y.J. Copolymers containing two types of reactive groups: New compatibilizer for immiscible PLLA/PA11 polymer blends. *Polymer* **2019**, *177*, 139–148. [CrossRef]
28. Acikbas, Y.; Capan, R.; Erdogan, M.; Bulut, L.; Soykan, C. Optical characterization and swelling behavior of Langmuir-Blodgett thin films of a novel poly[(styrene (ST)-co-glycidyl methacrylate (GMA)]. *Sens. Actuators B* **2017**, *241*, 1111–1120. [CrossRef]
29. Lei, Z.; Gao, J.; Liu, X.; Liu, D.; Wang, Z. Poly(glycidyl methacrylate-co-2-hydroxyethyl methacrylate) brushes as peptide/protein microarray substrate for improving protein binding and functionality. *ACS Appl. Mater. Interfaces* **2016**, *8*, 10174–10182. [CrossRef]
30. Nastasovic, A.B.; Ekmescic, B.M.; Sandic, Z.P.; Randelovic, D.V.; Mozetic, M.; Vesel, A.; Onjia, A.E. Mechanism of Cu(II), Cd(II) and Pb(II) ions sorption from aqueous solution by macroporous poly(glycidyl methacrylate-co-ethylene glycol dimethacrylate). *Appl. Surf. Sci.* **2016**, *385*, 605–615. [CrossRef]
31. Chauhan, G.S.; Guleria, L.; Sharma, R. Synthesis, characterization and metal ion sorption studies of graft copolymers of cellulose with glycidyl methacrylate and some comonomers. *Cellulose* **2005**, *12*, 97–110. [CrossRef]
32. Hus, S.; Kolar, M.; Krajnc, P. Separation of heavy metals from water by functionalized glycidyl methacrylate poly(high internal phase emulsions). *J. Chromatogr. A* **2016**, *1437*, 168–175. [CrossRef]
33. Li, X.; Xu, Z.; Yin, H.; Feng, Y.; Quan, H. Comparative Studies on Enhanced Oil Recovery: Thermoviscosifying Polymer Versus Polyacrylamide. *Energy Fuels* **2017**, *31*, 2479–2487. [CrossRef]

4. Li, X.; Yin, H.-Y.; Zhang, R.-S.; Cui, J.; Wu, J.-W.; Feng, Y.-J. A salt-induced viscosifying smart polymer for fracturing inter-salt shale oil reservoirs. *Petroleum Sci.* **2019**, *16*, 816–829. [CrossRef]
5. Su, X.; Feng, Y. Thermoviscosifying Smart Polymers for Oil and Gas Production: State of the Art. *Chem. Phys. Chem.* **2018**, *19*, 1941–1955. [CrossRef] [PubMed]
6. Ashraf, S.; Park, H.-K.; Park, H.; Lee, S.-H. Snapshot of phase transition in thermoresponsive hydrogel PNIPAM: Role in drug delivery and tissue engineering. *Macromol. Res.* **2016**, *24*, 297–304. [CrossRef]
7. Utrata-Wesolek, A.; Oleszko-Torbus, N.; Bochenek, M.; Kosowski, D.; Kowalczuk, A.; Trzebicka, B.; Dworak, A. Thermoresponsive polymer surfaces and their application in tissue engineering. *Polimery* **2018**, *5*, 325–406. [CrossRef]
8. Song, X.; Zhu, J.-L.; Wen, Y.; Zhao, F.; Zhang, Z.-X.; Li, J. Thermoresponsive supramolecular micellar drug delivery system based on star-linear pseudo-block polymer consisting of β-cyclodextrin-poly(*N*-isopropylacrylamide) and adamantyl-poly(ethylene glycol). *J. Colloid Interface Sci.* **2017**, *490*, 372–379. [CrossRef]
9. Gandhi, A.; Paul, A.; Sen, S.O.; Sen, K.K. Studies on thermoresponsive polymers: Phase behavior, drug delivery and biomedical applications. *Asian J. Pharm. Sci.* **2015**, *10*, 99–107. [CrossRef]
0. Gong, D.; Cao, T.; Han, S.-C.; Zhu, X.; Iqbal, A.; Liu, W.; Qin, W.; Guo, H. Fluorescence enhancement thermoresponsive polymer luminescent sensors based on BODIPY for intracellular temperature. *Sens. Actuators B* **2017**, *252*, 577–583. [CrossRef]
1. Chan, E.W.C.; Baek, P.; De la Rosa, V.R.; Barker, D.; Hoogenboom, R.; Travas-Sejdic, J. Thermoresponsive laterally-branched polythiophene phenylene derivative as water-soluble temperature sensor. *Polym. Chem.* **2017**, *8*, 4352–4359. [CrossRef]
2. Owusu-Nkwantabisah, S.; Gillmor, J.; Switalski, S.; Mis, M.R.; Bennett, G.; Moody, R.; Antalek, B.; Gutierrez, R.; Slater, G. Synergistic thermoresponsive optical properties of a composite self-healing hydrogel. *Macromolecules* **2017**, *50*, 3671–3679. [CrossRef]
3. Vidal, F.; Lin, H.; Morales, C.; Jakle, F. Polysiloxane/polystyrene thermos-responsive and self-healing polymer network via Lewis asic-Lewis base pair formation. *Molecules* **2018**, *23*, 405. [CrossRef]
4. Kim, J.-H.; Jung, Y.; Lee, D.; Jang, W.-D. Thermoresponsive polymer and fluorescent dye hybrids for tunable multicolour emission. *Adv. Mater.* **2016**, *28*, 3499–3503. [CrossRef]
5. Chin, S.M.; Synatschke, C.V.; Liu, S.; Nap, R.J.; Sather, N.A.; Wang, Q.; Álvarez, Z.; Edelbrock, A.N.; Fyrner, T.; Palmer, L.C.; et al. Covalent-supramolecular hybrid polymers as muscle-inspired anisotropic actuators. *Nat. Commun.* **2018**, *9*, 2395. [CrossRef]
6. Osváth, Z.; Iván, B. The Dependence of the Cloud Point, Clearing Point, and Hysteresis of Poly(*N*-isopropylacrylamide) on Experimental Conditions: The Need for Standardization of Thermoresponsive Transition Determinations. *Macromol. Chem. Phys.* **2017**, *218*, 1600470. [CrossRef]
7. Osváth, Z.; Tóth, T.; Iván, B. Synthesis, characterization, LCST-type behavior and unprecedented heating-cooling hysteresis of poly(*N*-isopropylacrylamide-co-3-(trimethoxysilyl) propyl methacrylate) copolymers. *Polymer* **2017**, *108*, 395–399. [CrossRef]
8. Osváth, Z.; Tóth, T.; Iván, B. Sustained Drug Release by Thermoresponsive Sol-Gel Hybrid Hydrogels of Poly(*N*-Isopropylacrylamide-co-3-(Trimethoxysilyl) Propyl Methacrylate) Copolymers. *Macromol. Rapid Commun.* **2017**, *38*, 1600724. [CrossRef] [PubMed]
9. Xu, X.; Liu, Y.; Fu, W.; Yao, M.; Ding, Z.; Xuan, J.; Li, D.; Wang, S.; Xia, Y.; Cao, M. Poly(*N*-isopropylacrylamide)-based thermoresponsive composite hydrogels for biomedical applications. *Polymers* **2020**, *12*, 580. [CrossRef] [PubMed]
0. Barsbay, M.; Güven, O. Modification of Polystyrene Cell-Culture-Dish Surfaces by Consecutive Grafting of Poly(acrylamide)/Poly (*N*-isopropylacrylamide) via Reversible Addition-Fragmentation Chain Transfer-Mediated Polymerization. *Eur. Polym. J.* **2021**, *147*, 110330. [CrossRef]
1. Luo, G.F.; Chen, W.H.; Zhang, X.Z. 100th Anniversary of Macromolecular Science Viewpoint: Poly(*N*-isopropylacrylamide)-Based Thermally Responsive Micelles. *ACS Macro Lett.* **2020**, *9*, 872–881. [CrossRef]
2. Hirayama, S.; Oohora, K.; Uchihashi, T.; Hayashi, T. Thermoresponsive micellar assembly constructed from a hexameric hemoprotein modified with poly(*N*-isopropylacrylamide) toward an artificial light-harvesting system. *J. Am. Chem. Soc.* **2020**, *142*, 1822–1831. [CrossRef] [PubMed]
3. Fischer, T.; Demco, D.E.; Fechete, R.; Möller, M.; Singh, S. Poly(vinylamine-co-*N*-isopropylacrylamide) linear polymer and hydrogels with tuned thermoresponsivity. *Soft Matter* **2020**, *16*, 6549–6562. [CrossRef]
4. Tang, L.; Wang, L.; Yang, X.; Feng, Y.; Li, Y.; Feng, W. Poly(*N*-isopropylacrylamide)-based smart hydrogels: Design, properties and applications. *Prog. Mater. Sci.* **2020**, 100702. [CrossRef]
5. Halperin, A.; Kröger, M.; Winnik, F.M. Poly(*N*-isopropylacrylamide) phase diagrams: Fifty years of research. *Angew. Chem. Int. Ed.* **2015**, *54*, 15342–15367. [CrossRef]
6. Virtanen, J.; Tenhu, H. Studies on copolymerization of *N*-isopropylacrylamide and glycidyl methacrylate. *J. Polym. Sci. Part A Polym. Chem.* **2001**, *39*, 3716–3725. [CrossRef]
7. Venugopal, B.; Shenoy, S.J.; Mohan, S.; Anil Kumar, P.R.; Kumary, T.V. Bioengineered corneal epithelial cell sheet from mesenchymal stem cells—A functional alternative to limbal stem cells for ocular surface reconstruction. *J. Biomed. Mater. Res. Part B* **2020**, *108*, 1033–1045. [CrossRef]
8. Pourjavadi, A.; Kohestanian, M.; Streb, C. pH and thermal dual-responsive poly(NIPAM-co-GMA)-coated magnetic nanoparticles via surface-initiated RAFT polymerization for controlled drug delivery. *Mater. Sci. Eng. C* **2020**, *108*, 110418. [CrossRef]
9. Li, P.; Xu, R.; Wang, W.; Li, X.; Xu, Z.; Yeung, K.W.; Chu, P.K. Thermosensitive poly(*N*-isopropylacrylamide-co-glycidyl methacrylate) microgels for controlled drug release. *Colloids Surf. B* **2013**, *101*, 251–255. [CrossRef]

60. Jiang, X.; Xiong, D.A.; An, Y.; Zheng, P.; Zhang, W.; Shi, L. Thermoresponsive hydrogel of poly(glycidyl methacrylate-co-N-isopropylacrylamide) as a nanoreactor of gold nanoparticles. *J. Polym. Sci. Part A Polym. Chem.* **2007**, *45*, 2812–2819. [CrossRef]
61. Chen, L.; Dong, J.; Ding, Y.; Han, W. Environmental responses of poly(*N*-isopropylacrylamide-co-glycidyl methacrylate derivatized dextran hydrogels. *J. Appl. Polym. Sci.* **2005**, *96*, 2435–2439. [CrossRef]
62. Nguyen, A.L.; Luong, J.H.T. Syntheses and applications of water-soluble reactive polymers for purification and immobilization of biomolecules. *Biotechnol. Bioeng.* **1989**, *34*, 1186–1190. [CrossRef]
63. Fang, S.-J.; Kawaguchi, H. A thermoresponsive amphoteric microsphere and its potential application as a biological carrier. *Colloid Polym. Sci.* **2002**, *280*, 984–989. [CrossRef]
64. Peng, H.; Rübsam, K.; Hu, C.; Jakob, F.; Schwaneberg, U.; Pich, A. Stimuli-Responsive Poly(N-Vinyllactams) with Glycidyl Side Groups: Synthesis, Characterization, and Conjugation with Enzymes. *Biomacromolecules* **2019**, *20*, 992–1006. [CrossRef]
65. Panayiotou, M.; Pöhner, C.; Vandevyver, C.; Wandrey, C.; Hilbrig, F.; Freitag, R. Synthesis and characterisation of thermo-responsive poly(*N,N'*-diethylacrylamide) microgels. *React. Funct. Polym.* **2007**, *67*, 807–819. [CrossRef]
66. Ida, S.; Harada, H.; Sakai, K.; Atsumi, K.; Tani, Y.; Tanimoto, S.; Hirokawa, Y. Shape and size regulation of gold nanoparticles by poly(*N,N*-diethylacrylamide) microgels. *Chem. Lett.* **2017**, *46*, 760–763. [CrossRef]
67. Isiklan, N.; Kazan, H. Thermoresponsive and biocompatible poly(vinyl alcohol)-graft-poly(*N,N*-diethylacrylamide) copolymer Microwave-assisted synthesis, characterization, and swelling behavior. *J. Appl. Polym. Sci.* **2018**, *135*, 45969. [CrossRef]
68. Zhang, X.; Yang, Z.; Xie, D.; Liu, D.; Chen, Z.; Li, K.; Li, Z.; Tichnell, B.; Liu, Z. Design and synthesis study of the thermo-sensitive poly(N-vinylpyrrolidone-b-*N,N*-diethylacrylamide). *Des. Monomers Polym.* **2018**, *21*, 43–54. [CrossRef] [PubMed]
69. Matsumoto, M.; Tada, T.; Asoh, T.A.; Shoji, T.; Nishiyama, T.; Horibe, H.; Katsumoto, Y.; Tsuboi, Y. Dynamics of the phase separation in a thermoresponsive polymer: Accelerated phase separation of stereocontrolled poly(*N,N*-diethylacrylamide) in water. *Langmuir* **2018**, *34*, 13690–13696. [CrossRef]
70. Havanur, S.; Farheenand, V.; JagadeeshBabu, P.E. Synthesis and optimization of poly(*N,N*-diethylacrylamide) hydrogel and evaluation of its anticancer drug doxorubicin's release behavior. *Iran. Polym. J.* **2019**, *28*, 99–112. [CrossRef]
71. Wu, M.; Zhang, H.; Liu, H. Study of phase separation behavior of poly(*N,N*-diethylacrylamide) in aqueous solution prepared by RAFT polymerization. *Polym. Bull.* **2019**, *76*, 825–848. [CrossRef]
72. Li, J.; Kikuchi, S.; Sato, S.I.; Chen, Y.; Xu, L.; Song, B.; Duan, Q.; Wang, Y.; Kakuchi, T.; Shen, X. Core-First Synthesis and Thermoresponsive Property of Three-, Four-, and Six-Arm Star-Shaped Poly(*N,N*-diethylacrylamide)s and Their Block Copolymers with Poly(*N,N*-dimethylacrylamide). *Macromolecules* **2019**, *52*, 7207–7217. [CrossRef]
73. Luo, G.; Lu, Y.; Wu, S.; Shen, X.; Zhu, M.; Li, S. Hierarchical polymer composites as smart reactor for formulating simple/tandem commutative catalytic ability. *J. Inorg. Organomet. Polym. Mater.* **2020**, *30*, 4394–4407. [CrossRef]
74. Zhang, L.; Xie, L.; Xu, S.; Kuchel, R.P.; Dai, Y.; Jung, K.; Boyer, C. Dual Role of Doxorubicin for Photopolymerization and Therapy. *Biomacromolecules* **2020**, *21*, 3887–3897. [CrossRef] [PubMed]
75. Paneysar, J.S.; Jain, S.; Ahmed, N.; Barton, S.; Ambre, P.; Coutinho, E. Novel smart composite materials for industrial wastewater treatment and reuse. *SN Appl. Sci.* **2020**, *2*, 1–12. [CrossRef]
76. Li, S.; Wang, F.; Yang, Z.; Xu, J.; Liu, H.; Zhang, L.; Xu, W. Emulsifying performance of near-infrared light responsive polydopamine-based silica particles to control drug release. *Powder Technol.* **2020**, *359*, 17–26. [CrossRef]
77. Baert, M.; Wicht, K.; Hou, Z.; Szucs, R.; Prez, F.D.; Lynen, F. Exploration of the selectivity and retention behavior of alternative polyacrylamides in temperature responsive liquid chromatography. *Anal. Chem.* **2020**, *92*, 9815–9822. [CrossRef]
78. Lee, C.H.; Bae, Y.C. Thermodynamic framework for switching the lower critical solution temperature of thermo-sensitive particle gels in aqueous solvent. *Polymer* **2020**, *195*, 122428. [CrossRef]
79. Ma, Y.; Li, M.; Shi, K.; Chen, Z.; Yang, B.; Rao, D.; Li, X.; Ma, W.; Hou, S.; Gou, G.; et al. Multiple stimuli-switchable electrocatalysis and logic gates of rutin based on semi-interpenetrating polymer network hydrogel films. *New J. Chem.* **2020**, *44*, 16045–16053. [CrossRef]
80. Ma, C.; Tchameni, A.P.; Pan, L.; Su, C.; Zhou, C. A thermothickening polymer as a novel flocculant for oily wastewater treatment. *Sep. Sci. Technol.* **2020**, *55*, 123–134. [CrossRef]
81. Ni, M.; Xu, Y.; Wang, C.; Zhao, P.; Yang, P.; Chen, C.; Zheng, K.; Wang, H.; Sun, X.; Lia, C.; et al. A novel thermo-controlled acetaminophen electrochemical sensor based on carboxylated multi-walled carbon nanotubes and thermosensitive polymer. *Diam. Relat. Mater.* **2020**, *107*, 107877. [CrossRef]
82. Zhang, X.; Burton, T.F.; In, M.; Begu, S.; Aubert-Pouëssel, A.; Robin, J.J.; Mongea, S.; Giani, O. Synthesis and behaviour of PEG-b-PDEAm block copolymers in aqueous solution. *Mater. Today Commun.* **2020**, *24*, 100987. [CrossRef]
83. Chen, L.; Chen, X.; Gong, Y.; Shao, T.; Zhang, X.; Chen, D.; Cai, Z. Study on Influence of Monobutyl Itaconate and *N,N*-diethylacrylamide on Acrylic Latex. *Prot. Met. Phys. Chem. Surf.* **2020**, *56*, 740–745. [CrossRef]
84. Hanyková, L.; Krakovský, I.; Šestáková, E.; Šťastná, J.; Labuta, J. Poly(*N,N'*-Diethylacrylamide)-Based Thermoresponsive Hydrogels with Double Network Structure. *Polymers* **2020**, *12*, 2502. [CrossRef]
85. Tuan, H.N.A.; Nhu, V.T.T. Synthesis and properties of pH-thermo dual responsive semi-IPN hydrogels based on *N,N'*-diethylacrylamide and itaconamic acid. *Polymers* **2020**, *12*, 1139. [CrossRef]
86. Yan, X.; Chu, Y.; Liu, B.; Ru, G.; Di, Y.; Feng, J. Dynamic mechanism of halide salts on the phase transition of protein models poly(*N*-isopropylacrylamide) and poly(*N,N*-diethylacrylamide). *Phys. Chem. Chem. Phys.* **2020**, *22*, 12644–12650. [CrossRef]

87. Zhang, J.; Wang, R.; Ou, X.; Zhang, X.; Liu, P.; Chen, Z.; Zhang, B.; Liu, C.; Zhao, S.; Chen, Z.; et al. Bio-inspired synthesis of thermo-responsive imprinted composite membranes for selective recognition and separation of ReO_4^-. *Sep. Purif. Technol.* **2021**, *259*, 118165. [CrossRef]
88. Drozdov, A.D.; Christiansen, J.D. Modulation of the volume phase transition temperature of thermo-responsive gels. *J. Mech. Behav. Biomed.* **2021**, *114*, 104215. [CrossRef] [PubMed]
89. Gao, H.; Mao, J.; Cai, Y.; Li, S.; Fu, Y.; Liu, X.; Liang, H.; Zhao, T.; Liu, M.; Jiang, L. Euryhaline Hydrogel with Constant Swelling and Salinity-Enhanced Mechanical Strength in a Wide Salinity Range. *Adv. Funct. Mater.* **2021**, *31*, 2007664. [CrossRef]
90. Xie, B.; Tchameni, A.P.; Luo, M.; Wen, J. A novel thermo-associating polymer as rheological control additive for bentonite drilling fluid in deep offshore drilling. *Mater. Lett.* **2021**, *284*, 128914. [CrossRef]
91. Li, K.; Liu, X.T.; Zhang, Y.F.; Liu, D.; Zhang, X.Y.; Ma, S.M.; Ruso, J.M.; Tang, Z.H.; Chen, Z.B.; Liu, Z. The engineering and immobilization of penicillin G acylase onto thermo-sensitive tri-block copolymer system. *Polym. Adv. Technol.* **2019**, *30*, 86–93. [CrossRef]
92. Li, K.; Shan, G.; Ma, X.; Zhang, X.; Chen, Z.; Tang, Z.; Liu, Z. Study of target spacing of thermo-sensitive carrier on the activity recovery of immobilized penicillin G acylase. *Colloids Surf. B.* **2019**, *179*, 153–160. [CrossRef] [PubMed]
93. Idziak, I.; Avoce, D.; Lessard, D.; Gravel, D.; Zhu, X.X. Thermosensitivity of Aqueous Solutions of Poly(*N*,*N*-diethylacrylamide). *Macromolecules* **1999**, *32*, 1260–1263. [CrossRef]
94. Lessard, D.G.; Ousalem, M.; Zhu, X.X. Effect of the molecular weight on the lower critical solution temperature of poly(*N*,*N*-diethylacrylamide) in aqueous solutions. *Can. J. Chem.* **2001**, *79*, 1870–1874. [CrossRef]
95. Lessard, D.G.; Ousalem, M.; Zhu, X.X.; Eisenberg, A.; Carreau, P.J. Study of the phase transition of poly(*N*,*N*-diethylacrylamide) in water by rheology and dynamic light scattering. *J. Polym. Sci. Part B Polym. Phys.* **2003**, *41*, 1627–1637. [CrossRef]
96. Zhou, K.; Lu, Y.; Li, J.; Shen, L.; Zhang, G.; Xie, Z.; Wu, C. The coil-to-globule-to-coil transition of linear polymer chains in dilute aqueous solutions: Effect of intrachain hydrogen bonding. *Macromolecules* **2008**, *41*, 8927–8931. [CrossRef]
97. Trzebicka, B.; Szweda, R.; Kosowski, D.; Szweda, D.; Otulakowski, Ł.; Haladjova, E.; Dworak, A. Thermoresponsive polymer-peptide/protein conjugates. *Prog. Polym. Sci.* **2017**, *68*, 35–76. [CrossRef]
98. Burridge, K.M.; Page, R.C.; Konkolewicz, D. Bioconjugates—From a specialized past to a diverse future. *Polymer* **2020**, *211*, 123062. [CrossRef]
99. Wright, T.A.; Page, R.C.; Konkolewicz, D. Polymer conjugation of proteins as a synthetic post-translational modification to impact their stability and activity. *Polym. Chem.* **2019**, *10*, 434–454. [CrossRef]
100. Chen, C.; Ng, D.Y.W.; Weil, T. Polymer bioconjugates: Modern design concepts toward precision hybrid materials. *Prog. Polym. Sci.* **2020**, *105*, 101241. [CrossRef]
101. Rodriguez-Abetxuko, A.; Sánchez-de Alcázar, D.; Muñumer, P.; Beloqui, A. Tunable polymeric scaffolds for enzyme immobilization. *Front. Bioeng. Biotechnol.* **2020**, *8*, 830. [CrossRef]
102. Wang, Y.; Wu, C. Site-specific conjugation of polymers to proteins. *Biomacromolecules* **2018**, *19*, 1804–1825. [CrossRef]
103. Messina, M.S.; Messina, K.M.; Bhattacharya, A.; Montgomery, H.R.; Maynard, H.D. Preparation of biomolecule-polymer conjugates by grafting-from using ATRP, RAFT, or ROMP. *Prog. Polym. Sci.* **2020**, *100*, 101186. [CrossRef] [PubMed]
104. Carmali, S.; Murata, H.; Cummings, C.; Matyjaszewski, K.; Russell, A.J. Polymer-Based Protein Engineering: Synthesis and Characterization of Armored, High Graft Density Polymer–Protein Conjugates. *Methods Enzymol.* **2017**, *590*, 347–380. [PubMed]
105. Baker, S.L.; Kaupbayeva, B.; Lathwal, S.; Das, S.R.; Russell, A.J.; Matyjaszewski, K. Atom transfer radical polymerization for biorelated hybrid materials. *Biomacromolecules* **2019**, *20*, 4272–4298. [CrossRef] [PubMed]
106. Cummings, C.; Murata, H.; Koepsel, R.; Russell, A.J. Dramatically increased pH and temperature stability of chymotrypsin using dual block polymer-based protein engineering. *Biomacromolecules* **2014**, *15*, 763–771. [CrossRef]
107. Kaupbayeva, B.; Russell, A.J. Polymer-enhanced biomacromolecules. *Prog. Polym. Sci.* **2020**, *101*, 101194. [CrossRef]
108. Shakya, A.K.; Nandakumar, K.S. An update on smart biocatalysts for industrial and biomedical applications. *J. R. Soc. Interface* **2018**, *15*, 20180062. [CrossRef]
109. Rottke, F.O.; Heyne, M.V.; Reinicke, S. Switching enzyme activity by a temperature responsive inhibitor modified polymer. *Chem. Commun.* **2020**, *56*, 2459–2462. [CrossRef]
110. Chapman, R.; Stenzel, M.H. All wrapped up: Stabilization of enzymes within single enzyme nanoparticles. *J. Am. Chem. Soc.* **2019**, *141*, 2754–2769. [CrossRef]
111. Yadavalli, N.S.; Borodinov, N.; Choudhury, C.K.; Quiñones-Ruiz, T.; Laradji, A.M.; Tu, S.; Lednev, I.K.; Kuksenok, O.; Luzinov, I.; Minko, S. Thermal stabilization of enzymes with molecular brushes. *ACS Catal.* **2017**, *7*, 8675–8684. [CrossRef]
112. Pasparakis, G.; Tsitsilianis, C. LCST polymers: Thermoresponsive nanostructured assemblies towards bioapplications. *Polymer* **2020**, *11*, 123146. [CrossRef]
113. Chado, G.R.; Holland, E.N.; Tice, A.K.; Stoykovich, M.P.; Kaar, J.L. Exploiting the benefits of homogeneous and heterogeneous biocatalysis: Tuning the molecular interaction of enzymes with solvents via polymer modification. *ACS Catal.* **2018**, *8*, 11579–11588. [CrossRef]
114. Bilici, Z.; Camli, S.T.; Unsal, E.; Tuncel, A. Activity behacior of a HPLC column including α-chymotrypsin immobilized monosized-porous particles. *Anal. Chim. Acta* **2004**, *516*, 125–133. [CrossRef]
115. Meller, K.; Pomastowski, P.; Grzywinski, D.; Szumski, M.; Buszewski, B. Preparation and evaluation of dual-enzyme microreactor with co-immobilized trypsin and chymotrypsin. *J. Chromatogr. A* **2016**, *1440*, 45–54. [CrossRef]

116. Munasinghe, A.; Baker, S.L.; Lin, P.; Russell, A.J.; Colina, C.M. Structure–function–dynamics of α-chymotrypsin based conjugates as a function of polymer charge. *Soft Matter* **2020**, *16*, 456–465. [CrossRef]
117. Mukhopadhayay, A.; Singh, D.; Sharma, K.P. Neat Ionic liquid and α-Chymotrypsin-Polymer Surfactant Conjugate-Based Biocatalytic Solvent. *Biomacromolecules* **2020**, *21*, 867–877. [CrossRef] [PubMed]
118. Murata, H.; Cummings, C.S.; Koepsel, R.R.; Russell, A.J. Rational tailoring of substrate and inhibitor affinity via ATRP polymer-based protein engineering. *Biomacromolecules* **2014**, *15*, 2817–2823. [CrossRef]
119. Hegedüs, I.; Nagy, E. Improvement of chymotrypsin enzyme stability as single enzyme nanoparticles. *Chem. Eng. Sci.* **2009**, *64*, 1053–1060. [CrossRef]
120. Hegedüs, I.; Vitai, M.; Jakab, M.; Nagy, E. Study of Prepared α-Chymotrypsin as Enzyme Nanoparticles and of Biocatalytic Membrane Reactor. *Catalysts* **2020**, *10*, 1454. [CrossRef]
121. Wang, X.; Yadavalli, N.S.; Laradji, A.M.; Minko, S. Grafting through method for implanting of lysozyme enzyme in molecular brush for improved biocatalytic activity and thermal stability. *Macromolecules* **2018**, *51*, 5039–5047. [CrossRef]
122. Rahman, M.S.; Brown, J.; Murphy, R.; Carnes, S.; Carey, B.; Averick, S.; Konkolewicz, D.; Page, R.C. Polymer Modification of Lipases, Substrate Interactions, and Potential Inhibition. *Biomacromolecules* **2021**, *22*, 309–318. [CrossRef] [PubMed]
123. Wirnt, R.; Bergmeyer, H.U. *Chymotrypsin, in Methods of Enzymatic Analysis*; Academic Press: New York, NY, USA, 1974; pp. 1009–1012.
124. Yin, X.; Stöver, H.D. Hydrogel microspheres by thermally induced coacervation of poly(*N*,*N*-dimethylacrylamide-co-glycidyl methacrylate) aqueous solutions. *Macromolecules* **2003**, *36*, 9817–9822. [CrossRef]
125. Ghaouar, N.; Elmissaoui, S.; Aschi, A.; Gharbi, A. Concentration regimes and denaturation effects on the conformational changes of α-chymotrypsin by viscosity and dynamic light scattering measurements. *Int. J. Biol. Macromol.* **2010**, *47*, 425–430. [CrossRef] [PubMed]

polymers

MDPI

Article

Modulating the Thermoresponse of Polymer-Protein Conjugates with Hydrogels for Controlled Release

Vincent Huynh [1], Natalie Ifraimov [2] and Ryan G. Wylie [1,2,*]

1 Department of Chemistry and Chemical Biology, McMaster University, Hamilton, ON L8S 4M1, Canada; huynhv2@mcmaster.ca
2 School of Biomedical Engineering, McMaster University, Hamilton, ON L8S 4M1, Canada; ifraimon@mcmaster.ca
* Correspondence: wylier@mcmaster.ca

Abstract: Sustained release is being explored to increase plasma and tissue residence times of polymer-protein therapeutics for improved efficacy. Recently, poly(oligo(ethylene glycol) methyl ether methacrylate) (PEGMA) polymers have been established as potential PEG alternatives to further decrease immunogenicity and introduce responsive or sieving properties. We developed a drug delivery system that locally depresses the lower critical solution temperature (LCST) of PEGMA-protein conjugates within zwitterionic hydrogels for controlled release. Inside the hydrogel the conjugates partially aggregate through PEGMA-PEGMA chain interactions to limit their release rates, whereas conjugates outside of the hydrogel are completely solubilized. Release can therefore be tuned by altering hydrogel components and the PEGMA's temperature sensitivity without the need for traditional controlled release mechanisms such as particle encapsulation or affinity interactions. Combining local LCST depression technology and degradable zwitterionic hydrogels, complete release of the conjugate was achieved over 13 days.

Keywords: polymer-protein conjugates; thermoresponsive; controlled release; temperature-sensitive polymers; hydrogels

Citation: Huynh, V.; Ifraimov, N.; Wylie, R.G. Modulating the Thermoresponse of Polymer-Protein Conjugates with Hydrogels for Controlled Release. *Polymers* **2021**, *13*, 2772. https://doi.org/10.3390/polym13162772

Academic Editor: Luminita Marin

Received: 9 July 2021
Accepted: 13 August 2021
Published: 18 August 2021

1. Introduction

Polymer conjugation to protein therapeutics is a common technique to enhance therapeutic efficacy by minimizing immunogenicity and increasing plasma and tissue residence times due to slower clearance rates [1,2]. Conjugation of poly(ethylene) glycol (PEG) is becoming a routine method to synthesize polymer-protein conjugates, with several used clinically [2,3]. To further improve the efficacy of polymer-protein conjugates, alternatives to PEG are being explored to maintain therapeutic effects while further minimizing antigenic or immunogenic effects [1,4]. PEG alternatives include poly(zwitterions), polysaccharides, poly(oligo(ethylene glycol) methyl ether methacrylate) (PEGMA), peptides, and many more [1,5].

PEGMA is an attractive polymer for protein conjugation due to low antigenicity, ease of polymerization and successful demonstrations in the production of polymer-protein conjugates [6–8]. For example, PEGMA conjugation to a GLP-1 agonist (exenatide) extended its plasma half-life, which enhanced and prolonged glucose suppression [7] Furthermore, PEGMA with side chains of only 2–3 ethylene glycol repeat units do not bind to anti-PEGs antibodies, potentially reducing immune recognition of the polymer [6]. PEGMA conjugation to proteins has also been shown to display molecular sieving effects, providing an additional strategy for increasing therapeutic half-life and decreasing immune recognition [9,10]. Therefore, PEGMA is well established as a potential alternative to PEG and warrants further exploration.

Despite the greater plasma and tissue half-lives provided upon polymer conjugation, repeated dosing of polymer-protein conjugates are often required to maintain therapeutic drug concentrations at disease sites. This has prompted research into stimuli responsive drug depots that slowly release polymer-protein conjugates to reduce the frequency of administration [11]. For example, temperature responsive elastin like peptides have been conjugated to protein drugs such as exenatide to form a sparingly soluble drug depot upon injection due to a temperature-sensitive phase transition [12,13]. The aggregated polymer-protein conjugate is then slowly released from the drug depot upon dissolution of the conjugate through phase transitions at the surface. This strategy resulted in prolonged glucose suppression in a diabetic mouse model with a single injection. In a similar manner, temperature-sensitive phase transitioning PEGMA conjugates can be used for controlled released applications.

To further enhance the controlled delivery of polymer-protein conjugates, we have developed an injectable hydrogel delivery vehicle that decreases the lower critical solution temperature (LCST) of encapsulated PEGMA-protein conjugates. The conjugates can therefore be designed to phase transition at body temperature within but not outside of the hydrogel, providing new a mechanism for the controlled release of PEGMA-protein conjugates without the need for traditional controlled release mechanisms such as particle formulations or hydrogel binding interactions. Therefore, controlled release can simply be achieved by locally injecting therapeutic PEGMA-protein conjugates with hydrogels simplifying formulations for potential clinical applications. By exploiting the influence of Hofmeister salt series in decreasing the LCST of PEGMA, we developed a zwitterionic hydrogel to influence PEGMA's LCST [14]. The zwitterionic hydrogel changes the local electrochemical environment, thereby lowering the LCST of an encapsulated PEGMA-protein conjugate, yielding a drug depot where release is governed by the dissolution of the formed aggregates and the hydrogel degradation rate. To ensure the PEGMA-protein conjugates form an aggregate depot within the hydrogel but not after their release from the hydrogel, the conjugate was designed to have an LCST above 37 °C in a buffered solution and below 37 °C in the hydrogel (Figure 1).

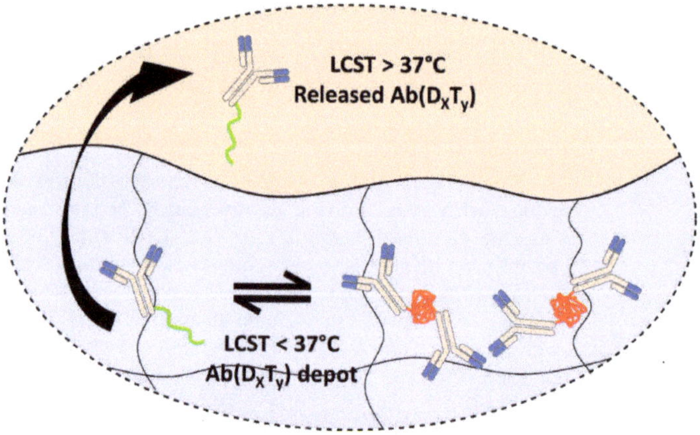

Figure 1. Local LCST depression of PEGMA-antibody conjugates ($Ab(D_xT_y)$) within a zwitterionic hydrogel for the controlled release of solubilized conjugates. Within the hydrogel, the $Ab(D_xT_y)$ aggregates form due to hydrogel promoted LCST depression and slowly dissolve for controlled release. PEGMA copolymers of diethylene glycol methyl ether methacrylate (D) and triethylene glycol methyl ether methacrylate (T) $P(D_xT_y)$ were synthesized to achieve temperature-sensitive polymers with suitable and hydrogel promoted phase transitions for $Ab(D_xT_y)$. The conjugates were designed to undergo phase transitions within but not outside of the hydrogel.

2. Materials and Methods

2.1. Materials

4-Cyano-4-(phenylcarbonothioylthio)pentanoic acid (CTA1), 4,4′-azobis(4-cyanovaleric acid) (ACVA), triethylamine, 2-(dodecylthiocarbonothioylthio)-2-methylpropionic acid 3-azido-1-propanol ester (CTA2), triethylene glycol methyl ether methacrylate, diethylene glycol methyl ether methacrylate, poly(ethylene glycol) methyl ether methacrylate (M_n 500) ($P(EG_9)MA$), 2-methacroloyloxyethyl phosphorylcholine (MPC), [2-(methacryloyloxy)ethyl] dimethyl-(3-sulfopropyl)ammonium hydroxide (SB), were purchased from Sigma Aldrich (Oakville, ON, Canada). N-(3-aminopropyl)methacrylamide hydrochloride (APMA) was purchased from Polysciences, Inc. (Warrington, PA, USA). Dibenzocyclooctyne-PEG_4-NHS (DBCO-PEG_4-NHS) was purchased from Click Chemistry Tools (Scottsdale, AZ, USA). 3-(3-dimethylaminopropyl)-1-ethyl-carbodiimide hydrochloride (EDC-HCl) was obtained from Chem-Impex International Inc. (Wood Dale, IL, USA) Alexa Fluor 647 NHS ester, Alexa Fluor 488 NHS Ester, Fetal bovine serum (FBS)DMEM, Alamar blue was purchased from Thermo Fisher Scientific (Burlington, ON, Canada). O-(6-azidohexyl)-O′-succinimidyl carbonate (NHS-azide), O-[1-(4-chlorophenylsulfonyl)-7-azido-2-heptyl]-O′-succinimidylcarbonate (NHS-AzCl) and carboxybetaine monomer were synthesized as previously described [15–17].

2.2. Methods

Poly(carboxybetaine) (pCB) hydrogel synthesis. pCB hydrogels were synthesized in a similar manner as previously described [15,17,18]. Copolymers of pCB-APMA, pCB-Azide (pCB-Az), pCB-AzCl and pCB-DBCO were synthesized as described below. Hydrogels were fabricated by mixing pCB-DBCO and pCB-Az at desired concentration.

pCB-APMA synthesis. CB monomer was synthesized according to previous protocols. CB monomer (1 g) and APMA-HCl (20 mg) was dissolved in 2.893 mL 1M sodium acetate buffer. Separately, CTA1 (5.5 mg) and ACVA (1.1) was dissolved in 579 μL dioxane. The reagents were added together resulting in 5:1 buffer: dioxane. The pH of the solution was then adjusted to pH 3–4 and was then placed in a Schlenk flask, subject to 3 cycles of freeze-pump-thaw and backfilled with N_2. The vessel was then submerged in an oil bath at 70 °C and reacted for 2 d. The polymerization was terminated by exposure to air and the solution was dialyzed against water (pH 3–4) for 3 d (MWCO 12–14k) and lyophilized yielding a pink powder. Polymer molecular weight and dispersity was determined with gel permeation chromatography using PEG standards.

pCB-Az, pCB-AzCl and pCB-DBCO synthesis. pCB-APMA random copolymers were synthesized according to previous procedures. pCB-APMA polymers were then functionalized with azide or DBCO moieties. We dissolved 500 mg of pCB-APMA in 5 mL of dry MeOH with 50 μL (0.50 mmol) of triethylamine. Subsequently, NHS-azide (14 mg, 0.05 mmol), NHS-DBCO (20 mg, 0.05 mmol) or NHS-AzCl (23 mg, 0.05 mmol) was added and reacted overnight under N_2 at room temperature. Polymers were then precipitated with 45 mL diethyl ether, dried and dissolved in 10–15 mL of deionized water. The aqueous solutions were then extracted twice with dichloromethane (DCM) and the aqueous layer was then dialyzed against water for 1 d (MWCO 12–14k). The polymers were then lyophilized yielding a white or pink powder. Polymer composition was determined by ^1H NMR upon integrating the methylene peak in the backbone with unique resonances for pCB-Az, pCB-AzCl (hydrogen adjacent to the carbamate bond) and pCB-DBCO (aromatic DBCO moieties; Figure S5).

$P(D_xT_y)$ Polymer Synthesis. We mixed 0.5 g of triethylene glycol methyl ether methacrylate (T) was mixed with di(ethylene glycol) methyl ether methacrylate (D) to yield PEGMA copolymers of varying D and T mole fractions. For T homopolymer and copolymers of $D_{10}T_{90}$, $D_{20}T_{80}$, $D_{30}T_{70}$, $D_{40}T_{60}$ and $D_{50}T_{50}$, 0, 45, 101, 174, 270 and 405 mg for D was added, respectively, to the 0.5 mg solution of T along with a 3 mL dioxane solution of CTA2 (8.2 mg) and initiator (1.2 mg). The reaction mixture was then transferred to a Schlenk flask and underwent three freeze-pump-thaw cycles, followed by a nitrogen backfill. The reaction was then submerged in a 70 °C oil bath and reacted for 16 h. The reaction was then dialyzed against water (MWCO 12–14k) for 3 d. The product was then lyophilized yielding a yellow oil.

P(EG_9)MA synthesis. Poly(ethylene glycol) methyl ether methacrylate (M_n 500) (1 g), CTA1 (5.6 mg) and ACVA (1.1 mg) was dissolved in 993 µL dioxane. The reaction mixture was then transferred to a Schlenk flask and underwent three freeze-pump-thaw cycles, followed by a nitrogen backfill. The reaction was then submerged in a 70 °C oil bath and reacted for 16 h. The reaction was then dialyzed against water at pH 3–4 (MWCO 12–14k) for 3 d and then lyophilized yielding a pink oil.

pMPC synthesis. MPC monomer (0.75 mg), CTA1 (4.2) and ACVA (1 mg) was dissolved in 1.787 mL of methanol. The reaction mixture was then sparged with nitrogen for 45 min, immersed in a 70 °C oil bath and reacted for 16 h. The polymer was then precipitated in acetone and dried overnight in the vacuum oven. The polymer was then dissolved in water and dialyzed against water at pH 3–4 (MWCO 12–14k) for 1 day. The product was then lyophilized, yielding a pink powder.

pSB synthesis. SB monomer (1 g), CTA1 (5.6 mg) and ACVA (1.1 mg) was dissolved in 2.380 mL 0.5 M NaCl. The reaction mixture was then transferred to a Schlenk flask and underwent three freeze-pump-thaw cycles, followed by a nitrogen backfill. The reaction was then submerged in a 70 °C oil bath and reacted for 16h. The reaction was then dialyzed against pH 3–4 water (MWCO 12–14k) for 3 d. The product was then lyophilized yielding a pink powder.

pCB synthesis. CB monomer (1 g), CTA1 (9.4 mg) and ACVA (1.9 mg) was dissolved in 3.375 mL 1 M acetate buffer (pH 5.2), the pH was the adjusted to pH 3–4. The reaction mixture was then transferred to a Schlenk flask and underwent three freeze-pump-thaw cycles and was backfilled with nitrogen. The reaction was then submerged in a 70 °C oil bath and reacted for 48 h. The reaction was then dialyzed against water at pH 3–4 (MWCO 12–14k) for 3 d. The product was then lyophilized yielding a pink powder.

Polymer LCST determination. D_xT_y polymers were dissolved in PBS (10 mg mL^{-1}) overnight at 4 °C. The polymers were then mixed with a solution of the modulating polymer (P(EG_9)MA, pMPC, pCB, pSB, pCB-APMA, pCB-Az, pCB-DBCO, PBS) resulting in a final concentration of 50 mg mL^{-1} of the modulating polymer and 5 mg mL^{-1} of the D_xT_y polymer in PBS. The sample was then transferred to a 1 mL quartz cuvette and the optical density at 550 nm (OD550) was measured using an Agilent UV Vis NIR spectrophotometer equipped with a Peltier accessory. Similarly, D_xT_y polymers were mixed with either pCB-DBCO and pCB-Az resulting in a final concentration of 100 mg mL^{-1} of pCB-DBCO/pCB-Az and 5 mg mL^{-1} D_xT_y polymer. The two mixtures were then combined in a 1 mL quartz cuvette and gelled for 30 min at room temperature and the optical density at 550 nm (OD550) was measured using an Agilent UV Vis NIR spectrophotometer equipped with a Peltier accessory. The sample was heated from 5 °C to 75 °C at a rate of 3 °C per minute and cooled from 75 °C to 5 °C at 3 °C per minute. OD550 measurements were taken every 0.5 °C.

Ab-DBCO and Ab-DBCO-488 synthesis. We added 33 µL of 10 mg mL^{-1} DBCO-PEG_4-NHS dissolved in DMSO to 1 mL of Avastin (10 mg mL^{-1}; Ab) in borax buffer (0.1 M borax, 0.15 M NaCl, pH 8.75) and reacted overnight at room temperature. Avastin-DBCO (Ab-DBCO) was purified by size exclusion chromatography using a GE Healthcare Superdex S200 column. Fractions were collected and concentrated using an Amicon Spin concentrator (MWCO 30k). DBCO conjugation was confirmed and quantified by UV Vis. Number of DBCO's conjugated to Av was determined using previously described methods [19].

Ab-DBCO-488 was then synthesized. Briefly, 2 µL of Alexa Fluor 488 NHS ester (10 mg mL^{-1}) was added to 300 µL of Ab-DBCO (5 mg mL^{-1}) in PBS and reacted at room temperature for 3 h in the dark. The reaction was then dialyzed against PBS (MWCO 12–14k) at 4 °C in the dark for 3 d. Ab-488 was synthesized as a control. Briefly, 2 µL of Alexa Fluor 488 was added to 500 µL of Ab in PBS (10 mg mL^{-1}) and incubated at room temperature for 3h in the dark. The solution was subsequently dialyzed against PBS (MWCO 12–14k) at 4 °C in the dark for 3 d.

Ab(D_xT_y) synthesis and validation. We added 125 μL of D_XT_y (25 mg mL^{-1}) dissolved in PBS to 125 μL of Av-DBCO (5 mg mL^{-1}) in PBS and reacted overnight at room temperature with gentle agitation. Ab(D_xT_y) was purified by size exclusion chromatography, using a GE Healthcare Superdex S200 column. Fractions were collected and concentrated with an Amicon Spin concentrator (MWCO 100k).

To ensure complete reaction of DBCO's conjugated Avastin, 4 μL of Ab-DBCO (5 mg mL^{-1}) in PBS was first added to 4 μL of D_xT_y (25 mg mL^{-1}) dissolved in PBS and reacted overnight at room temperature. We then added 2 μL of N_3-Cy5 dye in water and reacted at room temperature overnight in the dark. Samples were then loaded onto an SDS PAGE and run at 120 V for 45 min. The gel was imaged using a Biorad Chemidoc using the Cy5 filter and processed using ImageLab software. The appearance of fluorescent bands would indicate the presence of unreacted DBCO's on Av.

Hydrogel degradation. On the bottom of pre-weighted 2 mL Eppendorf tubes, 10 wt% AzCl hydrogels were formed. Hydrogels were then allowed to gel at room temperature for 30 min to ensure complete gelation. We carefully pipetted 1 mL of PBS over top the gels, which were then incubated at 37 °C. At specific time intervals, PBS was removed, and the surface of the gels were dabbed with a kimwipe. Wet gel weights were then measured and plotted against time.

Ab(D_xT_y) LCST determination. Ab(D_xT_y) conjugates (0.5 mg mL^{-1}) were transferred to a 100 μL quartz cuvette and OD550 was measured using an Agilent UV Vis NIR spectrophotometer equipped with a Peltier accessory. Samples were heated from 15 to 60 °C at a rate of 0.5 °C per minute and then cooled from 60 to 15 °C at 0.5 °C per minute. OD550 readings were taken every 0.5 °C.

Controlled release experiments of Ab(D_xT_y) from pCB hydrogels. Ab(D_xT_y)-488 in PBS (2 mg mL^{-1}) was incubated without or with D_xT_y polymers (0, 2.5 or 5 $mgmL^{-1}$) dissolved in PBS (40 mg mL^{-1}) overnight at 4 °C in the dark. The solution was then mixed with pCB-DBCO and pCB-Az forming 60 μL gels on the bottom of a black 96-well plate resulting in 10 wt% pCB hydrogels containing 0, 2.5 or 5 mg mL^{-1} D_xT_y polymers and 0.25 mg mL^{-1} Ab-DBCO-488. Gels were incubated at 37 °C for 1 h to ensure gelation. We pipetted 200 μL of warm (37 °C) PBS on top of the gels. Supernatants were taken at specific time intervals and replenished with warm PBS. Supernatant fluorescence was read using a Biotek Cytation 5 plate reader (Ex. 495 nm, Em. 519 nm) and amounts of released Ab(D_xT_y) were determined using a calibration curve.

Cytotoxicity assay. Human lung fibroblasts (HLF) (10,000 cells) were seeded onto a clear flat bottom 96-well plate with growth medium (DMEM, 10% FBS). D_xT_y polymers dissolved in PBS (1 mg mL^{-1}) were then added to each well. The cells were incubated at 37 °C 5% CO_2 for 24h. Alamar Blue (Invitrogen), was then added to each well and further incubated at 37 °C 5% CO_2 for 1 h. Fluorescence was then measured using a Cytation 5 plate reader (ex. 560 nm, em. 590 nm). Fluorescence readings were normalized to a control where PBS was added.

3. Results and Discussion

3.1. Selection of Polymer Components for Hydrogel and Antibody Conjugates

To identify suitable polymers for antibody conjugates and injectable hydrogels for LCST modulating delivery vehicles, we first synthesized azide terminated PEGMA random copolymers with various D:T ratios with MWs of 26–40k (P(D_xT_y); Figure 2a, Figure S1, Table S1), and thus LCSTs, for conjugation to antibodies. The LCST, midpoint (50%) of turbidity curves, of 0.5 wt% P(D_xT_y) solutions in PBS was determined in the presence of different hydrophilic polymers that are either zwitterionic (pCB, pSB or pMPC) or PEGMA with greater hydrophilicity (P(EG_9)MA) than P(D_xT_y) copolymers (Figure S2). As expected, hydrophilic polymers decreased the LCST of all P(D_xT_y) copolymers, with pCB resulting in the greatest (~3.2 °C) shift (Figure 2f, Figures S3 and S4), most likely due to pCB's greater ability to form a water shell under physiological conditions and know ability to promote hydrophobic interactions. This is similar to previous reports wher the hydration state of pCB enhanced hydrophobic interactions and displayed propertie similar to the Hofmeister salt series. The Hofmeister salt series has been shown to depres the LCST transition temperatures of PEGMA polymers; additionally, the hydration state o pCB polymers have been compared to Hofmeister salt series in promoting hydrophobi interactions [20,21]. pSB, pMPC and P(EG$_9$)MA decreased the P(D$_x$T$_y$) LCSTs by ~1.3, ~2. and ~1.5 °C, respectively (Figure S4). Therefore, pCB was chosen as the optimal materia for hydrogel fabrication as it had the greatest impact on P(D$_x$T$_y$) LCST.

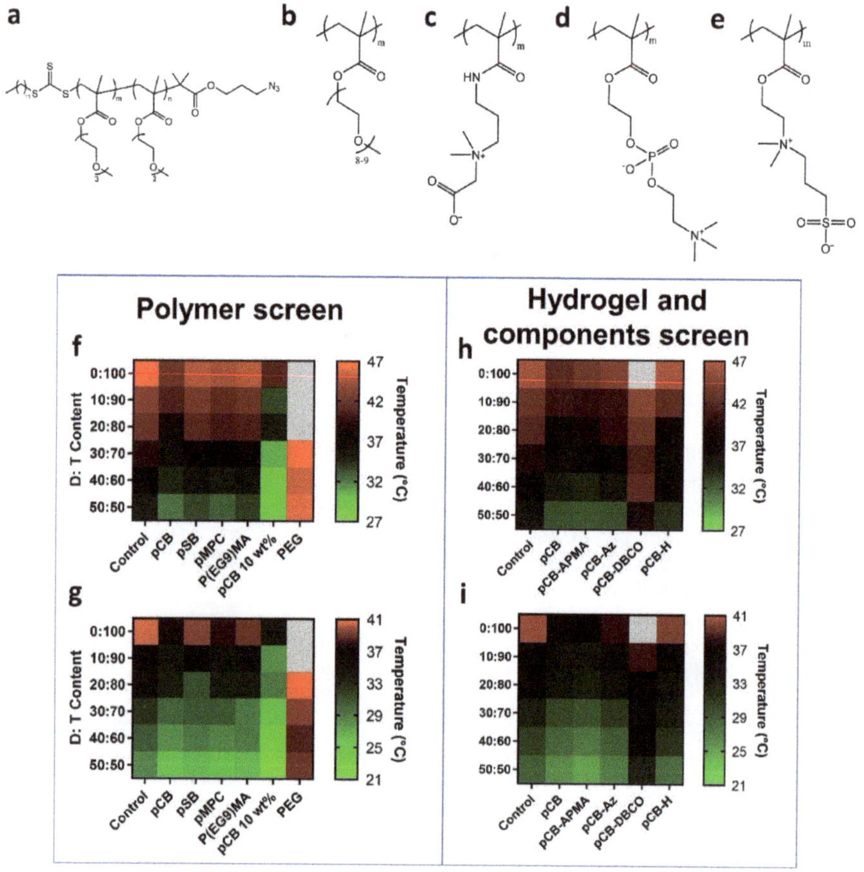

Figure 2. LCST screen of P(D$_x$T$_y$) copolymers with varying D:T content in the presence of hydrophilic polymers, hydrogel components and encapsulated in pCB hydrogels. Structures of polymers used in the study: (**a**) P(D$_x$T$_y$), (**b**) P(EG$_9$)MA, (**c**) pCB, (**d**) pMPC, (**e**) pSB. (**f,h**) Heat map of LCST transition temperatures (at 50% of maximum turbidity of heating curves) of different PEGMA copolymer compositions (10 mg/mL) in combination with modulating polymers (50 mg/mL unless specified). (**g,i**) Heat map of LCST cooling transition temperatures (at 50% of maximum turbidity of cooling curves) of different PEGMA polymer compositions (10 mg/mL) in combination with modulating polymers (50 mg/mL unless specified). Colored temperature scales are presented with 37 °C set as black.

As we have previously demonstrated that 10 wt% pCB hydrogels formed upon crosslinking copolymers of pCB-azide and pCB-DBCO that resulted in injectable, low fouling hydrogels, we synthesized the relevant pCB-azide, pCB-DBCO copolymers and precursors (Table S1 and Figure S5) and studied their influence on P(D$_x$T$_y$) LCSTs [17].

We synthesized pCB-Az and pCB-DBCO copolymers with 3.5 and 3.9 mol% of azide and DBCO moieties, respectively, resulting in a hydrogel with ~3.5 mol% of crosslinks, which was previously shown to be low-fouling and enabled protein diffusion. We first confirmed that the influence of pCB on $P(D_xT_y)$ LCST was concentration-dependent; 10 wt% pCB homopolymer shifted the LCST by 4.3 °C, whereas 5 wt% pCB shifted the LCST by 3.2 °C (Figure 2f, Figure S6). pCB copolymers were synthesized by polymerizing CB and APMA monomers to yield a copolymer with ~3.5 mol% APMA (Table S1). pCB-APMA was either modified with NHS-azide or NHS-DBCO to yield corresponding copolymers for hydrogel crosslinking (Figure S5). $P(D_xT_y)$ LCSTs were then screened in the presence of hydrogel components and compared to a homopolymer of pCB (Figure 2h,i, Figure S7). At 5 wt%, pCB-APMA and pCB-Az decreased $P(D_xT_y)$ LCSTs by an average of 2.6 and 1.8 °C, respectively, whereas pCB-DBCO increased the LCST by 1 °C due to the hydrophobicity of the DBCO moiety. The pCB hydrogel decreased the LCST of $P(D_xT_y)$ by 1.7 °C, on average. Therefore, the synthesized pCB hydrogels can decrease the LCST of $P(D_xT_y)$ polymers.

Ideally the $P(D_xT_y)$ copolymers for the antibody conjugate would demonstrate phase transitions below 37 °C within the hydrogel and above 37 °C outside of the hydrogel for the sustained release of fully solubilized antibody drug conjugates. The D:T copolymer ratio greatly influenced LCST temperatures when different modulating polymers were added (Figure S8). Consequently, a $P(D_xT_y)$ copolymer with correct D:T ratio was chosen. Only the copolymers with a D:T ratio of 30:70, $P(D_{30}T_{70})$, clearly met this criterion with LCSTs (midpoint of turbidity curve) of 36 °C and 39 °C in the hydrogel and PBS, respectively, and was therefore selected for further study. It should be noted that the LCST turbidity curve of $P(D_xT_y)$ within hydrogels were broadened compared to uncrosslinked polymers. The homopolymer of triethylene glycol methyl ether methacrylate, $P(T_{100})$, was also selected for comparison with $P(D_{30}T_{70})$ as its LCST within and outside of the hydrogel is >37 °C (Figure S7, Table S2). Additionally, both polymers showed no cytotoxicity against human lung fibroblasts (Figure S9), as expected. Therefore, they can be used and safely conjugated to therapeutic proteins.

3.2. *Synthesis and Characterization of PEGMA-Antibody Conjugates*

Avastin, an anti-VEGF humanized antibody, was selected as a model antibody because it is of interest for controlled delivery applications to the posterior segment of the eye and cancerous tumors. Avastin was first modified with DBCO-PEG_4-NHS, which resulted in an average of 4.1 DBCOs per antibody according to absorbance measurements [22]. Azide terminated $P(D_xT_y)$ polymers were then grafted onto the antibody until complete consumption of antibody DBCOs according to SDS-PAGE with a fluorescent azide tag (Figure 3a–c). The antibody conjugates with $P(D_{30}T_{70})$ and $P(T_{100})$, $AbP(D_{30}T_{70})$ and $AbP(T_{100})$, therefore had an average of 4.1 polymers per antibody. The conjugates were then purified and characterized by size-exclusion chromatography (Figure 3d). The conjugates had short retention times and broader peaks than unmodified antibody, indicating successful polymer grafting.

The temperature dependence of $P(D_{30}T_{70})$ and $P(T_{100})$ as free polymer and the conjugates, $AbP(D_{30}T_{70})$ and $AbP(T_{100})$, were then studied in PBS and the hydrogel by following temperature-dependent phase transition with OD550 turbidity measurements (Figure 4, Figure S10), polymer concentration in all samples were held at 0.5 mg mL^{-1}. Grafting polymers to the antibody shifted the turbidity curves lower with a larger effect on $P(T_{100})$'s LCST than $P(D_{30}T_{70})$. Compared to free polymers, the LCST of $AbP(D_{30}T_{70})$ and $AbP(T_{100})$ shifted by 1 °C and 6 °C at 50% turbidity, respectively (Figure 4a,b). Interestingly, grafting $P(T_{100})$ to the antibody influenced the LCST to a greater extent than $P(D_{30}T_{70})$. It has also been previously observed that grafting different PEGMA polymers to proteins had variable LCST shifts [8,23]. Encapsulating the conjugates in the hydrogel significantly broadened their phase transition curves, where even the turbidity curve begins to increase below 37 °C for the $P(T_{100})$ conjugate. There is a similar effect (but to a lesser extent) seen from free $P(D_{30}T_{70})$ and $P(T_{100})$ polymers in the pCB hydrogel. We hypothesize that this effect may be due to molecular crowding promoted by hydrogels, which can therefore induce phase transitions of stimuli responsive polymers at a lower temperature [24–27]. Furthermore, PEGMA phase transitions through multistep aggregation; therefore, the simultaneous effects of macromolecular crowding combined with reduced diffusivity, which influences aggregates size, can result in the broadening of phase transition temperatures curves [28–30]. Therefore, both the $P(D_{30}T_{70})$ and $P(T_{100})$ conjugates can be used for controlled release applications within the pCB hydrogel. As the LCST of the conjugates are above 37 °C outside of the hydrogel, any released conjugates will be completely solubilized, which is beneficial for therapeutic applications.

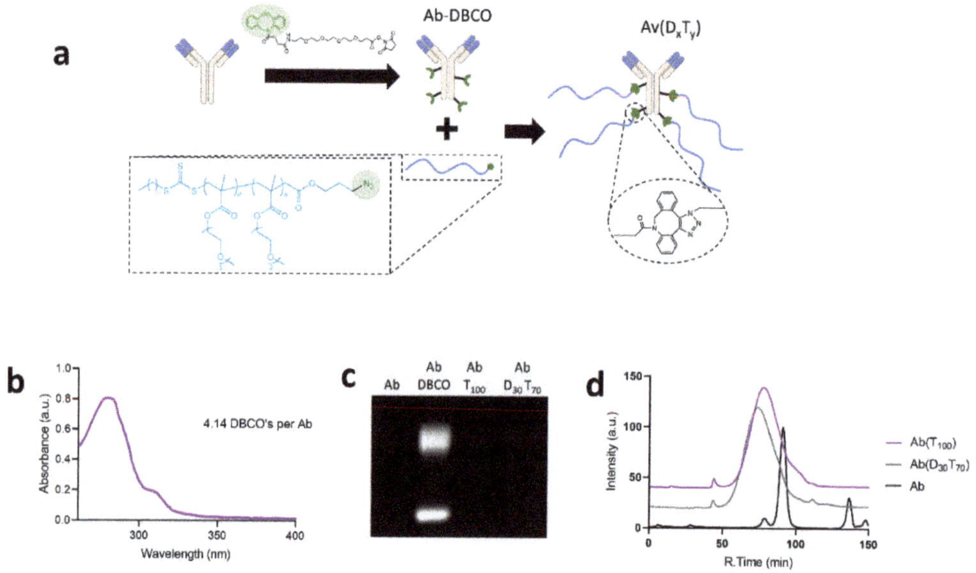

Figure 3. Synthesis and characterization of the $Ab(D_xT_y)$ conjugates. (**a**) Illustration of $Ab(D_xT_y)$ conjugate synthesis by grafting $P(D_xT_y)$ to Ab-DBCO. (**b**) Absorbance spectrum of Ab-DBCO for quantification of DBCO grafting degree. (**c**) Fluorescent SDS PAGE of antibody conjugates when reacted with Cy5-azide to detect reactive DBCO groups on the antibody. Loss of fluorescence in conjugates indicates complete consumption of DBCO (Fluorescent signal indicates free DBCO reacting with Cy5-azide). (**d**) Size exclusion chromatograms of antibody-polymer conjugates.

3.3. *Controlled Release of PEGMA-Protein Conjugates from pCB Hydrogels*

To demonstrate that controlled release is temperature-sensitive due to $P(D_xT_y)$ phase transitions and not solely diffusion limited, we first investigated the release of the conjugate with the lowest LCST, $Ab(D_{30}T_{70})$, from pCB hydrogels at 4 °C, which prevents aggregation (Figure 5a). As expected, complete release of the conjugate occurred after three to seven days, indicating that the conjugate can diffuse through the hydrogel if completely solubilized.

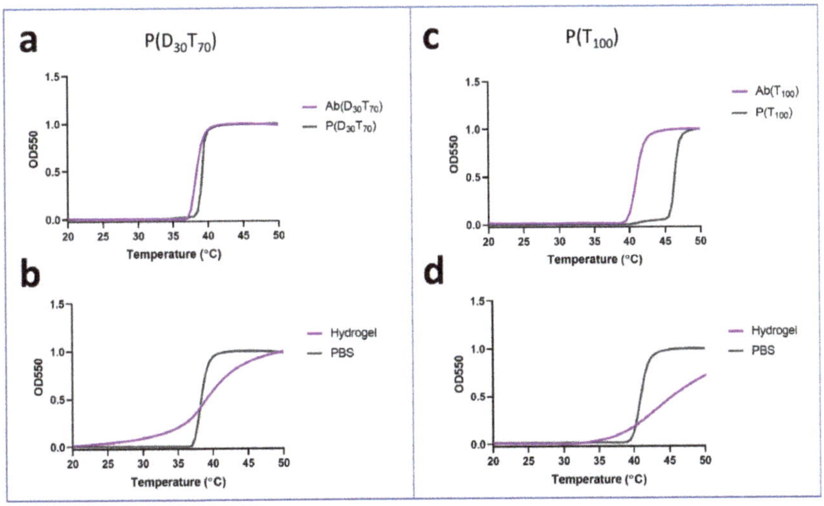

Figure 4. Temperature-sensitive turbidity curves of $P(D_xT_y)$ copolymers (5 mg mL^{-1}), $Ab(D_xT_y)$ (0.5 mg/mL^{-1}) in PBS and $Ab(D_xT_y)$ within the pCB hydrogel. (**a**) The copolymer $P(D_{30}T_{70})$ showed similar temperature responses as a free polymer and a conjugate, $Ab(D_{30}T_{70})$. (**b**) Compared the PBS, the phase transition profile of $Ab(D_{30}T_{70})$ was broadened when encapsulated within the pCB hydrogel. (**c**) The $Ab(T_{100})$ conjugate phase transitioned at lower temperatures than the corresponding free polymer, $P(T_{100})$. (**d**) Compared the PBS, the phase transition profile of $Ab(T_{100})$ was broadened when encapsulated within the pCB hydrogel.

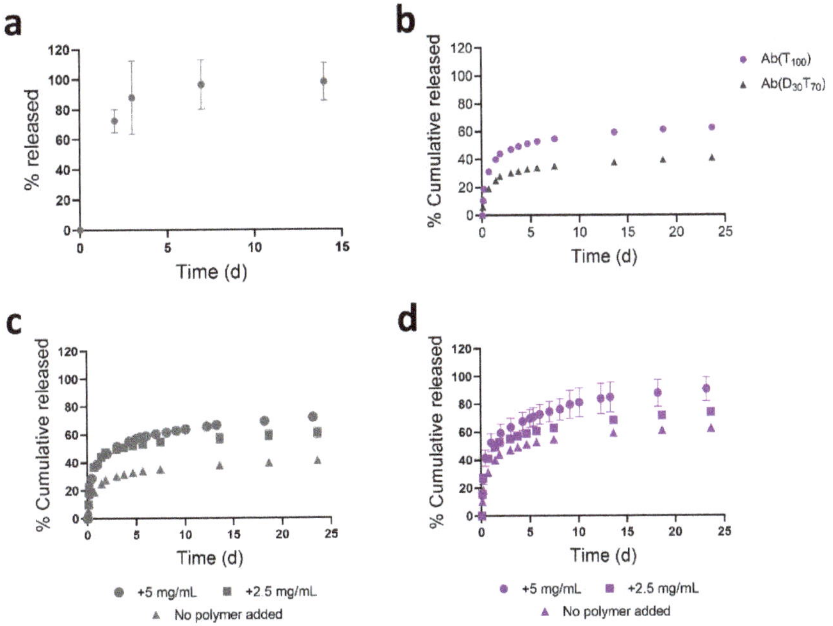

Figure 5. Controlled release of $Ab(D_xT_y)$ release from pCB hydrogels influenced by the type of thermoresponsive polymers. (**a**) Release of the $Ab(D_{30}T_{70})$ at 4 °C demonstrating that complete release occurs within ~3 days if the conjugate is completely solubilized. (**b**) Release curves for $Ab(D_{30}T_{70})$ and $Ab(T_{100})$ at 37 °C demonstrating that $P(D_{30}T_{70})$ lower LCST results in slower release. Release of (**c**) $Ab(D_{30}T_{70})$ and (**d**) $Ab(T_{100})$ at 37 °C in the presence of 2.5 and 5 $mgmL^{-1}$ of free $P(D_{30}T_{70})$ or $P(T_{100})$, respectively, demonstrating that release is due to conjugate phase transitions ($n = 3$, ± standard deviation).

At body temperature (37 °C), the release rate of the conjugates from the hydrogel agrees with LCST values, where partial phase transitions at lower temperatures resulted in slower released rates (Figure 5b). At three days, the conjugates differed in release by 17.7% where Ab($D_{30}T_{70}$) and Ab(T_{100}) released 31.6% and 49.3% of the initial loading. After three days, Ab($D_{30}T_{70}$) and Ab(T_{100}) sustained release rates of 73.2 ± 5.0 and 106.0 ± 8.6 ng per day, respectively (Figure S11). To further demonstrate that controlled release is due to polymer phase transition and aggregation, we conducted release experiments in the presence of free polymer corresponding to the conjugate (Figure 5c,d, Figure S11). Addition of free polymer will increase the amount of solubilized conjugate for release, which was observed in a concentration-dependent manner for both conjugates. Although the encapsulation of thermoresponsive polymer-protein conjugates in pCB hydrogels resulted in controlled delivery, the release rates plateaued at longer timepoints before complete release was achieved. Most of the difference in release profiles between Ab($D_{30}T_{70}$) and Ab(T_{100}) occurred during the first few hours and days, in a similar manner to many hydrogel affinity release systems [31,32].

To achieve sustained and complete release of temperature-sensitive Ab(D_xT_y) conjugates from pCB hydrogels, we combined the hydrogel promoted polymer aggregation technology with controlled hydrogel degradation. Hydrogel degradation will promote the solubilization of the conjugate for greater release rates and eventual complete release. The pCB-Az copolymer for crosslinking with pCB-DBCO was modified to contain β eliminative carbamate bonds with reported half-lives of 36 h (Figure 6a); the hydrogel crosslinks will therefore hydrolyze for complete hydrogel degradation [15,16,33]. To demonstrate degradability, the hydrogel was formed in a pre-weighed microcentrifuge tube and wet weight was followed over time (Figure 6b); wet weight increases at first due to hydrogel swelling. The Ab($D_{30}T_{70}$) conjugate was then encapsulated in the degradable pCB hydrogel and release was followed over 2 weeks. At first, release was similar to Ab($D_{30}T_{70}$) in a non-degradable gel, but release increased by day 5 with a substantial increase by day 7 due to hydrogel degradation (Figure 6c). Due to the differences in hydrogel volume and surface area between the degradation study (Figure 6b) and conjugate release (Figure 6c), hydrogel degradation occurred over a longer period for the release experiment (11 vs. 13 days). Therefore, the combination of temperature-sensitive conjugate with degradable hydrogels results in sustained and complete conjugate release.

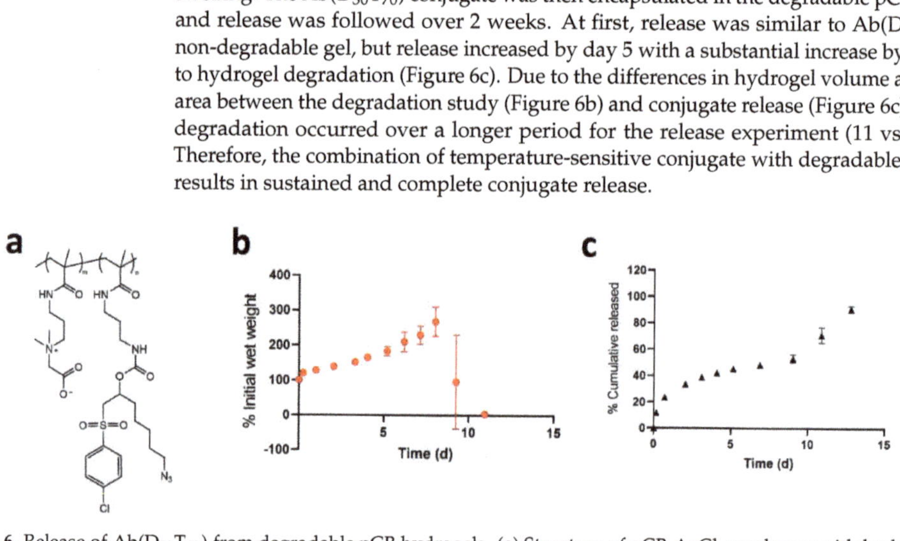

Figure 6. Release of Ab($D_{30}T_{70}$) from degradable pCB hydrogels. (**a**) Structure of pCB-AzCl copolymer with hydrolytic carbamate bond for hydrolysis of hydrogel crosslinks. (**b**) Degradation of pCB hydrogels over time by following the wet weight of hydrogels gelled in microcentrifuge tubes. (**c**) Complete and sustained release of Ab($D_{30}T_{70}$) from degradable pCB hydrogels where release is governed by the hydrogel promoted aggregation of Ab($D_{30}T_{70}$) and hydrogel degradation (n = 3, ± standard deviation).

4. Conclusions

Polymer-protein therapeutics, including PEGMA conjugates, continue to represent promising new therapeutics that may benefit from sustained release vehicles [2]. Here, we demonstrated that a zwitterionic hydrogel, pCB, can locally depress the LCST of thermoresponsive PEGMA-protein conjugates for their sustained release as solubilized molecules. In combination with a degradable pCB hydrogel, complete and sustained release of the conjugate occurred over 13 days. Therefore, the design of thermoresponsive polymers and hydrogels can yield drug delivery systems where polymer-protein conjugates aggregate within but not outside of hydrogels, representing a new mechanism for the controlled delivery of polymer conjugates.

Supplementary Materials: The following are available online at https://www.mdpi.com/article/10.3390/polym13162772/s1, Figure S1: ^{1}H NMR of $P(D_xT_y)$ in D_2O synthesized through RAFT polymerization, Figure S2: ^{1}H NRM spectra and chemical structures of the modulating polymers used, Figure S3: Temperature turbidity curves (OD550) of $P(D_xT_y)$ (5 mg mL^{-1}) with modulating polymers (50 mg mL^{-1}) in PBS, Figure S4: Decrease in heating and cooling LCST transition temperatures of $P(D_xT_y)$ polymers with he addition of zwitterionic polymer, Figure S5 Chemical structures of (a) pCB-APMA, (b) pCB-DBCO, (c) pCB-Az and their respective ^{1}H NMR spectra in D_2O (d), Figure S6: Temperature turbidity curves (OD550) of PEGMA polymers with pCB at different concentrations in PBS, Figure S7: Temperature turbidity curves (OD550) of PEGMA polymers with hydrogel precursor polymers and encapsulated within pCB hydrogel in PBS, Figure S8: Phase transition temperatures of PEGMA polymers when heating and cooling plotted with respect to triethylene glycol content with he addition of modulating polymers, different pCB concentrations and hydrogel precursors. Slopes of the aggregated data, Figure S9: Cytotoxicity of $P(T_{100})$ and $P(D_{30}T_{70})$ polymers incubated with 10,000 cells of human lung fibroblasts compared to PBS controls. No cytotoxicity from the polymers was observed, Figure S10: Temperature turbidity curves of protein polymer conjugates and their corresponding polymer in PBS, Figure S11: First order curves of release after day 3 of $Ab(D_{30}T_{70})$ and $Ab(T_{100})$, Table S1: MW of polymers synthesized obtained from GPC with PEG standards, Table S2: Phase transition temperatures of PEGMA polymers with the addition of modulating polymers and hydrogel precursor polymers when heating and cooling.

Author Contributions: Conceptualization, V.H. and R.G.W.; methodology, V.H.; validation, V.H.; formal analysis, V.H and N.I.; investigation, V.H. and N.I.; data curation, V.H. and N.I.; writing—original draft preparation, V.H. and R.G.W.; writing—review and editing, V.H. and R.G.W.; supervision, R.G.W.; funding acquisition, R.G.W. All authors have read and agreed to the published version of the manuscript.

Funding: This research was funded by the New Frontiers Research Fund, grant number NFRFE-2018-00943, Canada Foundation for Innovation: John R. Evans Leaders Fund, grant number 34107, and Ontario Research Fund-Research Infrastructure, grant number ORI-RI; 34107.

Institutional Review Board Statement: Not applicable.

Informed Consent Statement: Not applicable.

Data Availability Statement: The data presented in this study are available on request from the corresponding author.

Acknowledgments: V.H. would like to acknowledge NSERC for scholarship funding. We would also like to thank the Biointerfaces Institute.

Conflicts of Interest: The authors declare no conflict of interest.

References

1. Pelegri-O'Day, E.M.; Lin, E.-W.; Maynard, H.D. Therapeutic Protein–Polymer Conjugates: Advancing Beyond PEGylation. *J. Am. Chem. Soc.* **2014**, *136*, 14323–14332. [CrossRef]
2. Ekladious, I.; Colson, Y.L.; Grinstaff, M.W. Polymer–drug conjugate therapeutics: Advances, insights and prospects. *Nat. Rev. Drug Discov.* **2019**, *18*, 273–294. [CrossRef] [PubMed]
3. Cobo, I.; Li, M.; Sumerlin, B.S.; Perrier, S. Smart hybrid materials by conjugation of responsive polymers to biomacro-molecules. *Nat. Mater.* **2015**, *14*, 143–149. [CrossRef] [PubMed]
4. Zhang, P.; Sun, F.; Liu, S.; Jiang, S. Anti-PEG antibodies in the clinic: Current issues and beyond PEGylation. *J. Control. Release* **2016**, *244*, 184–193. [CrossRef]
5. Basu, A.; Kunduru, K.R.; Abtew, E.; Domb, A.J. Polysaccharide-Based Conjugates for Biomedical Applications. *Bioconjugate Chem.* **2015**, *26*, 1396–1412. [CrossRef]

6. Joh, D.Y.; Zimmers, Z.; Avlani, M.; Heggestad, J.T.; Aydin, H.B.; Ganson, N.; Kumar, S.; Fontes, C.M.; Achar, R.K.; Hershfield, M.S.; et al. Architectural Modification of Conformal PEG-Bottlebrush Coatings Minimizes Anti-PEG Antigenicity While Preserving Stealth Properties. *Adv. Health Mater.* **2019**, *8*, e1801177. [CrossRef]
7. Qi, Y.; Simakova, A.; Ganson, N.J.; Li, X.; Luginbuhl, K.M.; Ozer, I.; Liu, W.; Hershfield, M.S.; Matyjaszewski, K.; Chilkoti, A. A brush-polymer/exendin-4 conjugate reduces blood glucose levels for up to five days and eliminates poly(ethylene glycol) antigenicity. *Nat. Biomed. Eng.* **2016**, *1*, 1–12. [CrossRef] [PubMed]
8. Bebis, K.; Jones, M.W.; Haddleton, D.M.; Gibson, M.I. Thermoresponsive behaviour of poly[(oligo(ethyleneglycol methacrylate)]s and their protein conjugates: Importance of concentration and solvent system. *Polym. Chem.* **2011**, *2*, 975–982. [CrossRef]
9. Liu, M.; Johansen, P.; Zabel, F.; Leroux, J.C.; Gauthier, M.A. Semi-permeable coatings fabricated from comb-polymers efficiently protect proteins in Vivo. *Nat. Commun.* **2014**, *5*, 1–8. [CrossRef]
10. Liu, M.; Tirino, P.; Radivojevic, M.; Phillips, D.J.; Gibson, M.; Leroux, J.-C.; Gauthier, M.A. Molecular Sieving on the Surface of a Protein Provides Protection Without Loss of Activity. *Adv. Funct. Mater.* **2012**, *23*, 2007–2015. [CrossRef]
11. Jiang, L.; Bonde, J.S.; Ye, L. Temperature and pH Controlled Self-Assembly of a Protein-Polymer Biohybrid. *Macromol. Chem. Phys.* **2018**, *219*, 1700597. [CrossRef]
12. Gilroy, C.A.; Roberts, S.; Chilkoti, A. Fusion of fibroblast growth factor 21 to a thermally responsive biopolymer forms an injectable depot with sustained anti-diabetic action. *J. Control. Release* **2018**, *277*, 154–164. [CrossRef]
13. Luginbuhl, K.M.; Schaal, J.; Umstead, B.; Mastria, E.; Li, X.; Banskota, S.; Arnold, S.; Feinglos, M.; D'Alessio, D.; Chilkoti, A. One-week glucose control via zero-order release kinetics from an injectable depot of glucagon-like peptide-1 fused to a thermosensitive biopolymer. *Nat. Biomed. Eng.* **2017**, *1*, 1–14. [CrossRef]
14. Magnusson, J.P.; Khan, A.; Pasparakis, G.; Saeed, A.O.; Wang, W.; Alexander, C. Ion-sensitive 'isothermal' responsive polymers prepared in water. *J. Am. Chem. Soc.* **2008**, *130*, 10852–10853. [CrossRef] [PubMed]
15. Shoaib, M.M.; Huynh, V.; Shad, Y.; Ahmed, R.; Jesmer, A.H.; Melacini, G.; Wylie, R.G. Controlled degradation of low-fouling poly(oligo(ethylene glycol)methyl ether methacrylate) hydrogels. *RSC Adv.* **2019**, *9*, 18978–18988. [CrossRef]
16. Santi, D.V.; Schneider, E.; Reid, R.; Robinson, L.; Ashley, G.W. Predictable and tunable half-life extension of therapeutic agents by controlled chemical release from macromolecular conjugates. *Proc. Natl. Acad. Sci. USA* **2012**, *109*, 6211–6216. [CrossRef] [PubMed]
17. Huynh, V.; Jesmer, A.H.; Shoaib, M.M.; Wylie, R.G. Influence of Hydrophobic Cross-Linkers on Carboxybetaine Co-polymer Stimuli Response and Hydrogel Biological Properties. *Langmuir* **2019**, *35*, 1631–1641. [CrossRef] [PubMed]
18. Huynh, V.; Wylie, R.G. Displacement Affinity Release of Antibodies from Injectable Hydrogels. *ACS Appl. Mater. Interfaces* **2019**, *11*, 30648–30660. [CrossRef]
19. Huynh, V.; Wylie, R.G. Competitive Affinity Release for Long-Term Delivery of Antibodies from Hydrogels. *Angew. Chem. Int. Ed.* **2018**, *57*, 3406–3410. [CrossRef]
20. Keefe, A.J.; Jiang, S. Poly(zwitterionic)protein conjugates offer increased stability without sacrificing binding affinity or bioactivity. *Nat. Chem.* **2011**, *4*, 59–63. [CrossRef] [PubMed]
21. Erfani, A.; Seaberg, J.; Aichele, C.P.; Ramsey, J.D. Interactions between Biomolecules and Zwitterionic Moieties: A Review. *Biomacromolecules* **2020**, *21*, 2557–2573. [CrossRef]
22. Liu, D.; Yang, J.; Wang, H.-F.; Wang, Z.; Huang, X.; Wang, Z.; Niu, G.; Walker, A.R.H.; Chen, X. Glucose oxidase-catalyzed growth of gold nanoparticles enables quantitative detection of attomolar cancer biomarkers. *Anal. Chem.* **2014**, *86*, 5800–5806. [CrossRef]
23. Moatsou, D.; Li, J.; Ranji, A.; Pitto-Barry, A.; Ntai, I.; Jewett, M.C.; O'Reilly, R.K. Self-Assembly of Temperature-Responsive Protein–Polymer Bioconjugates. *Bioconjug. Chem.* **2015**, *26*, 1890–1899. [CrossRef]
24. Nandy, A.; Chakraborty, S.; Nandi, S.; Bhattacharyya, K.; Mukherjee, S. Structure, Activity, and Dynamics of Human Serum Albumin in a Crowded Pluronic F127 Hydrogel. *J. Phys. Chem. B* **2019**, *123*, 3397–3408. [CrossRef]
25. Xu, G.; Liu, K.; Xu, B.; Yao, Y.; Li, W.; Yan, J.; Zhang, A. Confined microenvironments from thermoresponsive dendronized polymers. *Macromol. Rapid Commun.* **2020**, *41*, 2000325. [CrossRef]
26. Park, S.; Barnes, R.; Lin, Y.; Jeon, B.-J.; Najafi, S.; Delaney, K.T.; Fredrickson, G.H.; Shea, J.-E.; Hwang, D.S.; Han, S. Dehydration entropy drives liquid-liquid phase separation by molecular crowding. *Commun. Chem.* **2020**, *3*, 1–12. [CrossRef]
27. Sakota, K.; Tabata, D.; Sekiya, H. Macromolecular Crowding Modifies the Impact of Specific Hofmeister Ions on the Coil–Globule Transition of PNIPAM. *J. Phys. Chem. B* **2015**, *119*, 10334–10340. [CrossRef]
28. Sun, S.; Wu, P. On the Thermally Reversible Dynamic Hydration Behavior of Oligo(ethylene glycol) Methacrylate-Based Polymers in Water. *Macromolecules* **2012**, *46*, 236–246. [CrossRef]
29. Zhang, B.; Tang, H.; Wu, P. In Depth Analysis on the Unusual Multistep Aggregation Process of Oligo(ethylene glycol) Methacrylate-Based Polymers in Water. *Macromolecules* **2014**, *47*, 4728–4737. [CrossRef]
30. Amsden, B. Solute Diffusion within Hydrogels. Mechanisms and Models. *Macromolecules* **1998**, *31*, 8382–8395. [CrossRef]
31. Soontornworajit, B.; Zhou, J.; Zhang, Z.; Wang, Y. Aptamer-Functionalized In Situ Injectable Hydrogel for Controlled Protein Release. *Biomacromolecules* **2010**, *11*, 2724–2730. [CrossRef] [PubMed]
32. Vulic, K.; Shoichet, M.S. Tunable Growth Factor Delivery from Injectable Hydrogels for Tissue Engineering. *J. Am. Chem. Soc.* **2011**, *134*, 882–885. [CrossRef] [PubMed]
33. Ashley, G.W.; Henise, J.; Reid, R.; Santi, D.V. Hydrogel drug delivery system with predictable and tunable drug release and degradation rates. *Proc. Natl. Acad. Sci. USA* **2013**, *110*, 2318–2323. [CrossRef] [PubMed]

polymers

MDPI

rticle

NIPAm-Based Modification of Poly(L-lysine): A pH-Dependent LCST-Type Thermo-Responsive Biodegradable Polymer

ggeliki Stamou [1], Hermis Iatrou [2] and Constantinos Tsitsilianis [1,*]

[1] Department of Chemical Engineering, University of Patras, 26500 Patras, Greece; aggestamou@hotmail.com
[2] Department of Chemistry, University of Athens, Panepistimiopolis, Zografou, 15771 Athens, Greece; iatrou@chem.uoa.gr
* Correspondence: ct@chemeng.upatras.gr

Abstract: Polylysine is a biocompatible, biodegradable, water soluble polypeptide. Thanks to the pendant primary amines it bears, it is susceptible to modification reactions. In this work Poly(L-lysine) (PLL) was partially modified via the effortless free-catalysed aza-Michael addition reaction at room temperature by grafting N-isopropylacrylamide (NIPAm) moieties onto the amines. The resulting PLL-g-NIPAm exhibited LCST-type thermosensitivity. The LCST can be tuned by the NIPAm content incorporated in the macromolecules. Importantly, depending on the NIPAm content, LCST is highly dependent on pH and ionic strength due to ionization capability of the remaining free lysine residues. PLL-g-NIPAm constitutes a novel biodegradable LCST polymer that could be used as "smart" block in block copolymers and/or terpolymers, of any macromolecular architecture, to design pH/Temperature-responsive self-assemblies (nanocarriers and/or networks) for potential bio-applications.

Keywords: poly(L-lysine); N-isopropylacrylamide; aza-Michael addition reaction; LCST; thermo-responsive; pH-responsive; biodegradable polymer

itation: Stamou, A.; Iatrou, H.; sitsilianis, C. NIPAm-Based Iodification of Poly(L-lysine): A H-Dependent LCST-Type hermo-Responsive Biodegradable olymer. *Polymers* **2022**, *14*, 802. ttps://doi.org/10.3390/ olym14040802

cademic Editors: Andreas Taubert nd Mattia Sponchioni

eceived: 19 January 2022
ccepted: 16 February 2022
ublished: 18 February 2022

1. Introduction

Polymeric "smart" materials can respond to external stimuli such as pH, light, temperature, enzymes, and electric or magnetic fields. Therefore, they are applicable in various scientific fields [1]. Thermosensitive polymers are a sub-category of smart polymers that attract great scientific interest, as they can respond to temperature and change their physicochemical properties. There are two main types of thermo-sensitive polymers. The first category includes polymers that exhibit lower critical temperature behaviour (LCST), while the second one consists of polymers that exhibit upper critical solution temperature (UCST) behaviour. The critical temperatures below and above which the polymer and the solvent become fully miscible are known as the LCST and UCST, respectively [2,3]. They change their conformation either upon heating (LCST) or cooling (UCST) from a rather random coil form to a collapsed, more globular one. Thermo-responsive polymers, integrated in block copolymers, can be exploited in a plethora of water dispersive self-assembling entities such as micelles, polymersomes, microcapsules and microgels, along with injectable hydrogels and thin films, all of them advantageous for biomedical applications such as controlled drug delivery, stem cell transplantation, wound healing, etc. [4].

Water soluble polymers that exhibit LCST are by far the most studied polymers. Among them, Poly(N-isopropylacrylamide) (PNIPAm) has attracted major attention thanks to its LCST of approximately 32 °C, which is close to and below body temperature (37 °C), which renders it suitable for biomedical applications [5–11]. However, provided that it is not entirely non-toxic and, more importantly, it is not biodegradable, PNIPAm does not seem to be acceptable for real biomedical applications and has mainly been used as a LCST thermo-responsive model polymer.

Polylysine is a biocompatible, biodegradable, water soluble polypeptide. These properties render polylysine and its derivatives ideal for a variety of biomedical applications [12,13]. Polylysine consists of either L-lysine or D-lysine. "L" and "D" refer to the chirality of central lysine carbon. Thus, two polypeptide isomers can be distinguished poly-L-lysine (PLL) and poly-D-lysine (PDL) [14]. In every case, it is a weak, cationic polyelectrolyte with a $pK_a = 10.5$, thanks to the pendant free amino group that can be charged by protonation at $pH < pK_a$. The polylysine conformational states in aqueous solutions, which determine its secondary polypeptide structure, have been extensively studied and are effected by a wide range of solution conditions such as pH, temperature and salt concentration [15]. In salt-free solutions and at a neutral pH, polylysine adopts a random coil conformation due to its protonated, charged state. At pH 11.5 the uncharged polylysine is entirely transformed into α-helix conformation and, upon heating, the α-helix structure is transformed into a β-sheet structure [16].

Another interesting property of poly(L-lysine) is that the free amino pendant groups it bears are susceptible to modification reactions. The objective of the present study is to combine the properties of poly(L-lysine) polypeptide with PNIPAm to form a biodegradable LCST-type thermo-responsive polymer, in order to ultimately utilize it in real biomedical applications. The free-catalysed aza-Michael addition reaction seems suitable for this purpose. This reaction constitutes a subcategory of Michael addition reactions, a very important reaction in organic chemistry. An amine group is a Michael donor, while α, β-unsaturated carbonyl compounds act as Michael acceptors, resulting in the formation of a Michael adduct [17]. Since the amino group can act as both a base and a nucleophile the aza-Michael addition reaction does not require the involvement of a base [18]. Michael donors can be both secondary amines and primary amines and the reaction activity depends, to a large extent, on the electronic and stereochemical environment of the amine When a Michael donor is added to a primary amine, bis-addition is possible, depending on the added amount [19]. The choice of solvents is critical for Michael addition reactions because aprotic solvents may cause incomplete alkylation of the amino group [17]. Michael addition reaction is an effective method for modifying an existing polymer. Li and Feng synthesized a series of highly sensitive stimuli-responsive polysiloxanes (SPSis) through a catalyst-free aza-Michael addition reaction. Particularly, Poly(aminopropylmethylsiloxane) was post-modified with Nipa monomers [20].

In the present communication, preliminary results dealing with the partial modification of PLL by NIPAm monomers, denoted as PLL-g-NIPAmX (where X is mol% of NIPAm with respect to LL monomer units), via the effortless aza-Michael reaction and their resulting thermo-responsiveness are demonstrated. The modified polypeptide exhibited LCST-type thermo-sensitivity. The LCST was found to be inextricably dependent on the amount of NIPAm grafted, along with the pH and salinity of the aqueous media.

2. Materials and Methods

2.1. Materials

PLL was synthesized by ring opening polymerization of the ε-tert-butyloxycarbonyl-L-lysine-NCA (NCA: N-carboxy anhydride) monomer, followed by acid hydrolysis to deprotect the amine groups, according to standard procedures. The molecular weight characteristics of the protected PLL, determined by gel permeation chromatography (GPC, Milford, MA, USA) are $M_w = 11{,}980$ Daltons, $M_n = 11{,}365$ Daltons (degree of polymerization 50) and molecular polydispersity, $M_w/M_n = 1.054$. Details of the synthesis and characterization of PLL and its NCA protected monomer are reported in the Supplementary Materials. NIPAm monomers (NIPAM, Fluorochem, Derbyshire, UK) were used as received.

2.2. Modification of PLL by Aza-Michael Addition

PLL was modified by grafting onto the NIPAm monomers, according to Scheme 1.

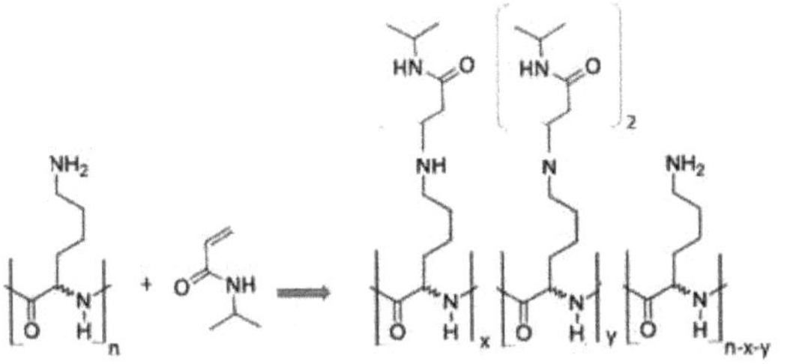

Scheme 1. Synthesis of PLL-g-NIPAmX via the free-catalysed aza-Michael addition reaction.

NIPAM monomers were added to the PLL aqueous solution (C_p~0.1 wt%), the pH of which was adjusted close to 12, by NaOH, ensuring non-protonation of the pendant primary amines. The reaction was performed at room temperature in a flask with a ground glass stopper in different NIPAm/Lysine mol ratios and various periods of time, as discussed. The next step was the purification of the solution from unreacted NIPAm monomers and from any other impurities, through dialysis membrane (MWCO 3500 Da) with 3D-H_2O. The final product was obtained in solid form by freeze drying. The NIPAm percentage (mol%) of the partially modified PLL-g-NIPAm samples was characterized by ^{1}H NMR in deuterated water (D_2O), using A Bruker Avance Iii Hd Prodigy Ascend Tm 600 MHz spectrometer (Billerica, MA, USA). A characteristic ^{1}H-NMR spectra is presented in Figure S1.

2.3. Preparation of Samples for Light Scattering

Aqueous solutions of the modified PLL-g-NIPAm were prepared at a concentration of 1% *w/v*. The pH value of the system was adjusted by adding HCl or NaOH solution (1 M). The study was performed at different pH, as PLL is pH responsive. NaCl was added to study salt effect. The concentration of the salt that forms by the addition of HCl or NaOH is negligible with respect to the concentration of added NaCl and was not considered in the final salt concentration.

2.4. Light Scattering

A physicochemical study of the modified PLL-g-NIPAm was performed by the static light scattering (SLS) technique. All measurements were carried out using a thermally regulated spectro-goniometer, Model BI-200SM (Brookhaven Holtsville, NY, USA), equipped with a He–Ne laser (632.8 nm). The aim of the study was to determine whether the modified polymers were thermo-sensitive and, in this case, to identify their critical temperature, T_{LCST}. More specifically, in a PLL-g-NIPAmX sample (transparent aqueous solution at low temperature), the light scattered intensity (I) was measured at a constant angle θ (90°) by increasing, stepwise, the temperature. At each temperature, sufficient time was left for equilibration before three repeating measurements were taken. The T_{LCST} was determined at T, above which an abrupt increase in LS intensity was observed. For the sake of comparison, the temperature-dependent optical density (OD) was also measured for one sample by UV visible to detect the cloud point T_{cp}, i.e., the temperature above which the solution turns turbid, using a Hitachi U-2001 UV–VIS spectrophotometer (Schaumburg, IL, USA).

3. Results and Discussion

The critical temperature of the aqueous solutions was determined by static light scattering which, in fact denotes, the onset of association (aggregation), T_{ass}, of the macromolecules, which is the consequence of their hydrophilic to hydrophobic transition following the phase separation of the solution upon heating. The T_{ass} coincides with the cloud point, T_{cp}, as observed by optical density measurements (Figure 1), since both are due to the intermolecular hydrophobic association. To emphasise that critical temperature occurs upon heating, we will refer to it hereafter as T_{LCST}.

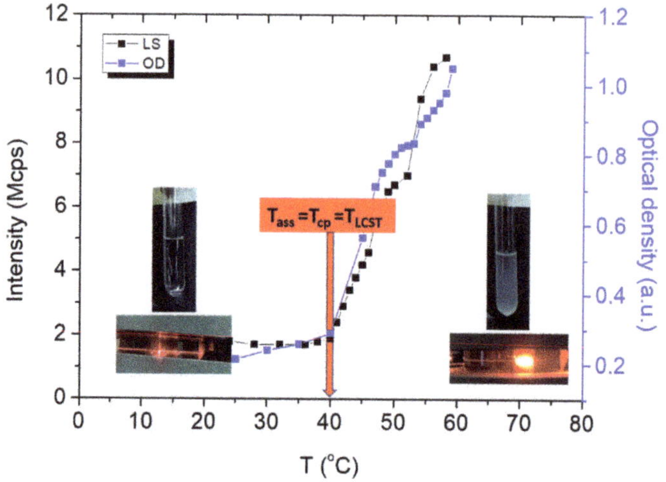

Figure 1. Example of temperature dependence of static light scattering intensity (LS) and optical density (OD) of the PLL-gNIPAM60 upon heating. The insets show digital photographs of the solution and the light scattering generated from a red laser, below and above the critical temperature. The cloud point T_{cp} (detected by OD) coincides with the onset of the polymer association T_{ass} (detected by LC).

Initially, the thermal behaviour of pure PLL in aqueous solutions was investigated at high pH (10.6), where the PLL is nearly uncharged (bearing non-protonated pendant primary amines), as it is shown by z-potential measurements [16]. As can be seen in Figure 2A, the light scattering intensity increases gradually above 45 °C, denoting the onset of association, and levels above 75 °C. On the contrary, no temperature effect is detected at pH 7. At pH 10.6, the 1 wt% polymer solution remains transparent in the entire temperature range investigated. However, a free-standing gel forms above 65 °C, as seen in Figure 2A. As it is well known, PLL adopts various secondary structures depending on pH and temperature [15,21]. At pH 10.6, the nearly uncharged PLL adopts mainly an α-helix secondary structure, i.e., about 80%, which becomes 100% above pH 11.5 [22]. Upon heating an α-helix-to-β-sheet conformational transition is expected above a critical temperature which depends on molecular weight, i.e., the lower the molecular weight the higher the transition temperature [23]. The thermo-induced formation of the transparent gel observed in Figure 2A is attributed to the α-helix-to-β-sheet transformation of the majority of the PLL chains, associated with the formation of a spanning 3D physical network. As reported, the peptides, when adopting a β-sheet secondary structure, self-assemble, forming entangled fibrils, which justifies the gel appearance [24]. The transition temperature was detected at about 45 °C, which seems reasonable for the molecular weight of PLL (Mn = 6500), as it is known that it depends on PLL molecular weight [25]. At pH 7, PLL does not exhibit thermosensitivity, since it exists as a random coil, due to its charged state.

Partial modification of PLL by conjugating NIPAm moieties onto primary amines via the aza-Michael addition reaction, with a feed molar ratio [NIPAm]/[LL] of 35 mL%, was accomplished first. The reaction was carried out at room temperature. The yield of the reaction was 82%, meaning that the percentage of NIPAm moieties incorporated into the PLL chain was 28.6 mol% (denoted as PLL-g-NIPAm29). The scattering intensity as a function of temperature in aqueous solutions of the partially modified PLL-g-NIPAm29 is shown in Figure 2B at various pH 7.8, 8.7 and 10.8. At pH = 7.8 and 8.7, the solutions are clear, and the intensity remains constant upon increasing temperature. The modified polymer does not show any thermo-sensitivity under these conditions. This behavior is attributed to the fact that PLL is positively charged, as the free primary amine groups are protonated (pH < pKa). At these pH conditions, the sample under study is characterized, to a large extent, by hydrophilicity, exhibiting very good solubility in the aqueous medium. Therefore, the grafting of NIPAm groups at 29%, does not appear to affect the thermal responsiveness of the samples, investigated in this temperature range, provided that the percentage of positively charged amine groups predominate. At pH 10.8, the solution shows thermosensitivity as the intensity of the light scattering increases abruptly above 54 °C. The presence of the NIPAm groups on the chain generate two new effects on the heat-induced solution behaviour. The scattering light intensity increases more sharply and the solution exhibits turbidity at higher temperatures. These effects are consistent with LCST-type thermo-sensitivity. Remarkably, at 60 °C the solution was transformed to a free-standing gel. It seems probable that the low percentage (29 mol%) of NIPAm attachment on the PLL chains does not entirely prevent the α-helix-to-β-sheet conformational transition, inducing gelation.

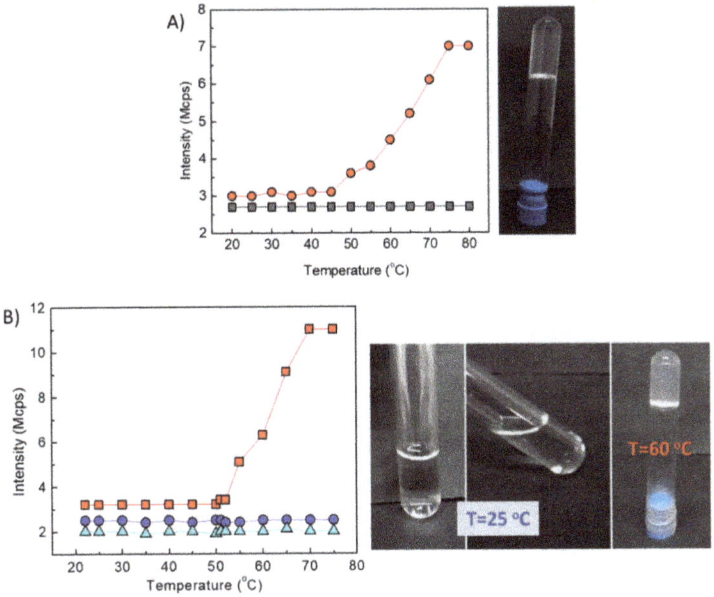

Figure 2. Light scattering intensity at 90° as a function of temperature of polymer aqueous solutions at various pH with C_p = 1 wt%: (**A**) PLL, (○) pH 10.6 and (□) pH 7, the photo illustrates a freestanding transparent gel of the solution at pH 10.6, and 65 °C, and (**B**) PLL-g-NIPAm29, (□) pH 10.8, (○) pH 8.7, (Δ) pH 7.4, the photo illustrates transparent solutions at pH 10.8, 25 °C and a turbid free-standing gel at pH 10.8 and 60 °C.

By increasing the [NIPAm]/[LL] feed ratio to 78%, the percentage of NIPAm modification reached 69 mol% (PLL-g-NIPAm69). The PLL-g-NIPAm69 aqueous solution at pH 11.8 exhibits a sharp increase in the light scattering intensity at 32 °C, above which the solution turns turbid, as shown in Figure 3A. More importantly, no gel formation was observed (Figure 3A, inset), implying that, at this NIPAm percentage, the formation of β-sheet responsible for the gel formation is prevented. Therefore, the observed thermal transition is clearly the LCST type. By decreasing pH at 9.5 and 8.3, the transition disappears at least up to 80 °C, due to ionization of the remaining free amine groups. Provided that the charged groups are highly hydrophilic, the T_{LCST} is likely shifted to higher temperatures.

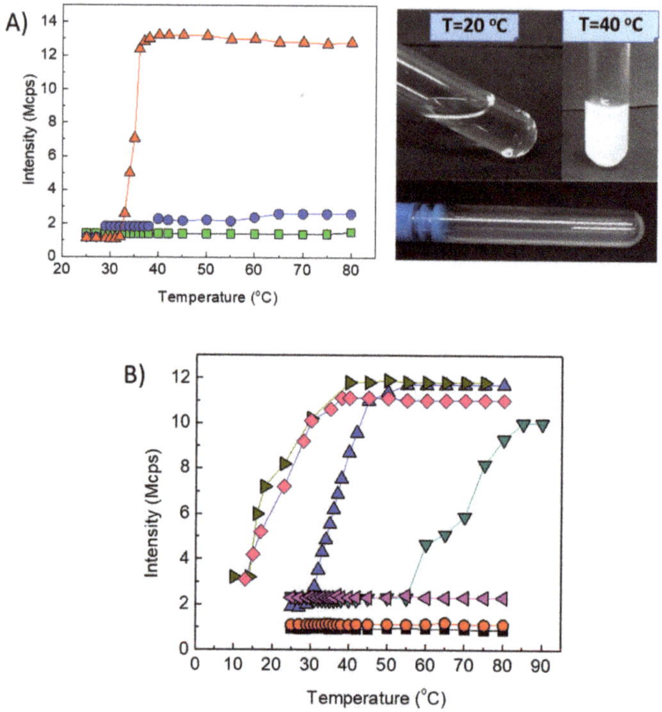

Figure 3. Light scattering intensity at 90° as a function of temperature of PLL-g-NIPAm69 in aqueous solutions of various pH: (**A**) (Δ) pH 11.8, (○) pH 9.5, (□) pH 8.3, the photo illustrates a transparent solution at pH 11.8, at 20 °C and a cloudy one at 40 °C, which flows as a liquid (lowest image), and (**B**) in the presence of NaCl at various pH: 1M NaCl, pH (□)7.5, (○) pH 8,5, (◁) pH 9.1, (Δ) pH 9.6, (▷) 9.8 and (∇) pH 9.0, 1.5M NaCl, (◊) pH 8.8, 2M NaCl.

The effect of ionic strength was explored by adding NaCl at various pHs, in an attempt to decrease the LCST. In Figure 3B, we can see the effect of pH in saline aqueous solutions of 1M NaCl. The thermosensitivity appears at T_{LCST} = 30 °C in pH 9.6, whereas it is not visible at lower pHs up to 80 °C. Compared to the absence of transition at pH 9.5 in salt free solutions (Figure 3A), the presence of 1M salt has a remarkable effect on the LCST shift, greater than 50 °C. At the slightly higher pH of 9.8 the critical temperature continues to decrease significantly to T_{LCST} = 14 °C. These intensive effects should be attributed to the partially substituted PLL with 69 mol% NIPAm which still bears a high number of free amines that are ionizable, depending on pH. Thus, the increase in ionic strength likely exerts two effects. It is known that hydrophilic comonomers increase the LCST, whereas hydrophobic ones have the opposite effect [4]. The presence of salts decreases the hydrophilicity of the protonated amines by electrostatic screening, thus inducing a decrease in LCST. Moreover, the well-known salting out effect of NIPAm moieties operates in the same direction [4,26]. Hence, both effects seem to be the reason for the observed LCST high shift at pH 9.5 for the PLL-g-NIPAm69 sample. At pH 9.0, more salt is needed, e.g., 1.5 M, for the appearance of thermosensitivity, which is detected at T_{LCST} = 55 °C. Finally, at pH 8.8/2M NaCl, the T_{LCST} is equivalent to that at pH 9.8/1M NaCl, clearly showing the combined effect of pH and ionic strength.

By increasing the percentage of NIPAm integration to 87 mol% (PLL-g-NIPAm87), the T_{LCST} is further shifted to lower temperatures, as seen in Figure 4. For the uncharged state (high pH) the T_{LCST} drops to 27 °C versus the 32 °C of 69 mol% content. At pH 9 and salt free solution it appears at 50 °C in contrast to the PLL-g-NIPAm69 sample, in which thermosensitivity is not visible, even up to 80 °C (Figure 3B). At the lower pH 8, a weak thermosensitivity appears at 65 °C.

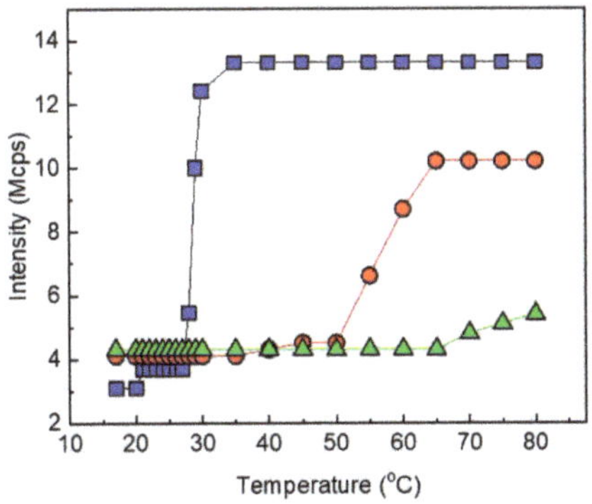

Figure 4. Light scattering intensity at 90° as a function of temperature of PLL-g-NIPAm87, $C_p = 1$ wt%, in aqueous solutions at various pH: (□) pH 11.6, (○) pH 9, (Δ) pH 8.

For the LCST to approach physiological conditions, namely pH 7.4, 37 °C, the following "in situ" experiment was designed (Scheme 2). To an aqueous solution of PLL-g-NIPAm87, a successive addition of NIPAm monomers was applied. Specifically, to 20 mg (0.095 mmol) of PLL-g-NIPAm87 an amount of 0.057 mmol of NIPAm was added to 2 mL aqueous solutions. NaOH (1M) was also added to adjust the pH of the mixture at 12. The solution was left under stirring for 48 h, and product 1 was characterized in terms of thermosensitivity at various pH. The same procedure was repeated sequentially three times, as demonstrated in Scheme 2. The final product was characterized by ^{1}H-NMR (i.e., PLL-g-NIPAm116). The successive addition of NIPAm resulted in an ongoing insertion of this moiety into the remaining free amines of PLL, reaching, after four steps, a total NIPAm percentage of 116 mol%, which is higher than 100%. This is attributed to the fact that the secondary amine remaining after the first addition is more nucleophilic than the primary one and bis-addition is unavoidable [20]. As can be seen in Figure 5, for the uncharged samples (pH > 11, red symbols), the consequence of NIPAm integration imposes a continuous T_{LCST} shift to lower temperature, about 2 or 3 degrees after each NIPAm addition, finally reaching 17 °C. More importantly, the T_{LCST} appears at lower pHs as the percentage of the remaining ionizable free amines decreases. For instance, in the PLL-g-NIPAm116 (Figure 5, product 4), the T_{LCST} appears at pH 7.4 at 39 °C, while it is not visible up to 55 °C when the pH decreased to 6.9, indicating that a considerable number of free amines remains intact, and their ionization capability still affects the LCST. It can also be observed that bis-addition is greater than 16%.

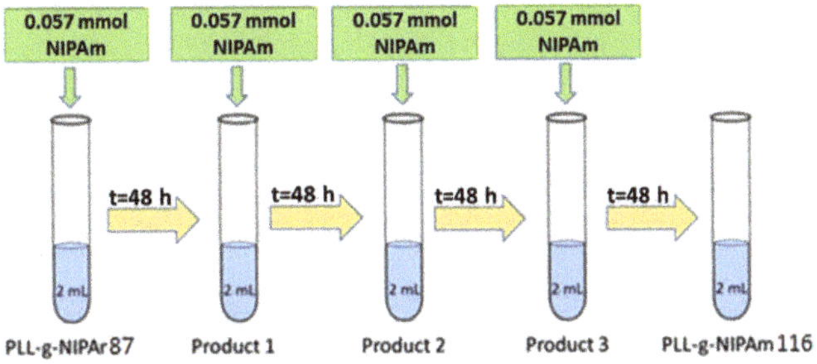

Scheme 2. Schematic representation of the "in situ" enrichment of PLL with NIPAm.

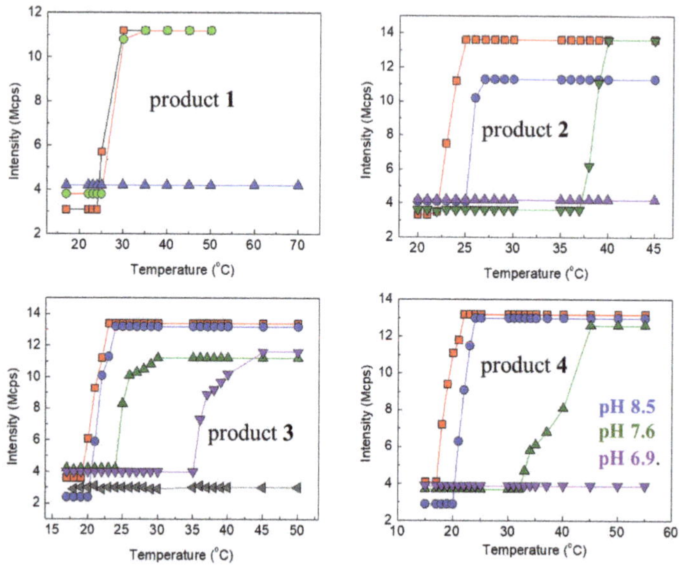

Figure 5. Light scattering intensity as a function of temperature of PLL-g-NIPAm87 aqueous solutions (C_p = 1 wt%) after in situ modification and at various pH: (product **1**) (□) pH 11.7, (○) pH 10, (Δ) pH 8 and 7.4, (product **2**) (□) pH 11.2, (○) pH 10.2, (∇) pH 8.3, (Δ) pH 7.4, (product **3**) (□) pH 11.6, (○) pH 9, (Δ) pH 8.4, (∇) pH 7.7, (◁) pH 7.25, (product **4**) PLL-g-NIPAm116 (□) pH 11.3, (○) pH 8.5, (Δ) pH 7.6, (∇) pH 6.9.

By attempting to increase the NIPAm content in a single experiment, the aza-Michael addition reaction was accomplished in NIPAm/LL ratio 2/1, prolonging the time at room temperature. The content of NIPAm reached 98 mol%, likely showing steric hindrance effects. It seems that the successive addition of NIPAm of the "in situ" experiment could enrich the PLL with more NIPAm (116 mol%). The temperature dependence of the light scattering intensity of the 1% PLL-g-NIPAm98 aqueous solutions at various pH and salt concentrations is depicted in Figure 6a. In the uncharged state of LL moieties (pH 11), the T_{LCST} was observed at 18 °C, which is shifted slightly at 20 °C when the pH dropped to 9. However, at pH 7.5, the intensity did not change upon the increase in temperature, at least up to 60 °C, due to ionization of the free remaining amines. Upon adding 0.15 M NaCl, the T_{LCST} appeared again at 52 °C, which further decreased with increasing NaCl concentration (Figure 6b). Finally, when 0.6 M NaCl was added, the T_{LCST} appeared at 37 °C.

By extrapolating to zero salt concentration, the T_{LCST} should appear at 57 °C in pH 7.5, which is not valid, as shown by the light scattering data in Figure 6a. Moreover, the T_{LCST} shift with salt concentration (ΔT more than 20 °C) is much more pronounced compared with the behaviour of pure PNIPAm (ΔT about 4 °C) in the presence of 0.6 M NaCl, attributed to the salting out effect [26]. Therefore, the strong salt effect is clearly due to the additional electrostatic screening of the charged amine groups that decrease their hydrophilicity, as discussed previously.

The T_{LCST} data of the various PLL-g-NIPAm samples is summarized in Figure 7. The influence of the grafted NIPAm percentage at high pH > 11, where the remaining free amines are in an uncharged state, is depicted in Figure 7a. The data follow a decreasing linear dependence of T_{LCST} with the mol% NIPAm, since the continuous decrease in the remaining amount of the hydrophilic primary amine groups also decreases the LCST.

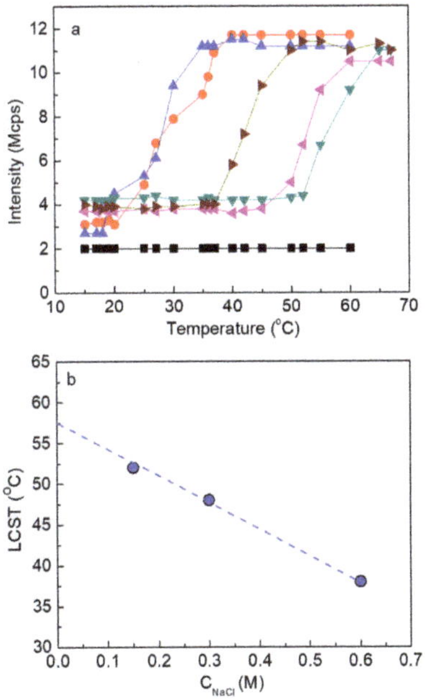

Figure 6. (**a**) Light scattering intensity at 90° as a function of temperature of PLL-g-NIPAm98 in aqueous solutions: (□) pH 7.5, C_p = 1 wt%, (○) pH 9, C_p = 1 wt%, (Δ) pH 11, C_p = 1 wt%, (∇) pH 7.7, 0.15M NaCl, C_p = 1 wt%, (◁) pH 7.5, 0.3M NaCl, C_p = 1 wt%, (▷) pH 7.5, 0.6 M NaCl, C_p = 1 wt%. (**b**) T_{LCST} at pH 7.5 as a function of salt concentration (in M).

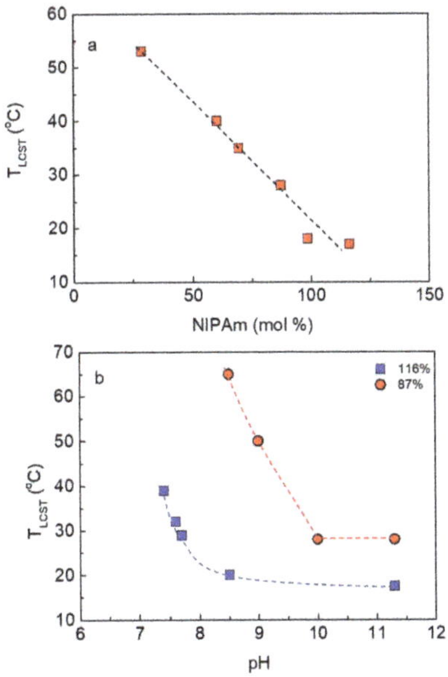

Figure 7. (**a**) T_{LCST} as a function of NIPAm mol% for the PLL-g-NIPAm in the uncharged state of the free amines (pH > 11). The line represents the linear fitting of the data and LCST as a function of pH (**b**) for various PLL-g-NIPAm of different mol% NIPAm (as indicated). The lines represent guide for the eyes.

More importantly, for the partially grafted samples, T_{LCST} is strongly pH-dependent. As observed in Figure 7b, the T_{LCST} decreases with increasing pH and levels off at high pH. Moreover, the degree of NIPAm grafted (mol%) remarkably influences this dependence, that is, the higher NIPAm content the lower T_{LCST}. For instance, by increasing the NIPAm content from 87 to 116 mol% at pH 9, the LCST is shifted by more than 30 °C at lower values. Therefore, the PLL-g-NIPAm is in fact a pH/thermo dual-responsive polymer.

4. Conclusions

The target of the present work was to prepare an LCST-type thermosensitive macromolecule, based on PLL polypeptide to ensure biodegradability. For this purpose, NIPAm moieties were conjugated onto the free amine pendants of PLL, via the effortless free-catalysed aza-Michael addition reaction in aqueous media at room temperature. The reaction yield was estimated at about 80%. A series of partially modified PLL was prepared. The addition reaction gave mono- and bis-addition products with free remaining primary amines (Scheme 1).

The partially modified PLL-g-NIPAmX exhibited LCST-type thermo-responsiveness, which is highly tuneable by the content of the incorporated NIPAm moieties, i.e., the LCST decreased with increasing NIPAm content of the polymer. More importantly, thanks to the presence of the remaining ionizable free primary amines, the LCST of the partially modified PLL is highly sensitive to pH and ionic strength. For the given NIPAM content, the LCST decreased with pH and salt concentration. The aza-Michael reaction between amines and double bonds can be applied to any LCST-type vinylic monomer, e.g., oligo(ethylene glycol) methacrylate, and polypeptides bearing primary and/or secondary amines, e.g., poly(L-histidine), towards biodegradable LCST-thermo-responsive polymers. It should also be mentioned that similar thermo-responsive modified polypeptides have been reported previously using different reactions [27–31]. The advantage of the present work relies on the simplest free-catalysed aza-Michael addition reaction, forming stable covalent bonds that can be accomplished at room temperature without any by-products and can be applied to any peptide (and/or protein) comprising amino acid residues bearing pendant amines.

The protected PLL precursor, bearing N-termini, can be integrated to block-type segmented polymers, e.g., block and/or graft copolymers, prior to deprotection and modification. These copolymers are expected to form pH/thermo dual-responsive "smart" self-assemblies that could be used as nano carriers and/or 3D networks (scaffolds). This constitutes our ongoing research, and the perspectives towards bio-applications seem promising.

Supplementary Materials: The following are available online at https://www.mdpi.com/article/10.3390/polym14040802/s1: Synthesis of PLL, Characterization of PLL-g-NIPAm by ^{1}H-NMR, Figure S1: Example of ^{1}H-NMR spectrum of PLL-g-NIPAmX.

Author Contributions: A.S., investigation, writing original draft; H.I., investigation, C.T., supervision, methodology, writing original draft, review and editing. All authors have read and agreed to the published version of the manuscript.

Funding: This research received no external funding.

Institutional Review Board Statement: Not applicable.

Informed Consent Statement: Not applicable.

Data Availability Statement: The data presented in this study are available on request from the corresponding author.

Acknowledgments: The authors acknowledge Maria-Helena Karga and Marilena Kargaki for the assistance in the confirmation experiments.

Conflicts of Interest: The authors declare no conflict of interest.

References

1. Stuart, M.A.C.; Huck, W.T.S.; Genzer, J.; Müller, M.; Ober, C.; Stamm, M.; Sukhorukov, G.B.; Szleifer, I.; Tsukruk, V.V.; Urban, M.; et al. Emerging applications of stimuli-responsive polymer materials. *Nat. Mater.* **2010**, *9*, 101–113. [CrossRef]
2. Wei, M.; Gao, Y.; Li, X.; Serpe, M.J. Stimuli-responsive polymers and their applications. *Polym. Chem.* **2017**, *8*, 127–143. [CrossRef]
3. Schild, H.G.; Tirrell, D.A. Microcalorimetric detection of lower critical solution temperatures in aqueous polymer solutions. *J. Phys. Chem.* **1990**, *94*, 4352–4356. [CrossRef]
4. Pasparakis, G.; Tsitsilianis, C. LCST polymers: Thermoresponsive nanostructured assemblies towards bioapplications. *Polymer* **2020**, *211*, 123146. [CrossRef]
5. Klouda, L. Thermoresponsive hydrogels in biomedical applications A seven-year update. *Eur. J. Pharm. Biopharm.* **2015**, *97*, 338–349. [CrossRef] [PubMed]
6. Dharmasiri, M.B.; Mudiyanselage, T.K. Thermo-responsive poly(N-isopropyl acrylamide) hydrogel with increased response rate. *Polym. Bull.* **2020**, *78*, 3183–3198. [CrossRef]
7. Halperin, A.; Kröger, M.; Winnik, F.M. Poly(N-isopropylacrylamide) phase diagrams: Fifty years of research. *Angew. Chem. Int. Ed.* **2015**, *54*, 15342–15367. [CrossRef]
8. Lanzalaco, S.; Armelin, E. Poly(N-isopropylacrylamide) and copolymers: A review on recent progresses in biomedical applications. *Gels* **2017**, *3*, 36. [CrossRef]
9. Inoue, M.; Hayashi, T.; Hikiri, S.; Ikeguchi, M.; Kinoshita, M. Mechanism of globule-to-coil transition of poly(N-isopropylacrylamide) in water: Relevance to cold denaturation of a protein. *J. Mol. Liq.* **2019**, *292*, 111374. [CrossRef]
10. Wu, C.; Wang, X. Globule-to-coil transition of a single homopolymer chain in solution. *Phys. Rev. Lett.* **1998**, *80*, 4092–4094. [CrossRef]
11. Hogan, K.J.; Mikos, A.G. Biodegradable thermoresponsive polymers: Applications in drug delivery and tissue engineering. *Polymer* **2020**, *211*, 123063. [CrossRef]
12. Wang, Y.; Chang, Y.C. Synthesis and conformational transition of surface-tethered polypeptide: Poly(L-lysine). *Macromolecules* **2003**, *36*, 6511–6518. [CrossRef]
13. Zheng, M.; Pan, M.; Zhang, W.; Lin, H.; Wu, S.; Lu, C.; Tang, S.; Liu, D.; Cai, J. Poly(α-L-lysine)-based nanomaterials for versatile biomedical applications: Current advances and perspectives. *Bioact. Mater.* **2021**, *6*, 1878–1909. [CrossRef] [PubMed]
14. Campbell, J.; Vikulina, A.S. Layer-by-layer assemblies of biopolymers: Build-up, mechanical stability and molecular dynamics. *Polymers* **2020**, *12*, 1949. [CrossRef]
15. Grigsby, J.J.; Blanch, H.W.; Prausnitz, J.M. Effect of secondary structure on the potential of mean force for poly-L-lysine in the α-helix and β-sheet conformations. *Biophys. Chem.* **2002**, *99*, 107–116. [CrossRef]
16. Naassaoui, I.; Aschi, A. Evaluation of properties and structural transitions of poly-L-lysine: Effects of pH and temperature. *J. Macromol. Sci. Part B Phys.* **2019**, *58*, 673–688. [CrossRef]
17. Genest, A.; Portinha, D.; Fleury, E.; Ganachaud, F. The aza-Michael reaction as an alternative strategy to generate advanced silicon-based (macro)molecules and materials. *Prog. Polym. Sci.* **2017**, *72*, 61–110. [CrossRef]
18. Narayanan, A.; Maiti, B.; De, P. Exploring the post-polymerization modification of side-chain amino acid containing polymers via Michael addition reactions. *React. Funct. Polym.* **2015**, *91–92*, 35–42. [CrossRef]
19. Bosica, G.; Abdilla, R. Aza-Michael mono-addition using acidic alumina under solventless conditions. *Molecules* **2016**, *21*, 815. [CrossRef] [PubMed]
20. Li, S.; Feng, S. High-sensitivity stimuli-responsive polysiloxane synthesized via catalyst-free aza-Michael addition for ibuprofen loading and controlled release. *RSC Adv.* **2016**, *6*, 99414–99421. [CrossRef]
21. Macovec, T. Poly-L-glutamic acid and poly-L-Lysine: Model substances for studying secondary structures of proteins. *Biochem. Mol. Biol. Educ.* **2000**, *28*, 244–247.
22. Matsumoto, K.; Kawamura, A.; Miyata, T. Structural transition of pH-responsive poly (L-lysine) hydrogel prepared via chemical crosslinking. *Chem. Lett.* **2015**, *44*, 1284–1286. [CrossRef]
23. Dzwolak, W.; Muraki, T.; Kato, M.; Taniguchi, T. Chain-length dependence of α-helix to β-sheet transition in polylysine: Model of protein aggregation studied by temperature-tuned FTIR spectroscopy. *Biopolymers* **2004**, *73*, 463–469. [CrossRef] [PubMed]
24. Popescu, M.-T.; Liontos, G.; Avgeropoulos, A.; Tsitsilianis, C. Stimuli responsive fibrous hydrogels from hierarchical self-assembly of a triblock copolypeptide. *Soft Matter* **2015**, *11*, 331–342. [CrossRef]
25. Iatrou, H.; Frielinghaus, H.; Hanski, S.; Ferderigos, N.; Ruokolainen, J.; Ikkala, O.; Richter, D.; Mays, J.; Hadjichristidis, N. Architecturally induced multiresponsive vesicles from well-defined polypeptides. Formation of gene vehicles. *Biomacromolecules* **2007**, *8*, 2173–2181. [CrossRef]
26. Zhang, Y.; Furyk, S.; Bergbreiter, D.E.; Cremer, P.S. Specific ion effects on the water solubility of macromolecules: PNIPAM and the Hofmeister series. *J. Am. Chem. Soc.* **2005**, *127*, 14505–14510. [CrossRef]
27. Shen, Y.; Fu, X.; Fu, W.; Li, Z. Biodegradable stimuli-responsive polypeptide materials prepared by ring opening polymerization. *Chem. Soc. Rev.* **2015**, *44*, 612–622. [CrossRef]
28. Chen, C.; Wang, Z.; Li, Z. Thermoresponsive polypeptides from pegylated poly-L-glutamates. *Biomacromolecules* **2011**, *12*, 2859–2863. [CrossRef]
29. Fu, X.; Shen, Y.; Fu, W.; Li, Z. Thermoresponsive oligo(ethylene glycol) functionalized poly-L-cysteine. *Macromolecules* **2013**, *46*, 3753–3760. [CrossRef]

30. Xiao, Y.; Tang, C.; Chen, Y.; Lang, M. Dual stimuli-responsive polypeptide prepared by thiol-ene click reaction of poly(L-cysteine) and N, N-dimethylaminoethyl acrylate. *Biopolymers* **2019**, *110*, e23318. [CrossRef]
31. Chopko, C.M.; Lowden, E.L.; Engler, A.C.; Griffith, L.G.; Hammond, P.T. Dual responsiveness of a tunable thermosensitive polypeptide. *ACS Macro Lett.* **2012**, *1*, 727–731. [CrossRef] [PubMed]

polymers

MDPI

rticle

Controlling Growth of Poly (Triethylene Glycol Acrylate-*Co*-Spiropyran Acrylate) Copolymer Liquid Films on a Hydrophilic Surface by Light and Temperature

ziz Ben-Miled [1], Afshin Nabiyan [2], Katrin Wondraczek [3], Felix H. Schacher [2,4] and Lothar Wondraczek [1,*]

1 Otto Schott Institute of Materials Research (OSIM), Friedrich Schiller University Jena, D-07743 Jena, Germany; aziz.ben.miled@uni-jena.de
2 Institute of Organic Chemistry and Macromolecular Chemistry (IOMC), Friedrich Schiller University Jena, D-07743 Jena, Germany; afshin.nabiyan@uni-jena.de (A.N.); felix.schacher@uni-jena.de (F.H.S.)
3 Leibniz Institute of Photonic Technology (Leibniz IPHT), D-07745 Jena, Germany; katrin.wondraczek@leibniz-ipht.de
4 Jena Center for Soft Matter (JCSM), Friedrich Schiller University Jena, D-07743 Jena, Germany
* Correspondence: lothar.wondraczek@uni-jena.de; Tel.: +49-3641-9-48500

Abstract: A quartz crystal microbalance with dissipation monitoring (QCM-D) was employed for in situ investigations of the effect of temperature and light on the conformational changes of a poly (triethylene glycol acrylate-*co*-spiropyran acrylate) (P (TEGA-*co*-SPA)) copolymer containing 12–14% of spiropyran at the silica–water interface. By monitoring shifts in resonance frequency and in acoustic dissipation as a function of temperature and illumination conditions, we investigated the evolution of viscoelastic properties of the P (TEGA-*co*-SPA)-rich wetting layer growing on the sensor, from which we deduced the characteristic coil-to-globule transition temperature, corresponding to the lower critical solution temperature (LCST) of the PTEGA part. We show that the coil-to-globule transition of the adsorbed copolymer being exposed to visible or UV light shifts to lower LCST as compared to the bulk solution: the transition temperature determined acoustically on the surface is 4 to 8 K lower than the cloud point temperature reported by UV/VIS spectroscopy in aqueous solution. We attribute our findings to non-equilibrium effects caused by confinement of the copolymer chains on the surface. Thermal stimuli and light can be used to manipulate the film formation process and the film's conformational state, which affects its subsequent response behavior.

Keywords: dual-stimuli-responsive materials; thin films; out-of-equilibrium

itation: Ben-Miled, A.; Nabiyan, A.; ondraczek, K.; Schacher, F.H.; ondraczek, L. Controlling Growth Poly (Triethylene Glycol crylate-*Co*-Spiropyran Acrylate) opolymer Liquid Films on a ydrophilic Surface by Light and mperature. *Polymers* **2021**, *13*, 1633. tps://doi.org/10.3390/ olym13101633

cademic Editor: Mattia Sponchioni

eceived: 14 April 2021
ccepted: 12 May 2021
ublished: 18 May 2021

ublisher's Note: MDPI stays neutral ith regard to jurisdictional claims in ublished maps and institutional affiltions.

1. Introduction

Controlling and understanding polymer adsorption at solid–liquid interfaces is of key importance in, e.g., coating [1], lubrication [2], surface adhesion [3], or colloid stabilization [4]. Polymer adsorption on a surface may occur in two general ways: by chemisorption or by physisorption. Chemisorption happens when polymers attach to a solid surface through a covalent bond. This type of adsorption is irreversible, and it is employed in many applications, such as repellant surface layers [5] or other types of functional coatings [6]. Alternatively, physisorption takes place as a result of physical attractive forces between polymer segments and the surface [7]. Physisorbed chains may consist of loops, tails and trains [8]. In general, physisorption of polymers from a bulk liquid on a solid surface can be either irreversible or reversible [9]. Irreversibility is usually achieved using hydrogen bonding or other dipolar forces, dispersive forces, or attractions between charged groups along the polymer backbone and the surface [10]. It typically occurs on metals, semiconductors, inorganic glasses, or sol-gel layers such as polydimethylsiloxane (PDMS), for example, when surface oxygens of the substrate form strong hydrogen bonds with the polymer [11]. Similarly, various macromolecules (polymers, proteins, DNA, etc.) are prone to adsorb strongly on oxide glass surfaces through hydrogen bonds or other physical forces (electrostatic attractions, hydrophobic interactions in the solvent) [1,12,13]. On the other hand, physisorption from a solution is reversible when the polymer binds weakly to the surface and has only few conformational restrictions.

In order to tailor surface adhesion, stimuli responsive polymers have attracted great attention in the last decades due to their ability to respond to external triggers, including temperature [14], light [15], pH [16], ionic strength [17], or combinations of thereof [18,19]. Layers formed from such polymers are expected to enable switchable surfaces which may change their properties in controllable and programmable ways [20]. Sometimes such switching can be through combinations of multiple stimuli (e.g., light and temperature, [21]), what enables logic gate operations. Thermo-responsive polymers are the most studied stimulus responsive materials. In aqueous solution, they usually display a fully reversible hydrophilic–hydrophobic transition (Figure 1), characterized by a lower critical solution temperature (LCST) [22]. Below the LCST, the polymer swells with a random coil conformation, while above LCST, the polymer collapses into a globular state and undergoes a liquid–liquid phase separation. This transformation from coil to globule is based on hydrogen bonds that are present between the polymer chains and the surrounding water molecules at temperatures below the LCST [23]. At higher temperatures the hydrogen bonds become weaker, leading to the dehydration of the polymer chains. Some prominent examples for this behavior are microgels [24,25], poly (*N*-isopropyl acrylamide, PNIPAM [23,26], acrylamides [27,28], poly (2-oxazolines) [29,30], poly (propylene glycol) [31,32], and poly (oligo (ethylene glycol) acrylates [33–35].

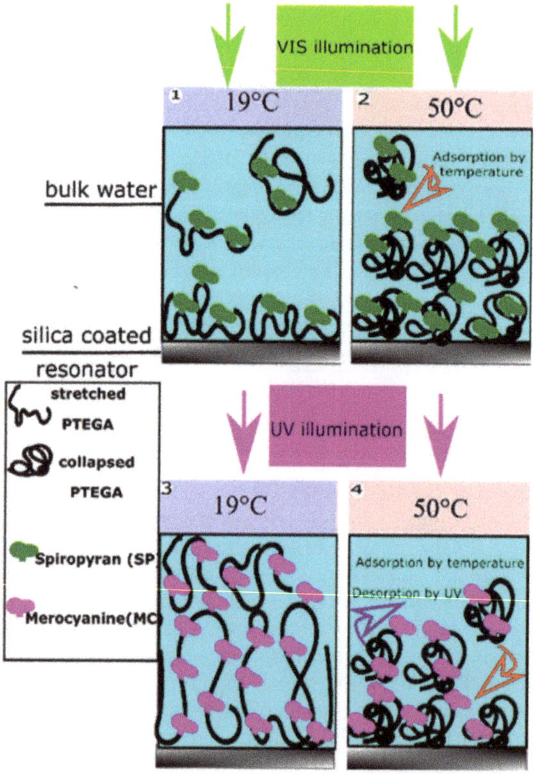

Figure 1. Coil to globule transition at different temperatures shown by way of example for the dual light and temperature responsive P (TEGA-co-SP/MC) copolymer during in situ observation of adsorption and film formation on a silica surface by QCM-D.

Incorporation of photochromic moieties into thermosensitive polymer backbones is a practical way to control their solubility in aqueous solutions by changing the temperature at which the phase transition happens through an optical stimulus [36]. Organic photochromic compounds that can be used for this purpose include spiropyran (SP) [37], azobenzene [38], and diarylethene [39]. These compounds are responsive to light irradiation through reversible or irreversible isomerization between two states of variable polarity. Isomerization reactions can be detected through observation of color changes due to photon absorption. [40] In case of SP-MC, the deep purple color of a liquid thin film upon UV irradiation originates from the absorption of UV photons causing a breakage of the C-O spiro bond in an excited singlet state, see example shown in Figure 2.

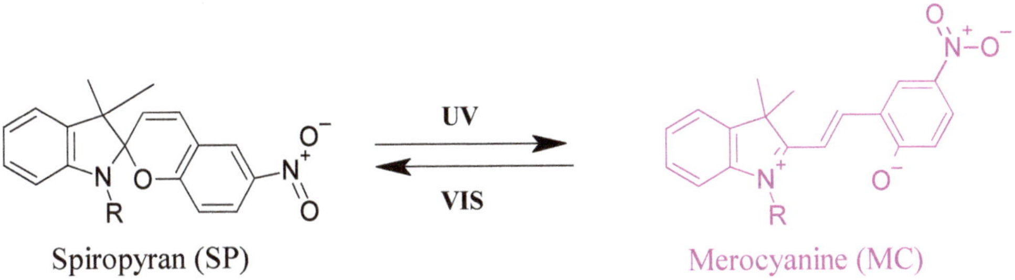

Figure 2. Photo switching between the spiropyran (SP) (left) and merocyanine (MC) form (right).

SP is one of the chromophores that is not only a photo-switch but also responds to other stimuli such as temperature, solvent, metal ions, and pH [41]. In response to UV light (λ = 365 nm), the closed nonpolar and colorless spiro form "SP" is transformed into the open, polar, colored and zwitterionic merocyanine form "MC". Irradiation with visible light (λ = 550 nm) causes ring closure and return to the initial state. The UV-light induced reversible isomerization of SP between nonpolar and polar states can be used to tune the phase separation of thermo-responsive polymers since the (UV-induced) polarity change affects the interaction between the polymer and the solvent. By combination with different types of backbone polymers this enabled, e.g., controlled foaming or bubble formation using light irradiation of spiropyran sulfonate surfactants [42], rewritable optical storage in spiropyran-doped liquid crystal polymer film [43], or controlling the enzymatic activity on orthogonally functionalized glycidyl methacrylate with spiropyran [44]. SP-incorporating poly (oligo (ethylene glycol) acrylate)-based copolymers have been synthesized by nitroxide mediated polymerization with varying amounts of SP (from 0 to 16 mol%). The visible light irradiation of the copolymer dissolved in pH 8 TRIS buffer resulted in a decrease in its cloud point temperature by 30 K at 16 mol% SP content, as previously detected by UV/Vis spectroscopy [45].

A quartz crystal microbalance with dissipation monitoring (QCM-D) is a highly sensitive technique for characterizing adsorption and desorption phenomena at the solid–liquid interface. Numerous experimental investigations and modeling studies have been carried out on the viscoelastic properties of adsorbed polymer films and their solid–liquid interfacial properties using QCM-D [46–48]. As an exemplary case, the adsorption of PNIPAAm on modified gold and silica surfaces was studied due to its conveniently accessible LCST of ~32 °C, and also for its potential relevance in biomedical applications [49]. These studies showed different behaviors of the adsorbed polymer depending on its state of adsorption, e.g., whether chemisorbed [50,51] or physisorbed [52,53]. The adsorption of thermosensitive block copolymers based on PNIPAAm on a gold surface was also investigated by QCM-D [54]. For example, the adsorption mechanism of a pentablock terpolymer poly (*N*-isopropylacrylamide)$_x$-*block*-poly (ethylene oxide)$_{20}$-*block*-poly (propylene oxide)$_{70}$-*block*-poly (ethylene oxide)$_{20}$-*block*-poly (*N*-isopropylacrylamide)$_x$ (PNIPAAm$_x$-*b*-PEO$_{20}$-*b*-PPO$_{70}$-*b*-PEO$_{20}$-*b*-PNIPAAm$_x$) on gold was found to be affected by several parameters including concentration, relative block length, temperature, and the substrate's physical properties. Furthermore, adsorption properties of pH sensitive cationic polyelectrolytes, e.g., poly (diallyl dimethyl ammonium chloride) or poly (allyl amine hydro-chloride) (PAH) on gold and silica surfaces were studied using QCM-D [55]. It was found that the adsorption property of the polyelectrolyte depends on the solid surface, solution concentration, and solution pH. As another example, QCM-D was employed to study the adsorption of polyelectrolyte monolayers of anionic poly (styrene sulfonate) (PSS) on amino-functionalized silica, as well as cationic PAH and poly-L-lysine (PLL) on bare silica [12]. In this example, the thickness of the polyelectrolyte monolayers increased when increasing the ionic strength (salt concentration) and the polyelectrolyte concentration.

Interestingly, also the light-induced swelling behavior of spin-coated thin layers of P (NIPAM-*co*-SPA) copolymers was described on the basis of QCM-D investigations [56]. However, although the employed deposition method is technologically important for the fabrication of thin films on solid surfaces, it also has the limitation of making the film prone to delamination once the solvent wets the substrate [57]. Nevertheless, studying adsorption of such copolymers appears very interesting from a physical point of view; they can adopt different conformations, which can be tuned by light irradiation and temperature.

In this paper, we report on the conformational change of the dual light and temperature responsive copolymer P (TEGA-*co*-SPA) in solution and confined at the silica-water interface using QCM-D measurements. We monitor the simultaneous effect of UV light irradiation and temperature changes on the co-polymer's adsorption behavior. Optical irradiation of the copolymer solution while undergoing adsorption provided us with direct access to the question as to how light can be used to tailor the kinetics of film formation and film conformation below and above the LCST.

2. Materials and Methods

2.1. Materials

P (TEGA-*co*-SPA) synthesis was reproduced from reference [45]. More details are provided in the Supplementary Section. Using this method, spiropyran acrylate (SPA) was obtained as a yellow powder. Commercial TEGA monomer was copolymerized with 15 mol% SPA in the initial monomer mixture. The obtained copolymer was investigated via size exclusion chromatography with triple detection to obtain absolute molar masses and ^{1}H liquid NMR to determine the composition by comparing the signal of the SPA moiety (8.2 ppm, 2H) and the TEGA moiety (3.3 ppm, 3H). The fraction of SPA in the obtained copolymer was between 12–14 mol%, the molar mass M_n was about 33,000 g/mol with a dispersity index PDI = 1.7. An aqueous solution of 0.15 wt.% P (TEGA-*co*-SPA) was obtained by diluting the copolymer in deionized water. Deionization was done using a Thermo Scientific Barnstead MicroPure water purification system to a resistivity of 18.2 MΩ cm^{-1}.

2.2. Dynamic Light Scattering

DLS measurements were performed using an ALV Laser CGS3 Goniometer (ALV GmbH, Langen, Germany) equipped with an He-Ne laser (λ = 633 nm) and an ALV-7004/USB FAST correlator. All DLS measurements were performed at 25 to 77 °C. To determine the hydrodynamic radius, three measurements of 30 s each were performed at an angle of 90 ° The analysis of the obtained correlation functions was performed using the correlator software (Correlator 3.2 beta 1).

2.3. QCM-D Experiment

QCM-D measurements were performed using a window module mounted on the QCM sensor (Q-sense E1 Biolin Scientifc, Västra Frölunda, Sweden). The employed sapphire window had an optical transmittance of >80% in the wavelength range 300 to 400 nm, in which UV irradiation was conducted.

AT-cut quartz crystal sensors coated with a 50 nm silicon dioxide layer (fundamental resonance frequency of typically ~4.95 MHz, sensor area 1.54 cm^2) were purchased from

Biolin Scientific, Sweden. Prior to experiments, the quartz sensor was cleaned by soaking in a 2 vol% sodium dodecyl sulfate SDS solution for 30 min, rinsing with ultra-pure water, blow-drying with a gentle nitrogen flow and, finally, exposing to a UV/ozone cleaner for 15 min.

Several overtones were acquired, although the third overtone was generally selected for further analysis because of its level of energy trapping at this particular overtone when operated in liquids [58].

For studying the dual light and temperature induced conformational response of the P (TEGA-*co*-SPA) solutions, all experiments were performed in the liquid exchange mode by first purging with ultra-pure water for 30 min at 19 °C at a flow rate of 50 µL/min. To avoid the formation of bubbles that can oscillate or migrate over the quartz crystal surface, all solutions were degassed in an ultrasonic bath (Elmasonic S 80) for 10 min prior to injection. If not otherwise stated, irradiation of the sensor with the light source was started 20 min after equilibration and referencing under continuous water flow was completed. The diluted P (TEGA-*co*-SPA) aqueous solution was then introduced into the chamber at 30 min and at a temperature of 20 °C ± 0.02 °C. At this point, the flow rate was reduced to 20 µL/min. Temperature ramping was conducted from the starting temperature of 20 °C up to a maximum of 47 °C, applying a constant heating rate of 0.2 K/min.

In the isothermal irradiation study, the P (TEGA-*co*-SPA) aqueous solution was fed for 25 min through the window module at a constant temperature prior to irradiation.

Irradiation was done with a fluorescent lamp (visible light) or using an ultraviolet spotlight (365 nm, Opsytech, Ettlingen, Germany). The power of UV LED was fixed at 10% via an LED controller (with a maximum nominal power density of 25 W/cm^2); the sample-to-LED distance was maintained at 75 mm.

During each run, changes in the resonance curves of the third overtone were continuously monitored and evaluated. The two resonance parameters under investigation were the change in dissipation factor ΔD_3, and the shift in resonance frequency $\Delta f_3/3$ being related to the mass of the adsorbate and the dynamically coupled liquid. While the resonance frequency shift $\Delta f_3/3$ is more sensitive to the mass of the film, the variation of the dissipation factor ΔD_3 is related to viscous losses and interfacial sliding [47]. The acquired datasets were corrected for each sensor using a temperature sweep in pure water for reference, see also Supplementary Material (Figure S2 and Table S1). This temperature correction was carried out by subtracting the calibration curve (pure water on sensor) from the one obtained in the presence of the dissolved copolymer. Furthermore, irradiation of the quartz crystal with UV light induced an increase in $\Delta f_3/3$ by a few Hz. This behavior was previously attributed to photo-induced mechanical stress [59,60]. A further calibration was, therefore, done for UV-illumination by subtracting the effect of the UV light on the crystal for the non-isothermal measurements, see calibration curve in Supplementary Material (Figure S3).

2.4. Data Evaluation

QCM-D is an established, sensitive tool to study in situ the adsorption from a liquid in contact with the surface of a quartz crystal resonator [61–63]. The resonance frequency is defined as the frequency where the electrical conductance of the equivalent circuit is maximal. If a Lorentzian peak function is fitted to the conductance curve, two parameters are obtained describing the complex resonance frequency f_n^*, the resonance frequency f_n of the quartz as the real part and the half width at half maximum of the resonance peak, Γ_n representing the imaginary part. A thin layer or any loading on the quartz crystal surface generates a complex resonance frequency shift Δf_n^* compared to the empty state, which can again be separated into Δf_n (the real part) and $\Delta\Gamma_n$ (the imaginary part),

$$\Delta f_n^* = \Delta f_n + i\Delta\Gamma_n \quad (1)$$

The fundamental resonance frequency of AT cut quartz crystal resonators operated in shear mode is typically near 5 MHz. More resonances are observed at the odd harmonics of this fundamental frequency, where the subscript n refers to the nth harmonic (i.e., $n = 1$ for the fundamental resonance frequency of 5 MHz, and $n = 3$ for the third overtone partial at ~15 MHz). The adsorbed rigid mass can be quantified using the Sauerbrey equation [64] where the adsorbed areal mass density m_f correlates with Δf_n^* [58].

$$\frac{\Delta f_n^*}{f_1} = \frac{-2f}{Z_q} m_f \tag{2}$$

where f_1 is the fundamental frequency, f is the measured resonance frequency and $Z_q = 8.8 \times 10^6$ kg·m^{-2}·s^{-1} is the acoustic impedance of quartz. The Sauerbrey equation is strictly valid only for rigid films. For a viscoelastic film immersed in liquid environment a viscoelastic correction is required to account for viscous dissipation, whereby softness reduces the apparent rigid Sauerbrey thickness [58],

$$\frac{\Delta f_n^*}{f_1} = -\frac{\omega\, m_f}{\pi Z_q}\left(1 - \frac{Z_{liq}^2}{Z_{film}^2}\right) \tag{3}$$

where $\omega = 2\pi f$, $Z_{liq} = \sqrt{n2\pi i f_1 \rho_{liq}\eta_{liq}}$, $Z_{film} = \sqrt{(\rho_{film} G_{film})}$; Z_{liq} is the acoustic field impedance of the liquid, Z_{film} the acoustic field impedance of the film, ρ_{liq} is the density of the liquid, η_{liq} the dynamic viscosity of the liquid, ρ_{film} the density of the film and G_{film} the shear modulus of the film.

Aside mass or Sauerbrey thickness, QCM-D simultaneously monitors dissipation which can be expressed by the factor D,

$$D_n = \frac{2\,\Gamma_n}{f_n} \tag{4}$$

Viscoelasticity, but also further effects such as surface roughness cause a shift ΔD_n of the dissipation factor [65–67]. When the crystal is immersed in a Newtonian liquid [68] the resonance frequency and dissipation factor shifts are proportional to the square root of liquid density ρ_{liq} times the liquid dynamic viscosity η_{liq} according to Kanazawa–Gordon–Mason relation [69],

$$\frac{\Delta f_n}{f_1} = \frac{-1}{\pi Z_q}\sqrt{\omega\,\rho_{liq}\,\eta_{liq}} \tag{5}$$

$$\Delta D_n = \frac{2}{n\pi Z_q}\sqrt{\omega\,\rho_{liq}\,\eta_{liq}} \tag{6}$$

3. Results and Discussions

3.1. Phase Separation of P (TEGA-Co-SPA) in Dilute Aqueous Solution

DLS data shown in Figure 3 provide an initial view at the effect of temperature on aggregation in the P (TEGA-co-SPA) polymer solutions containing between 12 and 14 mol% of spiropyran in terms of the hydrodynamic radius. In order to reduce the effect of particle aggregation, we chose to work with a dilute concentration of 0.06 wt.% (optically clear at room temperature). This is below the concentration used for DLS studies of similar thermoresponsive copolymers [70]. The hydrodynamic radius observed by DLS shows a sudden transition at a temperature of ~66 °C. Below this temperature, the polymer chains exist as individually dissolved polymer chains with small hydrodynamic radius of approximately 4–6 nm. Above 66 °C, aggregates (mesoglobules) with larger hydrodynamic radius of around 100–200 nm are formed. These values are comparable in size to other known polymers with a LCST [71,72]. At temperatures below the LCST, the copolymer chains are well solvated through hydrogen bonds [73,74]. Above the LCST, these exhibit van der Waals character, e.g., such as reported for PNIPAAm [70,75]. Interestingly the observed transition temperature occurs ~23 K above the reported cloud point for the same copolymer composition diluted in pH 8 TRIS buffer, as detected by UV/VIS spectroscopy [45]. This observation is attributed to the effect of salts contained in the buffer on the electrostatic interactions between the copolymer and water as reported recently for various thermoresponsive polymers [76].

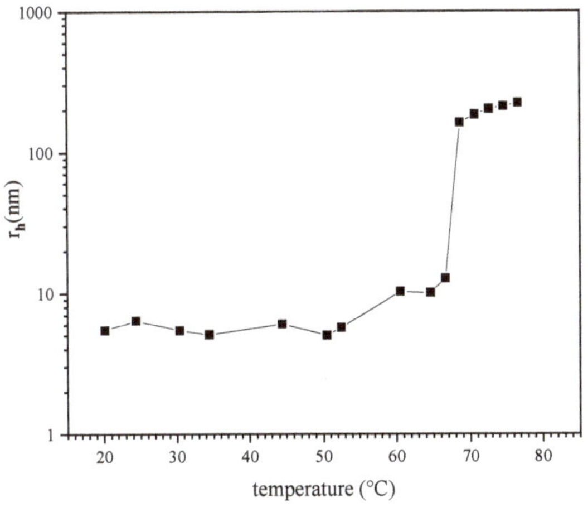

Figure 3. Hydrodynamic radius of a P (TEGA-co-SPA) copolymer in aqueous solution upon heating as determined from DLS measurements.

3.2. *Effect of UV-Irradiation on the Hydration of P (TEGA-Co-SPA) Films below and above the LCST*

The P (TEGA-co-SPA) liquid thin film adsorbed onto silica appears almost transparent under visible light, but switches to deep purple upon UV irradiation (Figure 4). As explained before, the deep purple color of the liquid thin film upon UV irradiation originates from the absorption of the UV photons causing a breakage of C-O spiro bonds in an excited singlet state yielding the colored MC form. Due to the physisorption of the copolymer in our case the chains of MC are enforced to rearrange in a way the ethylene oxide groups point to the solution that may stabilize the merocyanine form via hydrogen bonds.

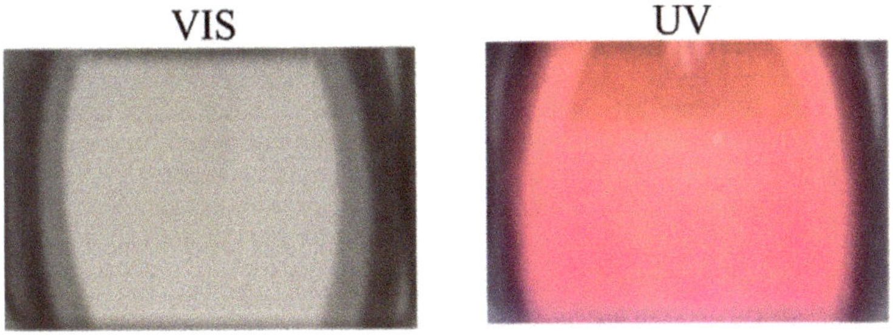

Figure 4. Effect of illumination on the P (TEGA-co-SPA) liquid film color. The photos were taken by a normal camera on the top of the QCM-D window cell.

By way of example, we selected different temperatures for isothermal treatment with and without illumination below and above the LCST when investigating with QCM-D. Figure 5a shows the effect of switching from visible to UV light irradiation on $\Delta f_3/3$ as a function of time at 19 °C, 35 °C, 45 °C, 50 °C; and 50 °C when the sensor was not irradiated with UV light, respectively. At 19 °C and 35 °C, the introduction of the copolymeric solution inside the window cell causes an initial frequency decrease (mass increase) followed by a slower frequency decrease as the system saturates at −31 Hz and −40 Hz, respectively.

Starting at 25 min, the sensor surface was irradiated with UV light, what caused a marginal increase in $\Delta f_3/3$ of a few Hz, followed by a linear decrease in the frequency in the next several minutes, see inset of Figure 5a. In comparison, when there is no light switch at 19 °C, $\Delta f_3/3$ and ΔD_3 signals do not show any significative change, see supporting information (Figure S5a,c). The spike of $\Delta f_3/3$ occurring immediately after illumination attributed to the effect of UV light on the crystal as described in the Materials and Methods section. The shallow linear decrease in the frequency shift is probably due to an increase in acoustic thickness as the copolymer chains swell. A similar result was observed in a previous study [56], where PNIPAAm-co-SPA thin films were illuminated with a UV lamp at 19 °C. In this material, the behavior was explained by a photoinduced hydration due to the photoisomerization of the rather hydrophobic spiropyran into the distinctly more hydrophilic merocyanine when the thermo-responsive part of the copolymer is sufficiently hydrophilic.

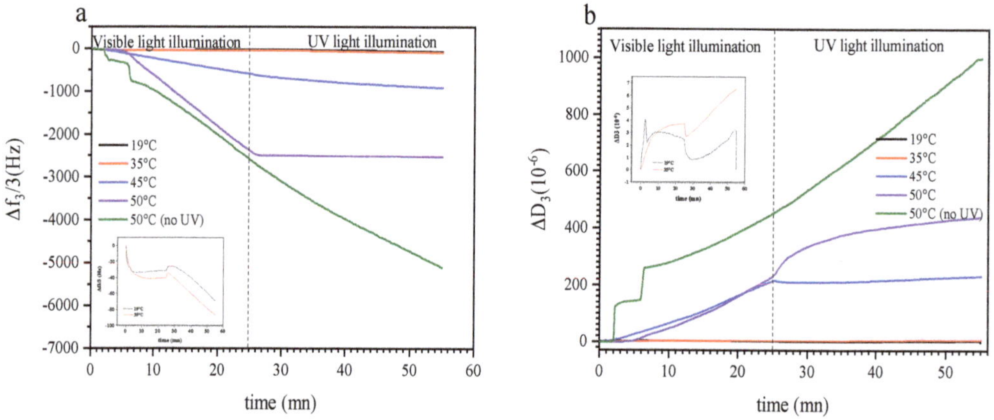

Figure 5. Variation of $\Delta f_3/3$ (**a**) and ΔD_3 (**b**) versus time of PTEGA-co-SPA at the interface silica-water at a constant temperature. The inset is a zoom at $\Delta f_3/3$ in the range of 19 °C to 35 °C.

At 45 °C and 50 °C, $\Delta f_3/3$ decreases linearly once the copolymer solution is in contact with the sensor. This decrease in $\Delta f_3/3$ is high in magnitude, reaching 0.58 and 2.38 kHz, respectively, after 25 min of continuous solution feed and visible light irradiation. Interestingly, UV light illumination affects $\Delta f_3/3$ differently at 45 °C and 50 °C. Although at 45 °C the rate of the observed decrease in $\Delta f_3/3$ slows down and causes a deviation from linearity, it stabilizes at a constant (but very low) value at 50 °C. Noteworthy, when continuing visible illumination and turning UV off beyond 25 min, the observed strong decrease in $\Delta f_3/3$ continues unaffected, indicating that indeed UV illumination (versus, e.g., some saturation effect) plays a role in the reaction observed at 50 °C (see also Figure S5b,d). We attribute this observation to a competition between PTEGA globule adsorption on the sensor surface and photoconversion of spiropyran to merocyanine. When there is no UV irradiation, surface adsorption is facilitated and the observed Sauerbrey thickness increases during prolonged solution injection. This process is interrupted by the conversion of the unipolar spiropyran to the polar merocyanine, which enhances the stability of the solution and thereby reduces the adsorption rate. Similar observation have been made for azobenzene surfactant adsorption and desorption at the air–water interface under UV irradiation [77].

Figure 5b shows the evolution of ΔD_3 corresponding to Figure 5a. At 19 °C and 35 °C, ΔD_3 shows low values in the first 25 min, suggesting that the film is forming a monolayer at the silica surface. Once the surface is irradiated with UV light, ΔD_3 increases linearly at both temperatures and reaches ~3 × 10^{-6} and 6 × 10^{-6}, respectively, at 19 °C and 35 °C after around 55 min. At the higher temperatures of 45 °C and 50 °C, ΔD_3 increases similarly (although at much higher rate) for as long as the sensor is irradiated with visible light. Once UV illumination is switched on at these temperatures, there is a very significant effect on dissipation: at 45 °C, ΔD_3 decreases slightly and subsequently reaches a plateau, while at 50 °C, ΔD_3 apparently evolves in a square root dependence on time, which could indicate some kind of diffusive process. Interestingly, the latter extends far beyond the time at which surface adsorption is interrupted (Figure 5a); we note that dissipation evolves as a convolution of swelling effects within the film, as well as adsorption from the solution, which are both affected by the two stimuli of temperature and light. When adsorption stops, conformational changes can still proceed within the film, but these would be significantly slower in their response rate due to the reduced film mobility as compared to the polymer in solution. The observed square root dependence on time corroborates this interpretation.

3.3. Dual Temperature and Light Effect on the Build-Up of P (TEGA-co-SPA) Layers on Silica Surfaces

Temperature ramping was carried out in order to investigate the concomitant effect of temperature and light on the conformational change of the P (TEGA-co-SPA) diluted solution during adsorption. We started by analyzing the behavior of a P (TEGA-co-SPA) thin film being formed on the QCM-D sensor surface.

Figure 6a shows the variation of the normalized resonance frequency shift $\Delta f_3/3$ over a temperature range of 20 °C to 47 °C, comparing the effects of visible light irradiation and UV irradiation (365 nm). Under UV exposure, we observe an initial, slow decrease in $\Delta f_3/3$ between 21 °C and 28 °C, which is less pronounced under visible light. This difference suggests that the sensed mass (load) increased with UV irradiation, which could be attributed to additional hydrodynamically coupled water inside the adsorbed film in this temperature range. Any masses as retrieved by QCM-D are non-specific, that is, both polymer and water (or solvent in general) bound in the adsorbed films are detected. For instance, in case of protein adsorption, an additional molecular weight increase of ~30% was reported, that was attributed to water bound to a protein molecule in solution [78]. In our present case, we believe that the photoisomerization of the spiropyran with UV irradiation results in a higher trapped amount of water inside the layer of P (TEGA-co-SPA) when it is sufficiently hydrophilic [56]. For visible light irradiation, we note a change in the slope of $\Delta f_3/3$ over T at ~28 °C; under UV irradiation, such a change is not observed until a much higher temperature of near ~47 °C. We attribute this change of the slope to a sudden increase in the amount of the adsorbed copolymer chains at the sensor surface. As we are approaching the LCST, one should expect that the copolymer is gradually collapsing and releasing water. This dehydration should express as increased $\Delta f_3/3$ values as reported, e.g., for PNIPAAm layers adsorbed on a hydrophobic gold surface [52].

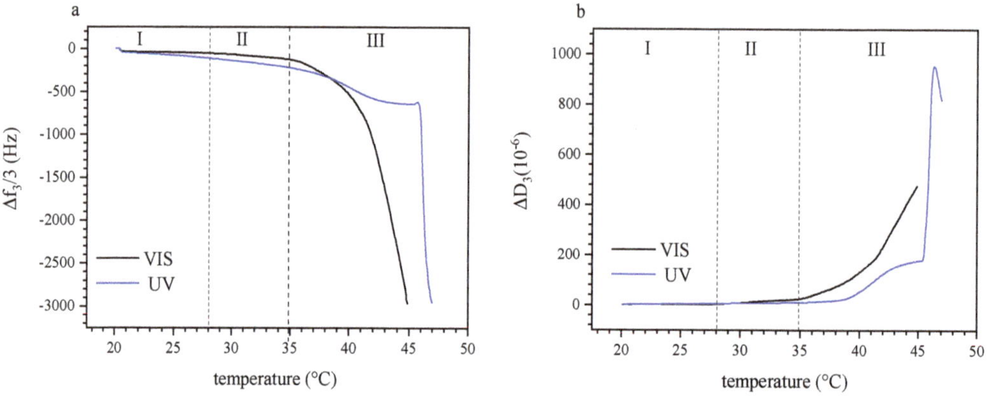

Figure 6. (**a**) Variation of normalized $\Delta f_3/3$ as a function of temperature upon irradiation of P (TEGA-co-SPA) copolymers at the silica–water interface, the copolymer was introduced at T = 20 °C, (**b**) variation of ΔD_3 as function of temperature of the same solution. Blue curves: Upon UV illumination, Black curves, upon illumination with visible light. The labels (I–III) mark the three regimes of adsorption and film response discussed in the text.

However, we must note again that we do not observe the properties of a preexisting film, but the process of a film being formed in situ from a photo-thermoresponsive solution. Thus, we argue that the observed decrease in $\Delta f_3/3$ (despite water release) is a result of polymer adsorption and film growth, which dominates over any water release reaction, in particular, as the hydrophilic coil to hydrophobic globule transition occurs already in solution, and only to a smaller extent within the film. At 35 °C, $\Delta f_3/3$ of the visible light irradiated sensor decreases drastically which we relate to the liquid–liquid phase separation. Interestingly, this large decrease in $\Delta f_3/3$ occurs at about 4–8 K lower than the reported cloud point temperature of the same copolymer in TRIS buffer solution when irradiated with 540 nm visible light [45]. This difference between the cloud point temperature detected by UV/Vis spectroscopy and the phase transition temperature determined acoustically on a surface suggests that the confinement affects the coil-to-globule transition of the copolymer at the interface. At 38.3 °C, we note a change in the the feature of $\Delta f_3/3$ over T of the UV irradiated solution, with an initial acceleration (higher negative slope) of adsorption, followed be a deceleration and a plateau reaching up to ~45 °C. This is in line with our isothermal observations summarized in Figure 5a, where UV irradiation at higher temperature decelerates film adsorption up to a certain extent. Here, the deceleration sets in just before LCST as would be occurring under visible illumination. At 45.7 °C, we observe a sudden, strong acceleration of the adsorption rate, with a sharp decrease in $\Delta f_3/3$. This is attributed to the retarded P (TEGA-co-SPA) LCST in the aqueous solution, shifted to a higher temperature due to the increase in the hydrophilicity of the polymer as induced by the 365 nm UV light irradiation. A similar temperature shift was also observed for the bulk material using UV-Vis spectroscopy, although the transition temperature happening at the interface silica-water was lowered by 2–3 K [45].

The dissipation data corresponding to the observed cases of $\Delta f_3/3$ is displayed in Figure 6b. Here, too, we distinguish the three regions of (I) T < 28 °C, (II) 28 °C < T < 35 °C and (III) T > 35 °C (marked as I–III in Figure 6). Again, an increase in dissipation correlates to the enhancement of coupling between water molecules and polymer chains due to the photo-induced hydration under UV illumination (region I). Moreover, there is a significant difference of the sensed masses on the sensor, depending on the type of irradiation. In the temperature range of 21 °C to 28 °C, ΔD_3 increases by a factor of about two, that is, from 1.6×10^{-6} to 3.3×10^{-6} and from 2.2×10^{-6} to 4.9×10^{-6} for the visible and UV irradiated film, respectively. In the temperature range of 28 °C to 35 °C, ΔD_3 increases more strongly for the solution exposed to visible light as compared to the one irradiated with

UV light. This observation corroborates our interpretation that the competition between dehydration and adsorption starts already several K below the commonly reported LCST. At 35 °C, ΔD_3 increases dramatically, in agreement with the resonance frequency data. According to the change of slope at ~38 °C, the dehydration happens gradually also under UV light. In this case, the retarded phase transition reflects in the over damping of the layer happening at ~46 °C, where the magnitude of ΔD_3 reaches 953×10^{-6}.

Examining the change of the energy dissipation as function of the negative frequency shift allows to eliminate the temperature as a variable and to focus on the effect of the light irradiation on the viscoelastic properties during layer build-up [79]. Figure 7a,b show the evolution of ΔD_3 as function of $-\Delta f_3/3$, respectively, for visible and UV irradiated P (TEGA-co-SPA) at the silica–water interface. Interestingly, both properties are not directly proportional; furthermore, the adsorbed film does not evolve in the same way whether it is irradiated with visible or UV light. Under Vis illumination, the change in dissipation underrepresents the change in resonance frequency whereas under UV light, it strongly exceeds the frequency change. A linear correlation between both properties is found only in the onset region of film formation, i.e., within 0 to 550 Hz (Vis) and 0–320 Hz (UV), where surface coverage of the layer is still low. In this range, the hydrodynamic thickness is expected to be small, and the number of polymer molecules adsorbed physically through trains, loops, and tails is negligible [80]. For as long as the dissipation values are low and ΔD_3 increases linearly with $-\Delta f_3/3$, we assume that the viscoelastic properties of the film remain unchanged and the parameter variations are solely due to continuous adsorption. The occurrence of such a region was similarly observed by QCM-D for different adsorbing systems, including polyelectrolytes [81] and homopolymers on gold [52].

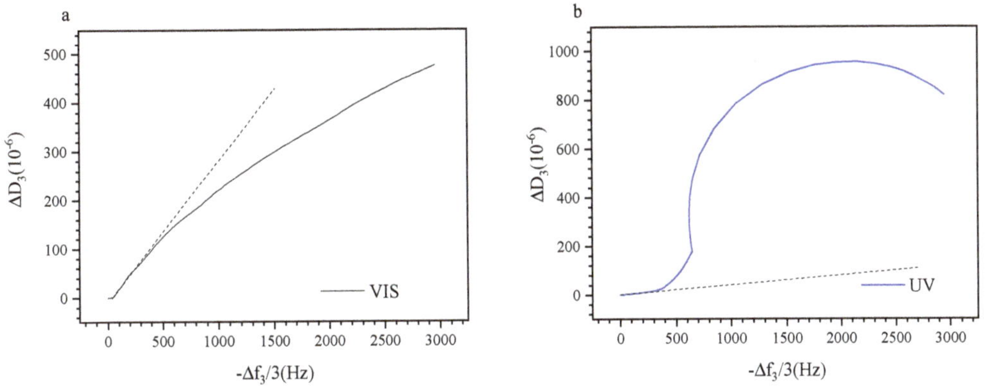

Figure 7. (**a**) Variation of ΔD_3 as function of $-\Delta f_3/3$ when the sensor is continuously exposed to visible light, (**b**) variation of ΔD_3 as function of $-\Delta f_3/3$ when the sensor is irradiated with UV light. Blue curves: upon UV illumination, black curves: upon visible illumination.

Beyond the linear onset regime, there are pronounced effects of temperature and illumination. For the visibly irradiated surface, we observe a decrease in dissipation as the coverage of the surface is increasing. This evolution can be explained by the densification of the film once the surface is saturated. For the UV irradiated layer, we observe a strong excess in dissipation which saturates at about 2000 Hz. The spiral shape is similar to previous observations made on polystyrene brushes in cyclohexane [82]. It indicates that the deposited film more pronouncedly interacts with the bulk solution, resulting in enhanced dissipation when the copolymer is in its polar (MC) state. In this case, we should expect a film with lower density as UV light leads to decelerated absorption and a polarity change, therefore the adsorbate has less time to rearrange itself as one its only irradiated with visible light.

The LCST of dilute P (TEGA-*co*-SPA) depends on illumination conditions. Adsorption kinetics and film growth at the silica–water interface can, therefore, be controlled through temperature and illumination conditions, relying on the thermally induced transition from hydrophilic coil arrangement to hydrophobic globules of the PTEGA components, and on the transition in polarity of the SPA-MC component controlled through illumination. Similarly, the film itself responds to thermal as well as optical stimuli through variable dissipation of acoustic excitation. Our results show that the coil-to-globule transition temperature is lower in diluted samples exposed to an adsorbing surface as compared to the solution. The conformational state of the adsorbed polymer chains is controlled by surface confinement and kinetics, whereby non-equilibrium conformational states could be frozen in for long times after adsorption [83]. A present assumption is that non-equilibrium effects originate from the polymer density and conformation at the interface of the adsorbing surface and the surrounding polymer solution or melt above its glass transition [84], which are kinetically frozen-in as a result of adsorption. For example, the slow rejuvenation of compressed polyethylene oxide PEO adsorbed on mica was found to be caused by the low mobility of the polymer chains in their adsorbed state [85]. Glassy dynamics of thermoresponsive, adsorbed polymers were investigated on solid substrates, e.g., latex particles in water. When the temperature was raised above the LCST temperature, PNIPAAm underwent a conformational transition from adsorbed loops to globules [86]. This transition process was slow: the relaxation time was found to vary between a few hundred to several thousand minutes [87]. Notwithstanding the difference in chemical structure between PNIPAAm and PTEGA, we assume that the difference between the bulk and surface LCST temperatures found here for P (TEGA-*co*-SPA) is likely due to similar kinetic considerations.

Another interesting aspect of the non-equilibrium nature of the adsorbed layer is related to the interplay between adsorption and wetting. The evolution of the frequency and dissipation shift as functions of temperature illustrate experimentally the surface-driven phase separation in polymer solutions, as predicted by Cahn [88]. Water and the P (TEGA-*co*-SPA) copolymer form one single solution phase at low temperatures. When the temperature of the system is increasing and, at the same time, the interaction between the solvent and the polymer is varied through an optical stimulus [89], we expect the system to first approach the wetting point at which the mixed and the de-mixed state of the binary mixture coexist. A further increase in temperature results in phase separation. Thereby, the phase with lower interfacial energy wets the silica surface [90]. Our QCM data supports this hypothesis, similar to previous observations on the adsorption of PNIPAAm on hydrophobic gold surfaces [52].

Although it is often claimed that thin hydrogel films are hydrophobic above their LCST, we show that SPA-copolymerization provides a means to circumvent this issue. For example [56], UV light exposure was found to not affect the hydration of PNIPAAm containing 2.5 mol% SPA when the temperature was above the LCST. This was explained by confinement of the chromophore within isopropyl groups, and the hydrophobic backbone of PNIPAAm. In our case, we found that UV light decelerated the growth of the wetting layer at 45 °C and 50 °C due to a competition between the copolymer globule adsorption and photoconversion of spiropyran to merocyanine facilitating desorption (Figure 1). In the absence of UV irradiation, the copolymer escapes from the solvent toward the silica surface, and thickness of the wetting layer increases for as long as the feeding solution is continuously injected. However, when illuminating with UV light, spiropyran rapidly converts to the polar merocyanine, leading to layer swelling and, eventually, globule desorption. The further difference between our observations and previous studies on PNIPAAm-SPA are attributed to different deposition techniques, major differences in the amount of the chromophore and even the difference in molar mass of the employed copolymer, which sets variable constraint on polymer conformation and deposition kinetics. A hydrogel film of PNIPAAm deposited by spin coating may delaminate from the surface due to osmotic stress caused by interaction with water molecules [91], even above LCST.

On the other hand, P (TEGA-*co*-SPA) surface rearrange both below and above the LCST; the isopropyl groups concentrate near air or other hydrophobic phases, whereas ethylene oxide groups rather orient towards water [92].

4. Conclusions

The conformational change of a thermal and light responsive copolymer layer of P (TEGA-*co*-SPA) on silica surfaces was investigated using quartz crystal microbalance with dissipation monitoring (QCM-D). First, we elucidate the effect of isothermal UV light illumination on the hydration state of the liquid film below and above its LCST. Second, we show that the phase separation temperature of the confined copolymer at the interface shifts to lower temperatures, namely 4–8 K lower compared to the cloud point temperatures as reported by UV/VIS spectroscopy in dilute aqueous solution. We attribute this difference to the formation of non-equilibrium adsorbed multilayers on the silica surface. Finally, we demonstrate that the built-up wetting layer displays variation of its viscoelastic properties with temperature and illumination conditions.

Supplementary Materials: The following are available online at https://www.mdpi.com/article/10.3390/polym13101633/s1, Figure S1: Synthetic route for 2-(3′,3′-Dimethyl-6-nitrospiro[chromene-2,2′-indolin]-1′-yl) ethyl Acrylate (SPA); Figure S2: Characterization of P (TEGA-co-SPA) copolymer: 1H-NMR in CDCl3; Figure S3: QCM-D water background used for data correction; Figure S4: Effect of UV illumination on bare (a) and water-wet sensor (b); Figure S5: Effect of continuous visible and UV light irradiation on the resonance frequency shift $\Delta f_3/3$ (a,b) and dissipation shift $\Delta D3$ (c,d) on the copolymer liquid film at 19 °C and 50 °C, respectively; Table S1: Fit parameters for the temperature correction as indicated in Figure S2.

Author Contributions: L.W. conceived of this study. A.N. and F.H.S. synthesized the PTEGA-SP co-polymer. A.B.-M. and K.W. conducted QCM-D analyses and, supported by L.W., evaluated the QCM-D data. All authors were involved in manuscript writing and draft revisions. All authors have read and agreed to the published version of the manuscript.

Funding: This project received funding from the European Research Council (ERC) under the European Union's Horizon 2020 research and innovation program (ERC grant UTOPES, grant agreement no. 681652), and was further supported by the Carl Zeiss Foundation (Durchbrüche 2019).

Institutional Review Board Statement: Not applicable.

Informed Consent Statement: Not applicable.

Data Availability Statement: All datasets reported in this study are available from the corresponding author on reasonable request.

Conflicts of Interest: The authors declare no conflict of interest.

References

1. Pirri, G.; Damin, F.; Chiari, M.; Bontempi, E.; Depero, L.E. Characterization of A Polymeric Adsorbed Coating for DNA Microarray Glass Slides. *Anal. Chem.* **2004**, *76*, 1352–1358. [CrossRef] [PubMed]
2. He, X.; Kim, S.H. Mechanochemistry of Physisorbed Molecules at Tribological Interfaces: Molecular Structure Dependence of Tribochemical Polymerization. *Langmuir* **2017**, *33*, 2717–2724. [CrossRef] [PubMed]
3. Jiang, N.; Sen, M.; Zeng, W.; Chen, Z.; Cheung, J.M.; Morimitsu, Y.; Endoh, M.K.; Koga, T.; Fukuto, M.; Yuan, G.; et al. Structure-Induced Switching of Interpolymer Adhesion at a Solid-Polymer Melt Interface. *Soft Matter* **2018**, *14*, 1108–1119. [CrossRef] [PubMed]
4. Tripp, C.P.; Hair, M.L. Measurement of Polymer Adsorption on Colloidal Silica by in Situ Transmission Fourier Transform Infrared Spectroscopy. *Langmuir* **1993**, *9*, 3523–3529. [CrossRef]
5. Blümmel, J.; Perschmann, N.; Aydin, D.; Drinjakovic, J.; Surrey, T.; Lopez-Garcia, M.; Kessler, H.; Spatz, J.P. Protein Repellent Properties of Covalently Attached PEG Coatings on Nanostructured SiO2-Based Interfaces. *Biomaterials* **2007**, *28*, 4739–4747. [CrossRef]
6. Hu, S.; Ren, X.; Bachman, M.; Sims, C.E.; Li, G.P.; Allbritton, N.L. Surface-Directed, Graft Polymerization within Microfluidic Channels. *Anal. Chem.* **2004**, *76*, 1865–1870. [CrossRef]
7. Källrot, N.; Dahlqvist, M.; Linse, P. Dynamics of Polymer Adsorption from Bulk Solution onto Planar Surfaces. *Macromolecules* **2009**, *42*, 3641–3649. [CrossRef]

8. Cohen Stuart, M.A.; Fleer, G.J. Adsorbed Polymer Layers in Nonequilibrium Situations. *Annu. Rev. Mater. Sci.* **1996**, *26*, 463–500. [CrossRef]
9. O'Shaughnessy, B.; Vavylonis, D. Irreversible Adsorption from Dilute Polymer Solutions. *Eur. Phys. J. E* **2003**, *11*, 213–230. [CrossRef]
10. Sims, R.A.; Harmer, S.L.; Quinton, J.S. The Role of Physisorption and Chemisorption in the Oscillatory Adsorption of Organosilanes on Aluminium Oxide. *Polymers* **2019**, *11*, 410. [CrossRef]
11. O'Shaughnessy, B.; Vavylonis, D. Non-Equilibrium in Adsorbed Polymer Layers. *J. Phys. Condens. Matter* **2005**, *17*. [CrossRef]
12. Porus, M.; Maroni, P.; Borkovec, M. Structure of Adsorbed Polyelectrolyte Monolayers Investigated by Combining Optical Reflectometry and Piezoelectric Techniques. *Langmuir* **2012**, *28*, 5642–5651. [CrossRef]
13. Roach, P.; Farrar, D.; Perry, C.C. Interpretation of Protein Adsorption: Surface-Induced Conformational Changes. *J. Am. Chem. Soc.* **2005**, *127*, 8168–8173. [CrossRef] [PubMed]
14. Taylor, M.; Tomlins, P.; Sahota, T. Thermoresponsive Gels. *Gels* **2017**, *3*, 4. [CrossRef]
15. Yu, L.; Schlaich, C.; Hou, Y.; Zhang, J.; Noeske, P.L.M.; Haag, R. Photoregulating Antifouling and Bioadhesion Functional Coating Surface Based on Spiropyran. *Chem. Eur. J.* **2018**, *24*, 7742–7748. [CrossRef] [PubMed]
16. Kocak, G.; Tuncer, C.; Bütün, V. PH-Responsive Polymers. *Polym. Chem.* **2017**, *8*, 144–176. [CrossRef]
17. Xiang, T.; Lu, T.; Zhao, W.F.; Zhao, C.S. Ionic-Strength Responsive Zwitterionic Copolymer Hydrogels with Tunable Swelling and Adsorption Behaviors. *Langmuir* **2019**, *35*, 1146–1155. [CrossRef]
18. Nabiyan, A.; Biehl, P.; Schacher, F.H. Crystallization vs Metal Chelation: Solution Self-Assembly of Dual Responsive Block Copolymers. *Macromolecules* **2020**, *53*, 5056–5067. [CrossRef]
19. Max, J.B.; Nabiyan, A.; Eichhorn, J.; Schacher, F.H. Triple-Responsive Polyampholytic Graft Copolymers as Smart Sensors with Varying Output. *Macromol. Rapid Commun.* **2020**, *2000671*, 1–5. [CrossRef]
20. Wondraczek, L.; Pohnert, G.; Schacher, F.H.; Köhler, A.; Gottschaldt, M.; Schubert, U.S.; Küsel, K.; Brakhage, A.A. Artificial Microbial Arenas: Materials for Observing and Manipulating Microbial Consortia. *Adv. Mater.* **2019**, *31*. [CrossRef]
21. Abdollahi, A.; Roghani-Mamaqani, H.; Razavi, B.; Salami-Kalajahi, M. The Light-Controlling of Temperature-Responsivity in Stimuli-Responsive Polymers. *Polym. Chem.* **2019**, *10*, 5686–5720. [CrossRef]
22. Zhang, Q.; Weber, C.; Schubert, U.S.; Hoogenboom, R. Thermoresponsive Polymers with Lower Critical Solution Temperature: From Fundamental Aspects and Measuring Techniques to Recommended Turbidimetry Conditions. *Mater. Horiz.* **2017**, *4*, 109–116. [CrossRef]
23. Heskins, M.; Guillet, J.E. Solution Properties of Poly(N-Isopropylacrylamide). *J. Macromol. Sci. Part Chem.* **1968**, *2*, 1441–1455. [CrossRef]
24. Sanson, N.; Rieger, J. Synthesis of Nanogels/Microgels by Conventional and Controlled Radical Crosslinking Copolymerization. *Polym. Chem.* **2010**, *1*, 965–977. [CrossRef]
25. Tavagnacco, L.; Chiessi, E.; Zanatta, M.; Orecchini, A.; Zaccarelli, E. Water-Polymer Coupling Induces a Dynamical Transition in Microgels. *J. Phys. Chem. Lett.* **2019**, *10*, 870–876. [CrossRef]
26. Halperin, A.; Krçger, M.; Winnik, F.M. Poly (N-Isopropylacrylamide) Phase Diagrams: Fifty Years of Research Angewandte. *Angew. Chem. Int. Ed.* **2015**, 15342–15367. [CrossRef]
27. De Oliveira, T.E.; Marques, C.M.; Netz, P.A. Molecular Dynamics Study of the LCST Transition in Aqueous Poly(N-n-Propylacrylamide). *Phys. Chem. Chem. Phys.* **2018**, *20*, 10100–10107. [CrossRef]
28. De Solorzano, I.O.; Bejagam, K.K.; An, Y.; Singh, S.K.; Deshmukh, S.A. Solvation Dynamics of N-Substituted Acrylamide Polymers and the Importance for Phase Transition Behavior. *Soft Matter* **2020**, *16*, 1582–1593. [CrossRef]
29. Hoogenboom, R.; Thijs, H.M.L.; Jochems, M.J.H.C.; Van Lankvelt, B.M.; Fijten, M.W.M.; Schubert, U.S. Tuning the LCST of Poly(2-Oxazoline)s by Varying Composition and Molecular Weight: Alternatives to Poly(N-Isopropylacrylamide)? *Chem. Commun.* **2008**, 5758–5760. [CrossRef]
30. Glassner, M.; Vergaelen, M.; Hoogenboom, R. Poly(2-Oxazoline)s: A Comprehensive Overview of Polymer Structures and Their Physical Properties. *Polym. Int.* **2018**, *67*, 32–45. [CrossRef]
31. Li, Z.; Zhang, Z.; Liu, K.L.; Ni, X.; Li, J. Biodegradable Hyperbranched Amphiphilic Polyurethane Multiblock Copolymers Consisting of Poly(Propylene Glycol), Poly(Ethylene Glycol), and Polycaprolactone as in Situ Thermogels. *Biomacromolecules* **2012**, *13*, 3977–3989. [CrossRef]
32. Su, X.; Tan, M.J.; Li, Z.; Wong, M.; Rajamani, L.; Lingam, G.; Loh, X.J. Recent Progress in Using Biomaterials as Vitreous Substitutes. *Biomacromolecules* **2015**, *16*, 3093–3102. [CrossRef] [PubMed]
33. Lutz, J.F.; Hoth, A. Preparation of Ideal PEG Analogues with a Tunable Thermosensitivity by Controlled Radical Copolymerization of 2-(2-Methoxyethoxy)Ethyl Methacrylate and Oligo(Ethylene Glycol) Methacrylate. *Macromolecules* **2006**, *39*, 893–896. [CrossRef]
34. Hedir, G.G.; Arno, M.C.; Langlais, M.; Husband, J.T.; O'Reilly, R.K.; Dove, A.P. Poly(Oligo(Ethylene Glycol) Vinyl Acetate)s: A Versatile Class of Thermoresponsive and Biocompatible Polymers. *Angew. Chem. Int. Ed.* **2017**, *56*, 9178–9182. [CrossRef]
35. Langer, M.; Brandt, J.; Lederer, A.; Goldmann, A.S.; Schacher, F.H.; Barner-Kowollik, C. Amphiphilic Block Copolymers Featuring a Reversible Hetero Diels-Alder Linkage. *Polym. Chem.* **2014**, *5*, 5330–5338. [CrossRef]
36. Feil, H.; Bae, Y.H.; Feijen, J.; Kim, S.W. Effect of Comonomer Hydrophilicity and Ionization on the Lower Critical Solution Temperature of N-Isopropylacrylamide Copolymers. *Macromolecules* **1993**, *26*, 2496–2500. [CrossRef]
37. Löwenbein, A.; Katz, W. Über Substituierte spiro-Dibenzopyrane. *Ber. Dtsch. Chem. Ges.* **1926**, *59*, 1377–1383. [CrossRef]

38. Hartley, G. The Cis Form of Azobenene. *Nature* **1937**, *14*, 281. [CrossRef]
39. Irie, M.; Mohri, M. Thermally Irreversible Photochromic Systems. Reversible Photocyclization of Diarylethene Derivatives. *J. Org. Chem.* **1988**, *53*, 803–808. [CrossRef]
40. Grimm, O.; Wendler, F.; Schacher, F.H. Micellization of Photo-Responsive Block Copolymers. *Polymers* **2017**, *9*, 396. [CrossRef]
41. Klajn, R. Spiropyran-Based Dynamic Materials. *Chem. Soc. Rev.* **2014**, *43*, 148–184. [CrossRef]
42. Schnurbus, M.; Kabat, M.; Jarek, E.; Krzan, M.; Warszynski, P.; Braunschweig, B. Spiropyran Sulfonates for Photo- And PH-Responsive Air-Water Interfaces and Aqueous Foam. *Langmuir* **2020**, *36*, 6871–6879. [CrossRef] [PubMed]
43. Petriashvili, G.; De Santo, M.P.; Devadze, L.; Zurabishvili, T.; Sepashvili, N.; Gary, R.; Barberi, R. Rewritable Optical Storage with a Spiropyran Doped Liquid Crystal Polymer Film. *Macromol. Rapid Commun.* **2016**, *37*, 500–505. [CrossRef]
44. Dübner, M.; Cadarso, V.J.; Gevrek, T.N.; Sanyal, A.; Spencer, N.D.; Padeste, C. Reversible Light-Switching of Enzymatic Activity on Orthogonally Functionalized Polymer Brushes. *ACS Appl. Mater. Interfaces* **2017**, *9*, 9245–9249. [CrossRef]
45. Grimm, O.; Maßmann, S.C.; Schacher, F.H. Synthesis and Solution Behaviour of Dual Light- and Temperature-Responsive Poly(Triethylene Glycol-:Co-Spiropyran) Copolymers and Block Copolymers. *Polym. Chem.* **2019**, *10*, 2674–2685. [CrossRef]
46. Johannsmann, D. Viscoelastic Analysis of Organic Thin Films on Quartz Resonators. *Macromol. Chem. Phys.* **1999**, *200*, 501–516. [CrossRef]
47. Voinova, M.V.; Jonson, M.; Kasemo, B. On Dissipation of Quartz Crystal Microbalance as a Mechanical Spectroscopy Tool. *Spectroscopy* **2004**, *18*, 537–544. [CrossRef]
48. Sadman, K.; Wiener, C.G.; Weiss, R.A.; White, C.C.; Shull, K.R.; Vogt, B.D. Quantitative Rheometry of Thin Soft Materials Using the Quartz Crystal Microbalance with Dissipation. *Anal. Chem.* **2018**, *90*, 4079–4088. [CrossRef]
49. Peppas, N.A.; Hilt, J.Z.; Khademhosseini, A.; Langer, R. Hydrogels in Biology and Medicine: From Molecular Principles to Bionanotechnology. *Adv. Mater.* **2006**, *18*, 1345–1360. [CrossRef]
50. Zhang, G. Study on Conformation Change of Thermally Sensitive Linear Grafted Poly(N-Isopropylacrylamide) Chains by Quartz Crystal Microbalance. *Macromolecules* **2004**, *37*, 6553–6557. [CrossRef]
51. Liu, G.; Zhang, G. Collapse and Swelling of Thermally Sensitive Poly (N-Isopropylacrylamide) Brushes Monitored with a Quartz Crystal Microbalance. *J. Phys. Chem. B* **2005**, *109*, 743–747. [CrossRef]
52. Plunkett, M.A.; Wang, Z.; Rutland, M.W.; Johannsmann, D. Adsorption of PNIPAM Layers on Hydrophobic Gold Surfaces, Measured in Situ by QCM and SPR. *Langmuir* **2003**, *19*, 6837–6844. [CrossRef]
53. Wu, K.; Wu, B.; Wang, P.; Hou, Y.; Zhang, G.; Zhu, D.M. Adsorption Isotherms and Dissipation of Adsorbed Poly(N- Isopropylacrylamide) in Its Swelling and Collapsed States. *J. Phys. Chem. B* **2007**, *111*, 8723–8727. [CrossRef] [PubMed]
54. Chen, T.; Lu, Y.; Chen, T.; Zhang, X.; Du, B. Adsorption of PNIPAmx-PEO20-PPO70-PEO 20-PNIPAmx Pentablock Terpolymer on Gold Surfaces: Effects of Concentration, Temperature, Block Length, and Surface Properties. *Phys. Chem. Chem. Phys.* **2014**, *16*, 5536–5544. [CrossRef] [PubMed]
55. Guo, Y.; Wang, D.; Yang, L.; Liu, S. Nanoscale Monolayer Adsorption of Polyelectrolytes at the Solid/Liquid Interface Observed by Quartz Crystal Microbalance. *Polym. J.* **2017**, *49*, 543–548. [CrossRef]
56. Ichi Edahiro, J.; Sumaru, K.; Takagi, T.; Shinbo, T.; Kanamori, T.; Sudoh, M. Analysis of Photo-Induced Hydration of a Photochromic Poly(N-Isopropylacrylamide)—Spiropyran Copolymer Thin Layer by Quartz Crystal Microbalance. *Eur. Polym. J.* **2008**, *44*, 300–307. [CrossRef]
57. Tay, A.; Bendejacq, D.; Monteux, C.; Lequeux, F. How Does Water Wet a Hydrosoluble Substrate? *Soft Matter* **2011**, *7*, 6953–6957. [CrossRef]
58. Johannsmann, D. Viscoelastic, Mechanical, and Dielectric Measurements on Complex Samples with the Quartz Crystal Microbalance. *Phys. Chem. Chem. Phys.* **2008**, *10*, 4516–4534. [CrossRef]
59. Benkoski, J.J.; Jesorka, A.; Kasemo, B.; Höök, F. Light-Activated Desorption of Photoactive Polyelectrolytes from Supported Lipid Bilayers. *Macromolecules* **2005**, *38*, 3852–3860. [CrossRef]
60. Heeb, R.; Bielecki, R.M.; Lee, S.; Spencer, N.D. Room-Temperature, Aqueous-Phase Fabrication of Poly(Methacrylic Acid) Brushes by UV-LED-Induced, Controlled Radical Polymerization with High Selectivity for Surface-Bound Species. *Macromolecules* **2009**, *42*, 9124–9132. [CrossRef]
61. Hook, F.F.; Vörös, J.; Rodahl, M.; Kurrat, R.; Böni, P.; Ramsden, J.J.; Textor, M.; Spencer, N.D.; Tengvall, P.; Gold, J.; et al. A Comparative Study of Protein Adsorption on Titanium Oxide Surfaces Using in Situ Ellipsometry, Optical Waveguide Lightmode Spectroscopy, and Quartz Crystal Microbalance/Dissipation. *Colloids Surf. B Biointerfaces* **2002**, *24*, 155–170. [CrossRef]
62. Qin, S.; Tang, X.; Zhu, L.; Wei, Y.; Du, X.; Zhu, D.M. Viscoelastic Signature of Physisorbed Macromolecules at the Solid-Liquid Interface. *J. Colloid Interface Sci.* **2012**, *383*, 208–214. [CrossRef]
63. Duarte, A.A.; Abegão, L.M.G.; Ribeiro, J.H.F.; Lourenço, J.P.; Ribeiro, P.A.; Raposo, M. Study of in Situ Adsorption Kinetics of Polyelectrolytes and Liposomes Using Quartz Crystal Microbalance: Influence of Experimental Layout. *Rev. Sci. Instrum.* **2015**, *86*. [CrossRef] [PubMed]
64. Sauerbrey, G. Verwendung von Schwingquarzen Zur Wägung Dünner Schichten Und Zur Mikrowägung. *Z. Phys.* **1959**, *155*, 206–222. [CrossRef]
65. Höök, F.; Kasemo, B.; Nylander, T.; Fant, C.; Sott, K.; Elwing, H. Variations in Coupled Water, Viscoelastic Properties, and Film Thickness of a Mefp-1 Protein Film during Adsorption and Cross-Linking: A Quartz Crystal Microbalance with Dissipation Monitoring, Ellipsometry, and Surface Plasmon Resonance Study. *Anal. Chem.* **2001**, *73*, 5796–5804. [CrossRef] [PubMed]

66. Wondraczek, K.; Bund, A.; Johannsmann, D. Acoustic Second Harmonic Generation from Rough Surfaces under Shear Excitation in Liquids. *Langmuir* **2004**, *20*, 10346–10350. [CrossRef]
67. Wehner, S.; Wondraczek, K.; Johannsmann, D.; Bund, A. Roughness-Induced Acoustic Second-Harmonic Generation during Electrochemical Metal Deposition on the Quartz-Crystal Microbalance. *Langmuir* **2004**, *20*, 2356–2360. [CrossRef]
68. Landau, L.D.; Lifshitz, E.M. *Fluid Mechanics: Landau and Lifshitz: Course of Theoretical Physics*; Pergamon: Oxford, UK, 1987.
69. Kanazawa, K.K.; Gordon, J.G. Frequency of a Quartz Microbalance in Contact with Liquid. *Anal. Chem.* **1985**. [CrossRef]
70. Lutz, J.F.; Weichenhan, K.; Akdemir, Ö.; Hoth, A. About the Phase Transitions in Aqueous Solutions of Thermoresponsive Copolymers and Hydrogels Based on 2-(2-Methoxyethoxy)Ethyl Methacrylate and Oligo(Ethylene Glycol) Methacrylate. *Macromolecules* **2007**, *40*, 2503–2508. [CrossRef]
71. Lessard, D.G.; Ousalem, M.; Zhu, X.X.; Eisenberg, A.; Carreau, P.J. Study of the Phase Transition of Poly(n,n-Diethylacrylamide) in Water by Rheology and Dynamic Light Scattering. *J. Polym. Sci. Part B Polym. Phys.* **2003**, *41*, 1627–1637. [CrossRef]
72. Kujawa, P.; Aseyev, V.; Tenhu, H.; Winnik, F.M. Temperature-Sensitive Properties of Poly(N-Isopropylacrylamide) Mesoglobules Formed in Dilute Aqueous Solutions Heated above Their Demixing Point. *Macromolecules* **2006**, *39*, 7686–7693. [CrossRef]
73. Israelachvili, J. Commentary The Different Faces of Poly (Ethylene Glycol). *Proc. Natl. Acad. Sci. USA* **1997**, *94*, 8378–8379. [CrossRef]
74. Begum, R.; Matsuura, H. Conformational Properties of Short Poly(Oxyethylene) Chains in Water Studied by IR Spectroscopy. *J. Chem. Soc. Faraday Trans.* **1997**, *93*, 3839–3848. [CrossRef]
75. Lutz, J.F.; Akdemir, Ö.; Hoth, A. Point by Point Comparison of Two Thermosensitive Polymers Exhibiting a Similar LCST: Is the Age of Poly(NIPAM) Over? *J. Am. Chem. Soc.* **2006**, *128*, 13046–13047. [CrossRef] [PubMed]
76. Judah, H.L.; Liu, P.; Zarbakhsh, A.; Resmini, M. Influence of Buffers, Ionic Strength, and PH on the Volume Phase Transition Behavior of Acrylamide-Based Nanogels. *Polymers* **2020**, *12*, 2590. [CrossRef]
77. Chevallier, E.; Mamane, A.; Stone, H.A.; Tribet, C.; Lequeux, F.; Monteux, C. Pumping-out Photo-Surfactants from an Air-Water Interface Using Light. *Soft Matter* **2011**, *7*, 7866–7874. [CrossRef]
78. Höök, F.; Rodahl, M.; Brzezinski, P.; Kasemo, B. Energy Dissipation Kinetics for Protein and Antibody-Antigen Adsorption under Shear Oscillation on a Quartz Crystal Microbalance. *Langmuir* **1998**, *14*, 729–734. [CrossRef]
79. Johannsmann, D.; Reviakine, I.; Richter, R.P. Dissipation in Films of Adsorbed Nanospheres Studied by Quartz Crystal Microbalance (QCM). *Anal. Chem.* **2009**, *81*, 8167–8176. [CrossRef]
80. Cohen Stuart, M.A.; Waajen, F.H.W.H.; Cosgrove, T.; Vincent, B.; Crowley, T.L. Hydrodynamic Thickness of Adsorbed Polymer Layers. *Macromolecules* **1984**, *17*, 1825–1830. [CrossRef]
81. Plunkett, M.A.; Claesson, P.M.; Rutland, M.W. Adsorption of a Cationic Polyelectrolyte Followed by Surfactant-Induced Swelling, Studied with a Quartz Crystal Microbalance. *Langmuir* **2002**, *18*, 1274–1280. [CrossRef]
82. Domack, A.; Prucker, O.; Rühe, J.; Johannsmann, D. Swelling of a Polymer Brush Probed with a Quartz Crystal Resonator. *Phys. Rev. E Stat. Phys. Plasmas Fluids Relat. Interdiscip. Top.* **1997**, *56*, 680–689. [CrossRef]
83. Chakraborty, A.K.; Shaffer, J.S.; Adriani, P.M. On the Existence of Quasi-Two-Dimensional Glasslike Structures at Strongly Interacting Polymer-Solid Interfaces. *Macromolecules* **1991**, *24*, 5226–5229. [CrossRef]
84. Kremer, K. Glassy States of Adsorbed Flexible Polymers and Spread Polymer "Monolayers". *J. Phys.* **1986**. [CrossRef]
85. Raviv, U.; Klein, J.; Witten, T.A. The Polymer Mat: Arrested Rebound of a Compressed Polymer Layer. *Eur. Phys. J. E* **2002**. [CrossRef] [PubMed]
86. Zhu, P.W.; Napper, D.H. Conformational Transitions of Poly(N-Isopropylacrylamide) Chains Loopily Absorbed at the Surfaces of Poly(N-Tert-Butylacrylamide) Latex Particles in Water. *J. Phys. Chem. B* **1997**, *101*, 3155–3160. [CrossRef]
87. Zhu, P.W.; Napper, D.H. Effects of Thermal History on the Dynamics of Relaxation of Poly([Formula Presented]-Isopropylacrylamide) Adsorbed at Latex Interfaces in Water. *Phys. Rev. E Stat. Phys. Plasmas Fluids Relat. Interdiscip. Top.* **1998**, *57*, 3101–3106. [CrossRef]
88. Cahn, J.W. Critical Point Wetting. *J. Chem. Phys.* **1977**, *66*, 3667–3672. [CrossRef]
89. Kawata, Y.; Yamamoto, T.; Kihara, H.; Yamamura, Y.; Saito, K.; Ohno, K. Unusual Photoresponses in the Upper Critical Solution Temperature of Polymer Solutions Mediated by Changes in Intermolecular Interactions in an Azo-Doped Liquid Crystalline Solvent. *Phys. Chem. Chem. Phys.* **2018**, *20*, 5850–5855. [CrossRef] [PubMed]
90. Jones, R.A.L.; Richards, R.W. *Polymers at Surfaces and Interfaces*; Cambridge University Press: New York, NY, USA, 1999.
91. Wiener, C.G.; Weiss, R.A.; Vogt, B.D. Overcoming Confinement Limited Swelling in Hydrogel Thin Films Using Supramolecular Interactions. *Soft Matter* **2014**. [CrossRef] [PubMed]
92. Pelton, R. Poly(N-Isopropylacrylamide) (PNIPAM) Is Never Hydrophobic. *J. Colloid Interface Sci.* **2010**. [CrossRef] [PubMed]

polymers

MDPI

rticle

'eatures of Solution Behavior of Polymer Stars with Arms of ?oly-2-alkyl-2-oxazolines Copolymers Grafted to the Upper Rim of Calix[8]arene

'atyana Kirila *, Alina Amirova, Alexey Blokhin, Andrey Tenkovtsev and Alexander Filippov

Institute of Macromolecular Compounds of the Russian Academy of Sciences, Bolshoy Pr. 31, 199004 Saint Petersburg, Russia; aliram.new@gmail.com (A.A.); 44stuff44@gmail.com (A.B.); avt@hq.macro.ru (A.T.); afil@imc.macro.ru (A.F.)

* Correspondence: tatyana_pyx@mail.ru; Tel.: +7-812-328-4102

Abstract: Star-shaped polymers with arms of block and gradient copolymers of 2-ethyl- and 2-isopropyl-2-oxazolines grafted to the upper rim of calix[8]arene were synthesized by the "grafting from" method. The ratio of 2-ethyl- and 2-isopropyl-2-oxazoline units was 1:1. Molar masses and hydrodynamic characteristics were measured using molecular hydrodynamics and optics methods in 2-nitropropane. The arms of the synthesized stars were short and the star-shaped macromolecules were characterized by compact dimensions and heightened intramolecular density. The influence of the arm structure on the conformation of star molecules was not observed. At low temperatures, the aqueous solutions of the studied stars were not molecular dispersed but individual molecules prevailed. One phase transition was detected for all solutions. The phase separation temperatures decreased with a growth of the content of more hydrophobic 2-isopropyl-2-oxazoline units. It was shown that the way of arms grafting to the calix[8]arene core affects the behavior of aqueous solutions of star-shaped poly-2-alkyl-2-oxazoline copolymers. In the case of upper rim functionalization, the shape of calix[8]arene resembles a plate. Accordingly, the core is less shielded from the solvent and the phase separation temperatures are lower than those for star-shaped poly-2-alkyl-2-oxazolines with lower rim functionalization of the calix[8]arene.

Keywords: synthesis; star-shaped macromolecules; calix[n]arene; block and gradient copolymers of poly-2-alkyl-2-oxazolines; conformation; thermoresponsibility; self-organization; phase separation

itation: Kirila, T.; Amirova, A.; ›okhin, A.; Tenkovtsev, A.; Filippov, . Features of Solution Behavior of ›lymer Stars with Arms of ›ly-2-alkyl-2-oxazolines ›opolymers Grafted to the Upper ›m of Calix[8]arene. *Polymers* **2021**, ', 2507. https://doi.org/10.3390/ ›lym13152507

cademic Editor: Mattia Sponchioni

eceived: 29 June 2021
ccepted: 26 July 2021
iblished: 29 July 2021

ublisher's Note: MDPI stays neutral ith regard to jurisdictional claims in ıblished maps and institutional affil- tions.

1. Introduction

New synthetic routes make it possible to obtain well-defined polymers with complex architecture [1–4] including multiarm stars [5–10]. Their behavior in solutions is determined by the chemical structure of the core and arms and the number and length of the latter. The use of copolymers as arms is a convenient way to control the properties of stimulus-sensitive star polymers. In the case of block copolymer arms, the sequence of block attachment affects both the phase separation temperatures and the dimensions of supramolecular structures present in solutions [11–14]. For example, the variation in solution behavior was observed for triblock copolymer stars with arms consisting of hydrophobic, hydrophilic, and thermosensitive blocks. When the thermosensitive block was located near a core, intramolecular aggregation took place and aggregates with a smaller diameter were formed in comparison with the supramolecular structures formed in solutions of macromolecules with a thermosensitive block in the outer layer [12]. From general considerations, it is clear that the stimulus-sensitivity of copolymer stars depends on the ratio of the components [11,15–19].

Thermoresponsive poly-2-alkyl-2-oxazolines (PAlOx) have been actively studied in recent decades due to the wide potential of their application [20–23]. One of the ways for using this class of polymers is medicine due to their biocompatibility and non-toxicity, stability in enzyme media [24], and lower critical solution temperature (LCST) near to the human body one. These are the reasons to enhance the investigation of star-shaped PAlOx properties depending on the molecule structure. The optimal conditions have been established, which make it possible to obtain PAlOx stars with a given number and length of arms and, accordingly, to regulate their conformational characteristics and behavior in water–salt solutions, including thermosensitivity and association with low molecular weight compounds. The study star-shaped block PAlOx copolymers revealed that a different sequence of block attachment to the core does not influence the phase separation temperatures, however, it determines the set and dimensions of scattering objects [13]. In the case of a polymer with a more hydrophilic outer block, aggregation processes prevail while for a polymer star with a more hydrophobic outer block in a wide temperature range the dominant process is aggregation [25]. The distribution of 2-alkyl-2-oxazoline units along the arm chains also affects the stimulus-sensitivity of copolymer PAlOx, in particular the temperature of the onset of phase separation T_1 for solutions of a stars with gradient arms is higher than the T_1 value for block copolymer stars [26].

Concerning the macromolecule core, its structure and size have a great effect on the properties of thermoresponsive star-shaped PAlOx. For example, competition between compaction and aggregation processes was observed upon heating of solutions of PAlOx stars with a massive hydrophobic dendrimer core, while aggregation dominates in solutions of PAlOx with a less hydrophobic calix[n]arene core [27]. The use of calixarene derivatives as the branching center of star-shaped polymers is due to bringing the unique ability of calix[n]arenes to complex formation with low molecular weight organic compounds [28–30]. Accordingly, a number of calix[n]arene derivatives with functionalization with low molecular weight fragments have been proposed for use in targeted drug delivery systems [31–36]. Calixarenes with polymer arms have also been obtained [25,37]. It is also important to point out that polymer stars with a calix[n]arene core are high macromolecular weight objects, components of which are selectively solvated by water. It provides another mechanism for regulating the characteristics of thermoresponsive supramolecular structures.

Note that most studies describe the results for stars in which polymer arms are grafted to the lower rim of calix[n]arenes. At the same time it is known that different positions of functional groups or polymer arms in calix[n]arene lead to a variation in physicochemical properties and self-organization of calix[n]arene derivatives [38,39]. It was shown that the grafting of the homopolymer PAlOx chains at the upper rim of calix[n]arene reduces the phase separation temperatures as compared to polymer stars with arm grafting to the lower rim [39]. The present work was aimed at the analysis of the influence of the arm structure and the configuration of the calix[8]arene (C8A) core on the molecular conformation, the solution behavior, and the self-organization of star-shaped PAlOx copolymers in aqueous solutions upon heating. To solve this problem, eight-arm polymer stars were synthesized and studied. Their arms were block copolymers of poly-2-ethyl-2-oxazoline (PEtOx) and poly-2-isopropyl-2-oxazoline (PiPrOx). These samples differed in the order in which the blocks were attached to the core. In the C8A-(PiPrOx-b-PEtOx) star, the inner block was PiPrOx, and in the C8A-(PEtOx-b-PiPrOx) copolymer, PEtOx was attached to the core. In addition, the star-shaped copolymer C8A-P(EtOx-grad-iPrOx) was studied, the arms of which were gradient copolymers of 2-ethyl-2-oxazoline (EtOx) and 2-isopropyl-2-oxazoline (iPrOx). In these stars, the content of EtOx units decreased with distance from the C8A core. For comparison, a star-shaped homopolymer C8A-PEtOx with PEtOx chains attached to the upper rim of calix[8]arene was studied. It is important that the synthesized star-shaped samples should have a similar arm length in order to avoid the influence of molar mass on the obtained characteristics.

2. Materials and Methods

2.1. Materials and Reagents

Dialysis bags "CellaSep", MWCO = 3000 Da (Orange Scientific, Braine-l'Alleud, Belgium) were used for the purification of polymer samples. Monomers, 2-ethyl- and 2-isopropyl-2-oxazolines (Sigma-Aldrich, St. Louis, MO, USA) were distilled over the calcium hydride. Sulfolane (Sigma-Aldrich, St. Louis, MO, USA) was purified by vacuum distillation. Pyrrolidine (Sigma-Aldrich, St. Louis, MO, USA) was distilled over calcium hydride.

2.2. Synthesis of Multicenter Macroiniciater

The octafunctional initiator, 5,11,17,23,29,35,41,47-octakis-(chlorosulfonyl)-49,50,51,52, 53,54,55,56-octakis-(methoxycarbonylmethoxy)-calix[8]arene, was synthesized following the described scheme [40].

2.2.1. 49,50,51,52,53,54,55,56-Octa(hydroxy)calix[8]arene

The mixture of 10 g (7.7×10^{-3} mol) of tert-butylcalix[8]arene, 5.8 g (0.062 mol) of phenol, 12.3 g (0.092 mol) of aluminum chloride, and 150 mL of toluene was stirred for 1 h at ambient temperature, after which it was poured into 170 mL of 0.2 M hydrochloric acid. The organic layer was separated and the solvent was distilled off. The precipitate was washed with 330 mL of methanol, acidified with a few drops of hydrochloric acid, and filtered out. The product was purified by chloroform extraction in Soxhlet apparatus during 24 h. Yield 5.5 g (84%). ^{1}H NMR (400 MHz, DMSO, 20 °C): δ (ppm) 6.0–7.0 (m, 24H), 3.5 (s 16H).

2.2.2. 49,50,51,52,53,54,55,56-Octa(methoxy(carbonylmethoxy)calix[8]arene

The mixture of 29 g (0.21 mol) of dry potassium carbonate, 11.2 g (0.067 mol) of dry potassium iodide, 3.45 g (4.1×10^{-3} mol) of calix[8]arene, 18 mL (0.21 mol) of methyl chloroacetate, and 180 mL of absolute acetonitril was heated at 80 °C during 24 h. The reaction mixture was poured into 300 mL of water. The product was extracted with diethyl ether (2 × 100 mL), washed with water, and dried ($MgSO_4$). After the evaporation product was recrystallized from methanol. Yield 2.1 g (36%). ^{1}H NMR (400 MHz, $CDCl_3$, 20 °C): δ (ppm) 6.9 (m, 24H), 4.27 (m, 16H), 4.10 (s, 16H), 3.7 (s, 24H).

2.2.3. 5,11,17,23,29,35,41,47-Octachlorosulfonyl-49,50,51,52,53,54,55,56-octa(methoxy(carbonylmethoxy)calix[8]arene

A solution of 2 g (1.41×10^{-3} mol) of octa(methoxy-(carbonylmethoxy)) calix[8]arene in 60 mL of chloroform was cooled to −10 °C and 20 mL (0.3 mol) of chlorosulfonic acid was added drop by drop. Then mixture was heated to 50 °C (at about 20 min) and left at this temperature for 20 min. After cooling to room temperature, the mixture was gradually poured into a mixture of 400 mL of ice water and 300 mL of petroleum ether and left for 30 min. The product was filtered off, washed with water, then with petroleum ether, and dried. The crude product was dissolved in a minimum amount of dichloromethane and reprecipitated into petroleum ether. This procedure was repeated twice. Yield: 1.5 g (49%). M.p 170 °C (with decomposition). ^{1}H NMR (400 MHz, CDCl3, 20 °C): δ (ppm) 7.6 (m, 16H), 4.2–4.7 (m, 32H), 3.7 (s, 24H). Elemental analysis: Calc. C 44.37%, H 3.72%, Cl 12.47%, S 11.28%. Found C 44.1%, H 4.0%, S 11.6%, Cl 12.8%.

2.3. Eight-Arm Star Poly(2-Ethyl-2-Oxazoline-Block-2-Isopropyl-2-Oxazoline) Copolymer Synthesis

The solution of 0.0931 g (0.0421 mmol) of the octafunctional initiator in 2 g (17 mmol) of sulfolane was prepared under the nitrogen atmosphere. The solution was mixed with 1 g (10.1 mmol) of 2-ethyl-2-oxazoline and sealed in a vial. The mixture was kept at 100 °C for 24 h. After that, 1.14 g (10.1 mmol) of 2-isopropyl-2-oxazoline was injected into the vial, which was sealed again and kept at 100 °C for 48 h. Then 1 mL (12.1 mmol) of pyrrolidine was added into the vial and the solution was stirred at 50 °C for 1 h. The polymer was purified by dialysis in 0.1 M sodium hydroxide aqueous solution followed by pure water, then dried at 20 °C, and finally evaporated from its chloroform solution (1.61 g, 72%). ^{1}H NMR (CDCl3, δ, ppm): 3.51 (m, 4H), 3.05–2.65 (m, 1H), 2.58–2.20 (m, 2H), 1.11 (m, 9H).

2.4. Eight-Arm Star Poly(2-Isopropyl-2-Oxazoline-Block-2-Ethyl-2-Oxazoline) Copolymer Synthesis

The same techniques of synthesis and purification were applied as previously discussed (1.68 g, 75%). ^{1}H NMR (CDCl3, δ, ppm): 3.52 (m, 4H), 3.05–2.60 (m, 1H), 2.58–2.20 (m, 2H), 1.11 (m, 9H).

2.5. Eight-Arm Star Poly(2-Ethyl-2-Oxazoline-Grad-2-Isopropyl-2-Oxazoline) Gradient Copolymer Synthesis

The solution of 0.1 g (0.0452 mmol) of the octafunctional initiator in 2.5 g (20 mmol) of sulfolane was prepared under the nitrogen atmosphere. Equimolar amounts (10.8 mmol) of 2-ethyl-2-oxazoline and 2-isopropyl-2-oxazoline were mixed and added to the solution, after that it was sealed in a vial. The mixture was kept at 100 °C for 72 h, then 1 mL (12.1 mmol) of pyrrolidine was added into the vial and the solution was stirred at 50 °C for 1 h. The same technique of purification was used as previously discussed (1.95 g, 82%). ^{1}H NMR (CDCl3, δ, ppm): 3.50 (m, 4H), 3.00–2.60 (m, 1H), 2.60–2.20 (m, 2H), 1.10 (m, 9H).

The synthesis and characterization of 8-arm of poly(2-ethyl-2-oxazoline) was described early [39].

2.6. Hydrolysis of Star-Shaped Polymers

A solution of 0.1 g of star-shaped poly-2-alkyl-2-oxazoline in 5 mL of 1 M hydrochloric acid was heated in a sealed ampoule at 100 °C during 24 h, after which it was evaporated to dryness. The residue was dissolved in 5 mL of ethyl alcohol, dialyzed against sodium bicarbonate (concentration 0.1 mol/L) using CellaSep dialysis bags with MWCO 3500 Da and freeze-dried. The product was dissolved in 15 mL of propionic anhydride, heated at 50 °C during 30 min, and evaporated under reduced pressure.

2.7. Characterization of Prepared Star Samples

UV–visible spectra were obtained using the SF-256 (LOMO-Photonika, Saint-Peterburg, Russia) spectrophotometer for ethanol solutions. The NMR spectra were measured on the Bruker AC400 (400 MHz) (Bruker, Billerica, MA, USA) spectrometer using chloroform solutions. Dialysis was conducted using dialysis sacks (CellaSep, Orange Scientific, Braine-l'Alleud, Belgium); MWCO, 3500 Da. Chromatographic analysis was performed on the Shimadzu LC-20AD chromatograph (Shimadzu Corporation, Nishinokyo Kuwabara, Japan) equipped with the TSKgel G5000HHR column (5 μm, 7.8 mm × 300 mm, Tosom-Bioscience, Tokyo, Japan) and light scattering and UV detectors. The mobile phase was a solution of LiBr (0.1 mol/L) in dimethylformamide at 60 °C. Polyethyleneglycol standards were chosen.

2.8. Investigation of Molecular-Dispersed Polymer Solutions

Molar mass and hydrodynamic characteristics of the synthesized polymers were obtained by molecular hydrodynamic and optics methods. Measurements were carried out in 2-nitropropane (dynamic viscosity η_0 = 0.72 cP, density ρ_0 = 0.982 g·cm^{-3}, and refractive index n_0 = 1.394) at 21 °C.

Dynamic and static light scattering was studied using the Photocor Complex setup (Photocor Instruments Inc., Moscow, Russia); the light source was a Photocor-DL diode laser with a wavelength λ = 658.7 nm. The correlation function of the scattered light intensity was obtained using the Photocor-PC2 correlator with 288 channels and processed using the DynaLS software (ver. 8.2.3, SoftScientific, Tirat Carmel, Israel).

The distribution of the light scattering intensity I over the hydrodynamic radii R_h of the particles present in the solutions was unimodal (Figure 1). Within the studied concentration range, radii $R_h(c)$ depended on concentration c. Therefore, to determine the hydrodynamic radius $R_{h\text{-}D}$ of macromolecules, the $R_h(c)$ values were extrapolated to zero concentration (Figure S1). The diffusion coefficients D_0 of macromolecules were calculated according to the Stokes–Einstein equation using the obtained values of $R_{h\text{-}D}$

$$D_0 = kT_a/(6\pi\eta_0 R_{h\text{-}D}) \quad (1)$$

where k is the Boltzmann constant and T_a is the absolute temperature.

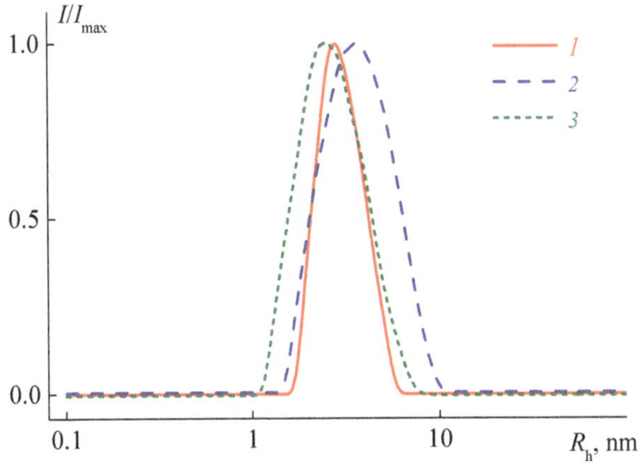

Figure 1. The dependencies of relative intensity I/I_{max} of scattered light on the hydrodynamic radii R_h of scattering species for 2-nitropropane solutions of C8A-(PiPrOx-b-PEtOx) at c = 0.040 g·cm^{-3} (1) C8A-(PEtOx-b-PiPrOx) at c =0.063 g·cm^{-3} (2) and C8A-P(EtOx-grad-iPrOx) at c = 0.069 g·cm^{-3} (3) I_{max} is the maximum value of light scattering intensity I at a given polymer concentration.

For all studied polymer solutions, there was no light scattering asymmetry; therefore, the weight average molar mass M_w and the second virial coefficient A_2 were found by the Debye method, taking measurements at a scattering angle 90°:

$$cH/I_{90} = 1/M_w + 2A_2c \quad (2)$$

where I_{90} is the light scattering intensity for an angle of 90° and c is the solution concentration. Optical constant H is calculated by the formula

$$H = 4\pi^2 n_0{}^2 (dn/dc)^2/N_A\lambda^4 \quad (3)$$

where dn/dc is the refractive index increment and N_A is the Avagadro number. Figure 2 shows the Debye dependencies for the studied star-shaped polymers. They are typical for dilute polymer solutions. The obtained values of M_w and A_2 are listed in Table 1. Note that positive values of the second virial coefficient indicate a good thermodynamic quality of 2-nitropropane for the studied polymer stars. The refractive index increment dn/dc was measured on the RA-620 refractometer (KEM, Tokyo, Japan). The dn/dc values (Table 1) were determined from the slope of the concentration dependence of the difference $dn = n - n_0$ in the refractive indices of solutions n and 2-nitropropane n_0 (Figure S2).

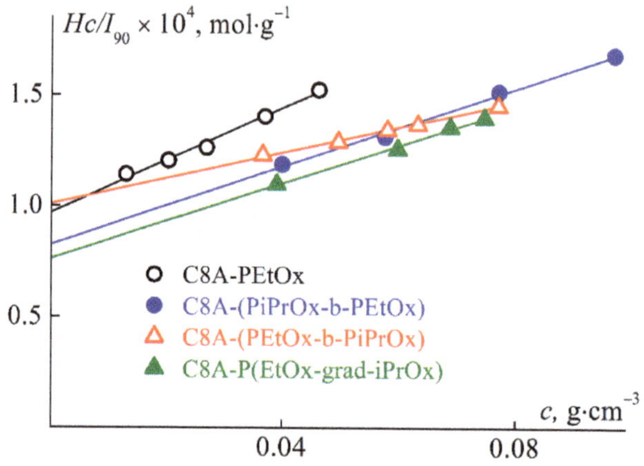

Figure 2. Concentration dependences of Hc/I_{90} for the CA8-PAlOx solutions in 2-nitropropane.

Table 1. Molar mass and hydrodynamic characteristics of star-shaped C8A-PAlOx.

Sample	M_w, g·mol^{-1}	Đ	$R_{h\text{-}D}$, nm	[η], cm^3·g^{-1}	$R_{h\text{-}[\eta]}$, nm	dn/dc, cm^3·g^{-1}	A_2·10^4, cm^3·mol·g^{-2}
C8A-PEtOx	10,300	1.38	2.6	8.2	2.4	0.1246	5.9
C8A-(PiPrOx-b-PEtOx)	12,100	1.21	2.4	5.9	2.2	0.1166	4.4
C8A-(PEtOx-b-PiPrOx)	10,000	1.35	2.0	4.9	2.0	0.1156	2.8
C8A-P(EtOx-grad-iPrOx)	13,200	1.41	2.9	7.7	2.5	0.1160	4.3

Ostwald-type glass viscometers (Cannon Instrument Company Inc., State College, PA USA) were used to measure intrinsic viscosity [η]. The solution temperature was regulated by a thermostat with a temperature control unit T-100 (Grant, Cambridge, UK). The solvent efflux time was 59.4 s. The concentration dependencies of the reduced viscosity η_{sp}/c (Figure S3) were analyzed using the Huggins equation:

$$\eta_{sp}/c = [\eta] + k'[\eta]^2 c \quad (4)$$

where k' is the Huggins constant. High k' values, from 1.4 to 2.4, were obtained for the studied polymers. Note that increased values of the Huggins constant are often reported for not very high molecular mass samples of polymers with increased intramolecular density [41,42]. Using the obtained values of the intrinsic viscosity [η], the so-called viscosity hydrodynamic radius $R_{h\text{-}\eta}$ of macromolecules were calculated by Einstein's formula:

$$R_{h\text{-}\eta} = (3M[\eta]/(10\pi N_A))^{1/3} \quad (5)$$

2.9. Investigation of Self-Organization in Aqueous Solutions

The thermosensitive behavior of aqueous solutions of CA8-PAlOx was studied by light scattering and turbidimetry methods using the Photocor Complex setup described above. The experiments were carried out in a wide range of concentrations and temperatures. The temperature T was changed discretely with a step from 0.5 to 5 °C; the value of T was regulated with an accuracy of 0.1 °C. The measurement procedure is described in detail in [43]. After the given temperature was established, the dependencies of the light scattering intensity I and the optical transmittance I^* on time t were obtained at a scattering angle of 90°. The hydrodynamic radii R_h of the scattering objects and their contribution S_i into total solution intensity I were determined when the values of I and I^* became constant in time. These measurements were carried out at scattering angles from 45° to 135° to confirm the diffusion nature of the modes and obtain the extrapolated values of

R_h and S_i. The laser power changes from 5 to 30 mV and/or optical filters placing on the photodetector allowed one to attenuate the light scattering signal to 1.5 MHz and maintain the linearity of the device regarding *I*.

Before static and dynamic light scattering experiments, the solutions, 2-nitropropane and calibration liquid, toluene, were filtered through the Millipore syringe filter (Merck, Germany) with a pore diameter of 0.20 μm. The water solutions were filtered through hydrophilic PTFE Millipore (Merck, Germany) membrane filters with a pore diameter of 0.45 μm.

3. Results and Discussion

3.1. Synthetic Approach

Star-shaped poly(2-alkyl-2-oxazoline) block copolymers were synthesized using a "grafting from" approach. At first, the octafunctional macrocyclic initiator was prepared based on the tert-butyl-calix[8]arene. The lower rim of macrocycle was modified with ester groups to increase solubility, whereas the upper rim was functionalized with initiating sulfonyl chloride moieties. The reaction scheme is presented in Figure 3. Since aromatic sulfonyl chlorides were shown to be effective initiators of oxazoline polymerization [44], this approach was successfully applied in the present paper. The kinetic studies of 2-ethyl-2-oxazoline polymerization initiated by the obtained abovementioned calixarene initiator was reported [45] and it showed that the initiation reaction is rapid and chain growth proceeds via the "living chain" mechanism.

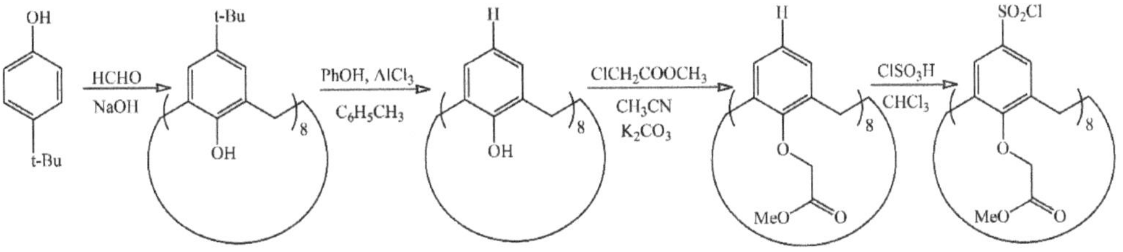

Figure 3. Synthesis of the octafunctional macrocyclic initiator.

3.2. Polymer Synthesis

Monomers of 2-ethyl- and 2-isopropyl-2-oxazolines with the optimal hydrophobic–hydrophilic balance were chosen to obtain thermosensitive polymers. Sulfolane was chosen as the solvent keeping in mind the high rate of oxazoline polymerization in this solvent [46] and the enough solubility of the initiator in sulpholane. The post-polymerization technique was applied to obtain 8-arm star-shaped block-copolymers. After the complete consumption of the first type monomer, the polymerization was reinitiated by injection of the second type monomer. Two samples of block-copolymers were synthesized with a different order of blocks, namely, CA8-(PEtOx-b-PiPrOx) and CA8-(PiPrOx-b-PEtOx) (Figure 4). Monomers were taken in equivalent molar amounts to obtain polymeric blocks with equal lengths. It was shown, that amine-type terminating agents are the most preferred for 2-oxazoline polymerization because of rapid termination on the 5-position of the oxazoline ring [47]. Therefore pyrrolidine was used as the termination agent.

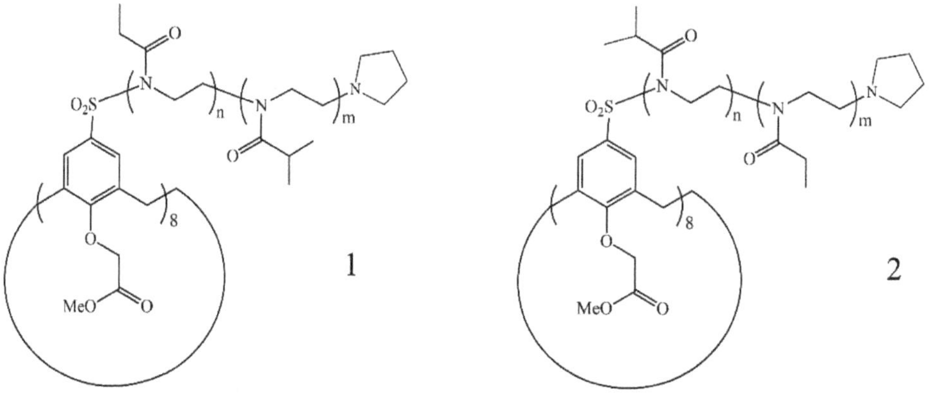

Figure 4. Structures of obtained star-shaped block-copolymers C8A-(PEtOx-b-PiPrOx) (**1**) and C8A-(PiPrOx-b-PEtOx) (**2**).

In order to obtain the statistical copolymer, equivalent amounts of 2-ethyl-2-oxazoline and 2-isopropyl-2-oxazoline were mixed in the polymerization vial. The proposed method of synthesis consisted of the simultaneous copolymerization of both monomers. It was found that the relative reactivity of EtOx and iPrOx in simultaneous copolymerization are equal to 0.79 and 1.78, respectively [48]. It can be assumed that the structure of the statistical copolymer formed in the reaction mixture would be the gradient most probably. Figure 5 shows the structures of C8A-P(EtOx-grad-iPrOx) and star-shaped homopolymer C8A-PEtOx.

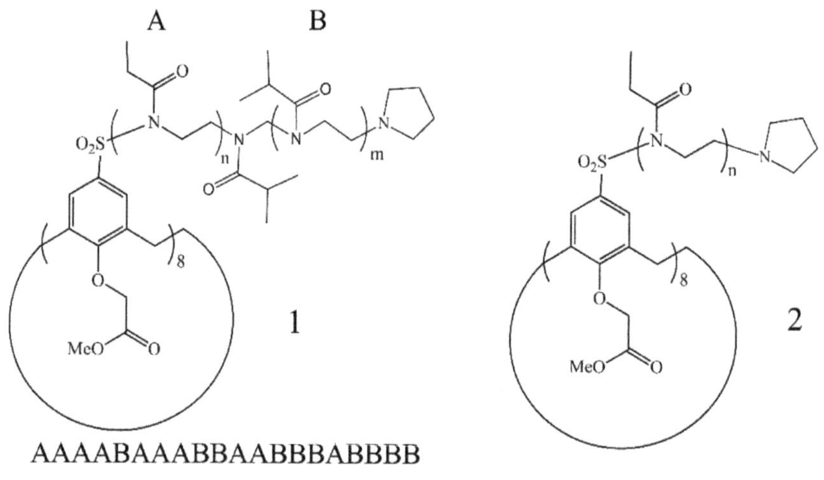

Figure 5. Structures of obtained star-shaped C8A-P(EtOx-grad-iPrOx) (**1**) and C8A-PEtOx (**2**).

It is well known that the only parent calix[8]arene exists in the stable cone conformation due to the intramolecular H-bonding of hydroxyl groups at the lower rim while any kinds of chemical transformation of these moieties leads to conformationally labile structures that are clearly visible in the NMR spectra. It was found that methylene protons of the macrocycle did not exhibit the AB quartet that is typical for the core conformer and had no more complicated signals that are typical for paco or the other conformers [40]. Broad singlet at about 4 ppm verified the quick rotation in macrocycle. On the other hand functional moieties at the lower rim really define the complexation ability of the macrocycle that keeps in mind that the probable biomedical applications is the goal of our research.

3.3. Characterization of Polymers

The spectra of all samples are completely similar and contain signals of both ethyl (2.50–2.20 ppm) and isopropyl (3.05–2.55 ppm) groups, which confirms the presence of both monomers in the polymer chains (Figure 6). Additionally, minor signals of the calix[8]arene core, attributed to the bridged methylene groups (4.46 ppm) were detected and ester methylene groups at the lower rim (4.23–3.97 ppm). According to the NMR data the integral intensities of proton signals at about 2.3 ppm ($C\underline{H}_2CH_3$ in ethyloxazoline) and doublet at about 2.5 and 2.8 ($C\underline{H}(CH_3)_2$ in isopropyloxazoline) are 2:1. Based on the integral intensities of proton signals, it was determined that the ratio of the monomer in the copolymer was at about 1:1 for all samples. Therefore, both blocks have a near equal degree of polymerization. The same ratio of components was calculated for the gradient copolymer sample. This conclusion is confirmed by the fact that the values of the refractive index increment dn/dc coincide within the experimental error for stars with copolymer arms (Table 1) but lower than dn/dc for C8A-PEtOx. This behavior was observed for PAlOx copolymers [25,39,49] and is explained by an increase in the refractive index on the passage from polymers with ethyl groups to samples containing isopropyl ones.

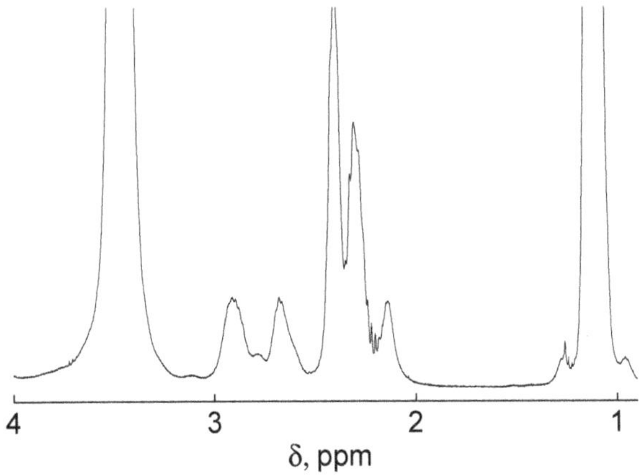

Figure 6. ^{1}H NMR spectrum of star-shaped C8A-(PEtOx-b-PiPrOx).

The presence of calix[8]arene cores in the synthesized copolymers was also confirmed by UV–visible spectroscopy (Figure S4). The typical absorption bands at about 250–290 cm^{-1} confirm the presence of the macrocycles core in the polymer structure.

The number of arms in a star-shaped polymer was determined by the selective destruction of the macromolecule without degradation of the poly-2-alkyl-2-oxazoline original length. For this purpose, it was applied the original procedure involving acid hydrolysis of sulfonamide groups to polyethylenimine followed by acylation with propionic anhydride and GPC analysis of the obtained oligomers. Using the MM of star-shaped polymers and their arms, the arm number fa was calculated. The fa values are at about 8, i.e., all polymers have an eight-arm structure.

The arms of the synthesized stars are short (Table 2). Their length L_a was calculated by the ratio:

$$L_a = N_a\lambda_a = \lambda_a(M_w - M_{C8A})/f_a M_{0\text{-}a} \quad (6)$$

where $f_a = 8$ is the arm number, N_a is the polymerization degree of arms, $\lambda_a = 0.378$ nm is the length of the monomer unit of poly-2-alkyl-2-oxazoline [50], and $M_{C8A} = 1928$ g·mol^{-1} is the molar mass of CA8. In the case of copolymers, the average molar mass (MM) of the arm monomer units was equal to 106 g·mol^{-1}, i.e., the average value of MM of 2-ethyl-2-oxazoline (99 g·mol^{-1}) and 2-isopropyl-2-oxazoline (113 g·mol^{-1}). Poly-2-ethyl and poly-2-isopropyl-2-oxazoline were comb-shaped with short side chains containing three valence bonds to the point of the most distant from the backbone chain and differing by one $-CH_3$ group. Systematic studies of various classes of comb-shaped polymers [51] showed that, with those insignificant structural variations, the conformational characteristics of polymers almost do not change. Therefore, it can be assumed that the Kuhn segment lengths for blocks of poly-2-ethyl- and poly-2-isopropyl-2-oxazoline are equal to $A = (1.4 - 1.8)$ nm [50,52]. Accordingly, using the average value $A = 1.6$ nm, we can estimate the number N^* of Kuhn segments in the arms of the studied polymer stars. From Table 2 it is seen that N^* did not exceed 3.

Table 2. Structure characteristics and contraction factors for star-shaped CA8-PAlOx.

Sample	M_w, g·mol^{-1}	M_w (arm), g·mol^{-1}	f_a	L_a, nm	N^*	g'	g	$A_0 \times 10^{10}$, erg·K^{-1}mol$^{-1/3}$
C8A-PEtOx	10,300	1100	8.0	4.0	2.4	0.45	0.37	2.7
C8A-(PiPrOx-b-PEtOx)	12,100	1300	8.0	4.5	2.8	0.29	0.23	2.7
C8A-(PEtOx-b-PiPrOx)	10,000	1050	8.1	3.6	2.3	0.27	0.21	2.9
C8A-P(EtOx-grad-iPrOx)	13,200	1400	7.6	4.7	2.9	0.34	0.27	2.7

3.4. *Hydrodynamic Characteristics and Conformation of Star-Shaped CA8-PAlOx-UR*

Gel permeation chromatography (Figure S5) shows that all samples are characterized by a monomodal molar mass distribution. This behavior is in qualitative agreement with the dynamic light scattering data obtained in molecularly dispersed solutions in a wide range of polymer concentrations. The polydispersity indexes $Đ = M_w/M_n$ of studied star samples are shown in Table 1.

It should also be noted that the molar masses of the synthesized star-shaped polymers differed insignificantly, the maximum difference was about 30%. Therefore, in further analysis and comparison of the results obtained, the influence of MM on the polymer characteristics could be neglected.

The hydrodynamic radii of CA8-PAlOx molecules are less than the arm lengths L_a (Tables 1 and 2). This indicates that the macromolecules were compact and the arms, despite their small length, were relatively strongly folded. The compact structure of CA8-PAlOx molecules was also confirmed by the low values of intrinsic viscosity [η]. Table 2 shows the values of the viscosity contraction factor.

$$g' = [\eta]_{star}/[\eta]_{lin} \quad (7)$$

where $[\eta]_{star}$ and $[\eta]_{lin}$ are the characteristic viscosities of star-shaped and linear polymers of the same MM. As the $[\eta]_{lin}$ values, we used the average values [η] for linear poly-2-ethyl 2-oxazoline, calculated from the Mark–Kuhn–Houwink–Sakurada equations obtained in thermodynamically good solvents [50,53]. Moreover, the conformation and hydrodynamic properties of linear PEtOx and PiPrOx could be assumed identical. This conclusion is supported by the results of studies of linear and star-shaped poly(2-ethyl-2-oxazine) [54]. It was shown that a change by one $-CH_2-$ group of the monomer unit of pseudo-polypeptoids does not lead to a change in the conformational characteristics of the polymer.

The behavior of contraction factor $g = (R_g)_{star}^2/(R_g)_{lin}^2$, determined from the ratio of the squared gyration radii of star-shaped $(R_g)_{star}$ and linear $(R_g)_{lin}$ polymers, has been theoretically analyzed in detail. Zimm and Stockmayer [55] showed polymer stars with long monodisperse arms:

$$g = (3f_a - 2)/f_a^2 \quad (8)$$

Therefore, $g = 0.34$ for a star with eight arms. In the case of polydisperse arms [56,57]:

$$g = 3f_a/(f_a + 1)^2, \quad (9)$$

where $g = 0.30$ for eight-arm stars. Daoud–Cotton theory [58] describes g for multiarm star-shaped polymers with short arms as

$$g = f_a^{-4/5}, \quad (10)$$

where the contraction factor is $g = 0.19$ at $f_a = 8$. For the studied CA8-PAlOx, the value of g can be calculated using the empirical equation [59]

$$g' = (1.104 - 0.104g^7)g^{0.906} \quad (11)$$

The g values for CA8-PAlOx with copolymer arms lie between the theoretical values of the contraction factor for star-shaped macromolecules with short and long arms (Table 2). This behavior is in agreement with the findings of the study of six-arm polypeptoids [49], which can be considered as long-arm molecules if the arms contain more than six Kuhn segments. A higher value of g was obtained for CA8-PEtOx, the reason for which remains unclear.

The low hydrodynamic invariant A_0 [51,60,61]:

$$A_0 = (\eta_0 D_0 (M[\eta]/100)^{1/3}/T_a \quad (12)$$

where it justifies the compact structure of the molecules of studied CA8-PAlOx. The obtained values of A_0 (Table 1) were less than 3.2×10^{-10} erg·K^{-1}·mol$^{-1/3}$, predicted theoretically for flexible chain polymers [51,60]. In particular, for linear poly-2-ethyl-2-oxazoline the average value of hydrodynamic invariant is 3.1×10^{-10} erg·K^{-1}·mol$^{-1/3}$ [58]. On the other hand, the A_0 values for the star-shaped CA8-PAlOx are noticeably larger than the hydrodynamic invariant for dendrimers and hyperbranched polymers [62–64], which are polymers with high intramolecular density. Similar values of A_0 were obtained previously for star-shaped polypeptoids with short arms [49]. Accordingly, analysis of the hydrodynamic invariant makes it possible to conclude that polymer stars occupy an intermediate position between linear flexible chain polymers and dendritic systems in terms of the intramolecular density.

3.5. *Self-Organization of C8A-PAlOx-UR Molecules in Aqueous Solutions*

For all studied solutions at low temperatures, three modes with hydrodynamic radii R_f (fast mode), R_m (middle mode), and R_s (slow mode) were observed (Figure 7). R_f values did not vary with concentration (Figure 8) and the concentration-average R_f value coincided with the size $R_{h\text{-}D}$ of macromolecules for each polymer and so the particles responsible for the fast mode are single macromolecules. The middle mode and slow mode reflect the diffusion of aggregates similar to those formed in solutions of thermoresponsive polymers [65–76], including star-shaped PAlOx [25,49]. The reason for the formation of aggregates is the interaction of hydrophobic cores.

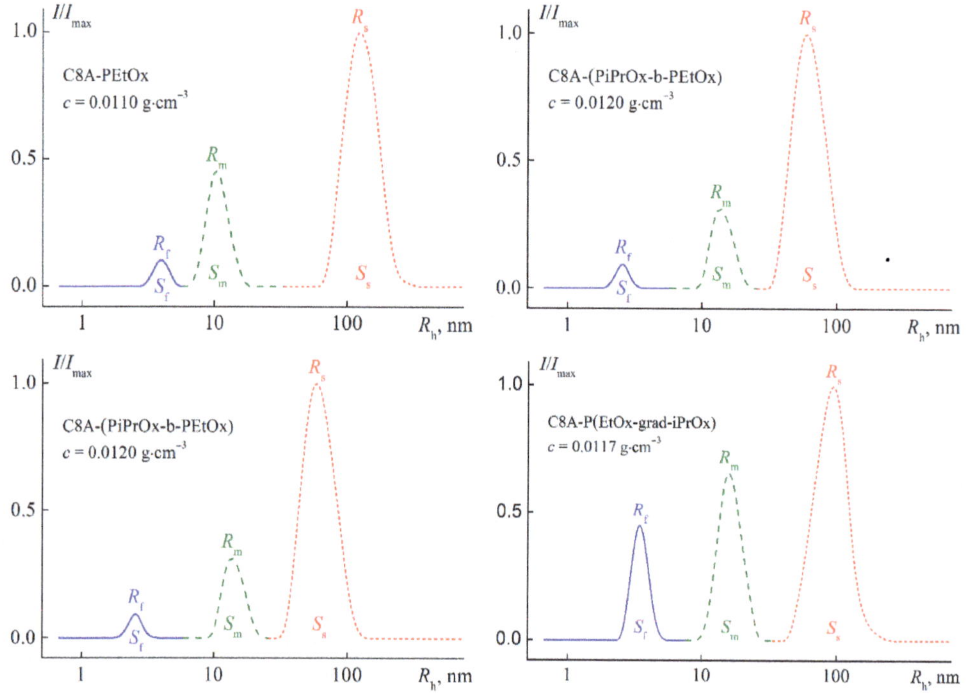

Figure 7. The dependences of relative intensity I/I_{max} of scattered light on the hydrodynamic radius R_h of scattering species for aqueous solutions of C8A-PAlOx at 21 °C. I_{max} is the maximum value of light scattering intensity for the given solution concentration.

The R_m and R_s values display an independence of concentrations up to $c \approx 0.015$ g·cm^{-3}, above which their slight growth is observed (Figure 8). As known, a change in the hydrodynamic radii can be caused both by a change in size of scattering species and by the concentration dependence of the diffusion coefficient D. The dependence $D(c)$ is determined by the values of the second virial coefficient, molar mass, concentration coefficient of sedimentation, and specific partial volume. Probably, these factors make up for each other at low concentrations, when the aggregate dimensions increase and the coefficient D reduces. Consequently, R_m and R_s were constant at $c < 0.015$ g·cm^{-3}.

One can see in Figure 7, the largest contribution S_s to light scattering is made by large aggregates with a radius R_s, while the contribution S_f of macromolecules is minimal. Nevertheless, the latter prevails in the solution. Indeed, the contribution $I_i = S_i I$ of ith set of particles to the total light scattering intensity I is described by the relation $I_i \sim c_i R_i^x$, where c_i and R_i are the weight concentration and radius of the ith particles, respectively [77,78]. The value of the exponent x depends on the shape of the scattering particles. The fraction of each type of particles in solutions of the studied polymers can be roughly estimated using the models of a hard sphere (molecules and aggregates with a radius R_m, $x = 3$) and a coil (large aggregates, $x = 2$). This approach is supported by the results of the conformational analysis of multiarm stars with short arms [58,79,80] and the studying micelle-like and large aggregates [71–75,81]. An estimation shows that the relative fraction c_f/c of molecules in solutions of the C8A-(PiPrOx-b-PEtOx) and C8A-(PEtOx-b-PiPrOx) stars with block copolymer arms was about 80% (c_f is the concentration of macromolecules in solution). The c_f/c ratio increased up to 87% for the homopolymer C8A-PEtOx and 98% for the gradient C8A-P(EtOx-grad-iPrOx). Note that, the weight fraction c_s/c of large aggregates did not exceed 10% for solutions of block copolymers and $c_S/c < 0.1\%$ for the two other stars (c_S is the concentration of large aggregates with a hydrodynamic radius R_S).

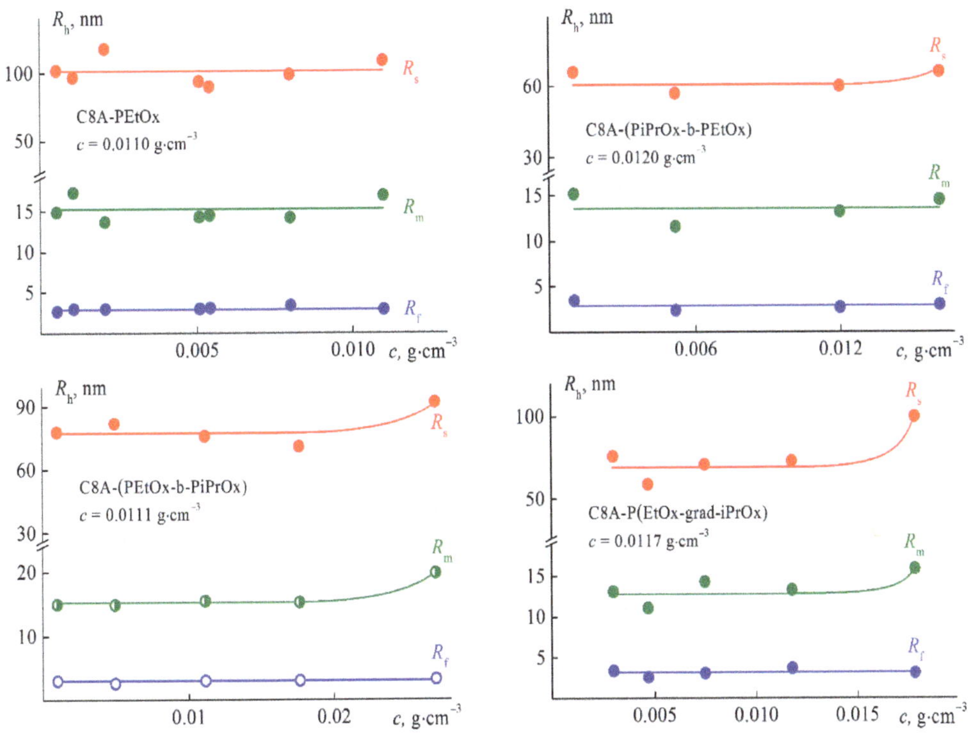

Figure 8. Concentration dependences of hydrodynamic radii R_h of scattering species for aqueous solutions of C8A-PAlOx at 21 °C.

Thus, the arm structure did not significantly affect the set of scattering objects and the hydrodynamic radii of aggregates, which is in opposition to results for star-shaped PAOx with copolymer arms grafted to a lower rim of the calix[8]arene [25,81]. This is probably due to both the more pleated loop conformation of C8A, functionalized along the upper rim [39], and the short arm length of the studied star samples. These factors lead to a decrease in the shielding of the core surface by arms, which promotes the aggregation. Probably, this is a reason why macromolecules disappear, or rather, have not been observed by the method of dynamic light scattering upon solution heating at relatively low temperatures, $T \leq 37$ °C, far from the phase separation interval (Figure 9).

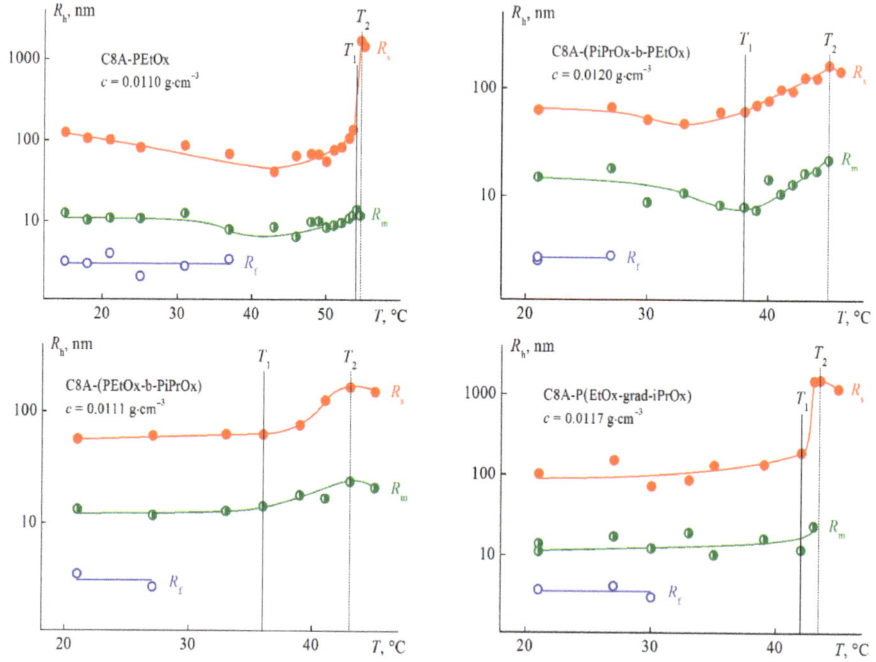

Figure 9. Temperature dependences of hydrodynamic radii R_h of scattering particles for aqueous solutions of C8A-PAlOx.

The temperatures of the onset T_1 and the finishing T_2 of the phase separation were determined by turbidimetry (Figure 10). Optical transmission I^* did not depend on the temperature up to T_1. On the contrary, the scattered light intensity I varies over the whole temperature range (Figure 10). At low temperatures, solution heating is accompanied by a slow I growth. The rate of change in intensity I dramatically increases at temperature T_1. Light scattering intensity reached a maximum value at T_2 and decreased slightly upon further heating for most of the studied solutions.

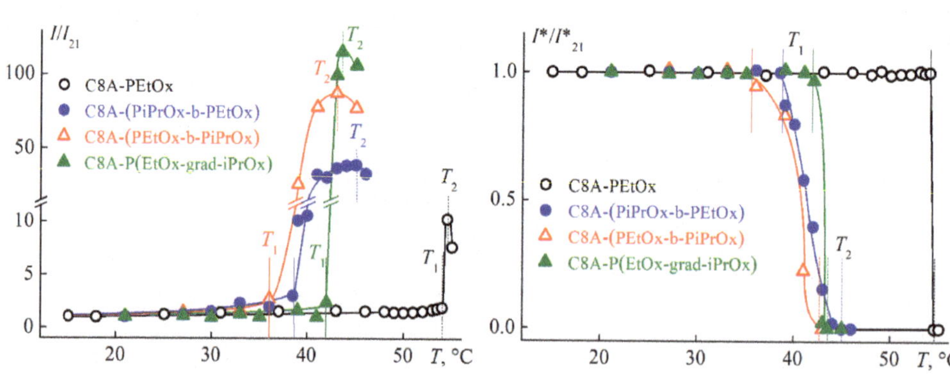

Figure 10. Dependencies of relative optical transmission I^*/I^*_{21} and relative light scattering intensity I/I_{21} on temperature T for investigated polymer stars. I^*_{21} and I_{21} are the optical transmission and light scattering intensity at 21 °C, respectively. The concentration of the solutions is the same as in Figures 7–9.

The $I(T)$ dependence was caused by changes in the size of the aggregates and their fraction in the solution with temperature. As mentioned above, macromolecules cease to be detected well below T_1, which indicates the aggregation of macromolecules at $T < T_1$, probably, caused by the dehydration of iPrOx units, which begins at relatively low temperatures [68]. As can be seen in Figure 9, for stars C8A-(PEtOx-b-PiPrOx) and C8A-P(EtOx-grad-iPrOx) on the periphery of which the more hydrophobic iPrOx units prevail, the hydrodynamic radii R_s of large aggregates increased with increasing temperature. At low temperatures, this change was very weak and greatly accelerated at temperature T_1 and the R_s values reached hundreds and even thousands of nanometers at T_2. In this case, the contribution S_s of large aggregates to the integral light scattering increased owing to a lowering in the contribution S_m of aggregates with a radius R_m. A similar behavior was observed earlier for four- and eight-pointed stars with PiPrOx arms grafting to the lower rim of calix[n]arene [79]. For solutions C8A-PEtOx and C8A-(PiPrOx-b-PEtOx) a decrease in R_s is observed at moderate temperatures (Figure 9). This change did not exceed 30% in comparisons with R_s value at 21 °C. A decrease in R_s, i.e., compaction of large aggregates, indicates the formation of inter- and intramolecular hydrogen bonds between dehydrated iPrOx units in one aggregate scale. Note that the R_s value decreased much more, sometimes more than ten times, for stars with a copolymer PAlOx arm grafted to the lower rim of C8A and for PiPrOx stars with carbosilane dendrimer cores [25,26,82]. The platter conformation of C8A and shorter arms in the studied polymer stars facilitated contacts between hydrophobic cores of different molecules, which should lead to aggregation. Indeed, the contributions S_s of large aggregates to light scattering increased with temperature for C8A-PEtOx and C8A-(PiPrOx-b-PEtOx). For example, for a C8A-PEtOx solution at a concentration of c = 0.0110 g·cm^{-3}, the S_s magnitude increased from 0.45 at 21 °C to 0.66 at a temperature of minimum value of R_s, and S_s varies from 0.81 to 0.94 for C8A-(PiPrOx-b-PEtOx) at c = 0.0120 g·cm^{-3}. Concerning smaller aggregates in solutions of the stars under discussion, the R_m value changes with temperature in the same way as the radius of R_s, but this change is within the experimental error (Figure 9). The dimensions of large aggregates reached a maximum value within the interval of phase separation and then decreased, reflecting the compaction of these particles. However, it should be taken into account that under these conditions, light scattering was multiple and a quantitative analysis of the results was impossible.

One phase transition was observed for all studied star-shaped polymers at all concentrations. Figure 11 shows the phase separation temperatures versus concentration. As expected, T_1 and T_2 depended on the arm structure. The highest values of T_1 and T_2 were obtained for C8A-PEtOx, which were 15–20 °C lower than T_1 and T_2 for the eight-arm star with the PEtOx arm grafted to the lower rim of C8A [83] at the same concentrations.

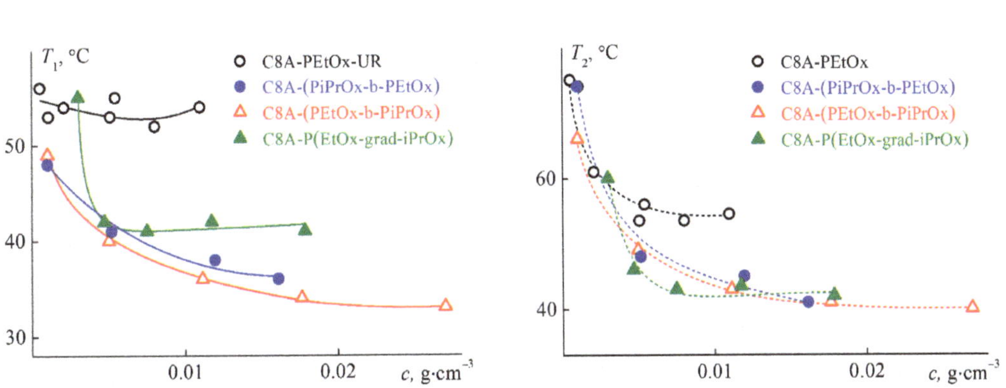

Figure 11. Phase transition temperatures vs. polymer concentration for investigated polymer solutions.

The introduction of a more hydrophobic block of PiPrOx into the star macromolecules led to a decrease in the T_1 and T_2 values for solutions of C8A-(PEtOx-b-PiPrOx), C8A-(PiPrOx-b-PEtOx) and C8A-P(EtOx-grad-iPrOx) as compared to C8A-PEtOx solutions. The temperatures T_1 of the phase separation onset for stars with copolymer arms differed and the temperatures T_2 of its finishing product almost coincided. Similar results were obtained for the copolymer PAlOx stars with the C8A core functionalized along the lower rim [25], for which, however, the values of T_1 and T_2 were significantly higher. The pattern of the concentration dependencies of the phase separation temperatures for the star-shaped polymers under consideration differed markedly. In particular, an LCST of about 41 °C can be determined reliably for the star C8A-P(EtOx-grad-iPrOx). The flattening of the $T_1(c)$ dependence for C8A-(PEtOx-b-PiPrOx) in the region of high concentrations allows suggesting that LCST was close to 32–33 °C, that is almost 10 °C lower than for C8A-P(EtOx-grad-iPrOx). The T_1 values for the block copolymer star with a more hydrophilic outer block changed quite strongly within the whole concentration range and it was not possible to estimate the LCST for C8A-(PiPrOx-b-PEtOx), but it was unlikely that it differed greatly from 30 °C.

At the same concentrations, the phase separation temperatures for the studied copolymer stars are several degrees higher than T_1 and T_2 for solutions of a star with a C8A core functionalized along the lower rim by PiPrOx chains [79]. It is expected given that the dehydration temperature of PEtOx is higher than for PiPrOx [68]. On the other hand, despite the presence of hydrophilic EtOx units in arms, the LCSTs for solutions of the studied stars with copolymer arms are near to the LCST of linear PiPrOx [65,84]. The latter can be explained by both the distinction in molecule architecture and hydrophobic C8A core and different MMs of the compared polymers.

All of the results discussed were obtained, when the characteristics of the solution reached constant values in time after a jump-like temperature change. The analysis of the processes of establishing the "equilibrium" state of the system, in particular, determining their duration depending on the chemical structure of the polymer and external conditions are an important task since the rate of change of the characteristics of the stimulus-sensitive polymer solution determines the application field of the material based on it. Nevertheless, the number of works devoted to solving this problem remains small [43,68,85–90] and many questions remain open.

Figure S6 shows the dependencies of the scattered light intensity I and optical transmission I* on time t after the discrete temperature change. The moment when the temperature reaches a given value is taken as $t = 0$. The intensity I and the transmission I^* reach a constant value in time t_{eq}.

The duration of the processes of establishing the "equilibrium" state was long for the studied star-shaped C8A-PAlOx and strongly depended on the temperature at each concentration (Figure 12). At low temperatures, the t_{eq} values ranged from 1200 to 3000 s, which is higher than the usual ones for linear stimulus-sensitive polymers [68,85,86]. Time t_{eq} sharply increased upon heating and reached the maximum t_{eq}^{max} value near the phase separation. At $T > T_1$, t_{eq} fell and was about 1000 s at $T \rightarrow T_2$. A similar $t_{eq}(T)$ dependence was observed earlier for PAlOx of a complex architecture [43,90–93].

No systematic variation of t_{eq}^{max} with concentration was found for the studied stars, and the t_{eq}^{max} values were in the range from 8000 to 15,000 s. Similar times for the reaching "equilibrium" values of solution characteristics were obtained for star-shaped PAlOx with copolymer arms grafted to the lower rim of the calix[8]arene [25,81]. In the case of star-shaped PiPrOx, the time t_{eq}^{max} strongly depends on the core configuration; the t_{eq}^{max} value did not exceed 20,000 s when arms grafted to the upper rim of C8A and reached 40,000 s when C8A core was functionalized along the lower rim [43]. Thus, the rate of self-organization processes in solutions of PAlOx stars depends both on the way of functionalization of the calix[n]arene core, which changes its configuration, and on the structure of PAlOx arms because the presence of EtOx units led to a decrease in values of t_{eq}^{max} as compared to star-shaped PiPrOx homopolymers.

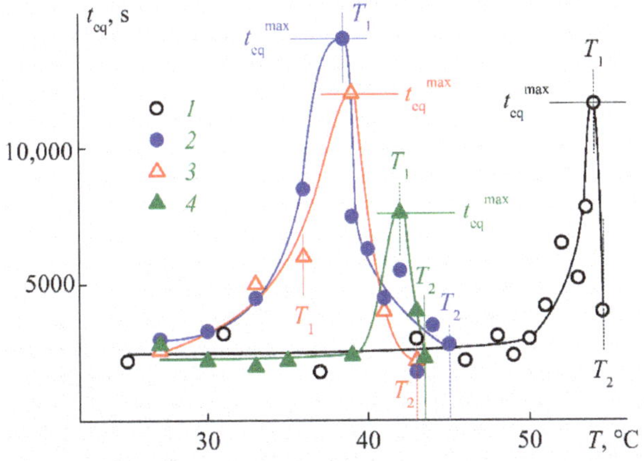

Figure 12. Dependences of time t_{eq} on temperature T for aqueous solutions of C8A-PEtOx (1) at concentration c = 0.0110 g·cm^{-3}, C8A-(PiPrOx-b-PEtOx) (2) at c = 0.0120 g·cm^{-3}, C8A-(PEtOx-b-PiPrOx) (3) at c = 0.0111 g·cm^{-3}, and C8A-P(EtOx-grad-iPrOx) (4) at c = 0.0117 g·cm^{-3}.

4. Conclusions

Star-shaped poly-2-alkyl-2-oxazolines with arms of block and gradient copolymers of 2-ethyl- and 2-isopropyl-2-oxazolines grafted to the upper rim of calix[8]arene were synthesized by the "grafting from" method. The ratio of 2-ethyl- and 2-isopropyl-2-oxazoline units was 1:1, which was confirmed by NMR spectroscopy and refractometry. The arms of the synthesized stars were relatively short, their length did not exceed 5 nm. The arm conformation was folded in an organic solvent, and the star-shaped C8A-PAlOx macromolecules were characterized by compact dimensions and heightened intramolecular density. The influence of the arm structure on the conformation of C8A-PAlOx molecules was not observed.

At low temperatures, the aqueous solutions of the studied C8A-PAlOx-UR were not molecular dispersed; there were aggregates formed due to the interaction of hydrophobic calix[8]arene cores. Nevertheless, single macromolecules prevailed, its relative fraction always exceeded 80%. No systematic changes in the set of scattering objects and their dimensions were found for studied stars.

Below the phase separation temperature, the heating of aqueous solutions of C8A-PAlOx with prevailing iPrOx units in the outer layer caused the aggregation of macromolecules because of an increase in the degree of dehydration of iPrOx units with temperature. For stars with an outer PEtOx layer, there was a rather wide temperature interval, in which compaction prevailed.

One phase transition was detected for all studied polymer stars. The temperatures of its onset T_1 and finish T_2 decreased with a growth of the content of more hydrophobic iPrOx units. The highest temperatures were obtained for the homopolymer C8A-PEtOx, while they decreased by 10–20 °C for stars with copolymer arms. The distinction between the phase separation temperatures for copolymer stars and a homopolymer with PiPrOx arms was moderate. Consequently, iPrOx dehydration determines the T_1 and T_2 values. The arm structure did not affect the value of T_2 but led to a change in the concentration dependence of T_1. Thus the highest LCST could be expected for a star with gradient copolymer arms, while LCST for C8A-(PEtOx-b-PiPrOx)-UR and C8A-(PiPrOx-b-PEtOx)-UR were 10 °C lower.

The molar masses of synthesized samples were similar and the effect of MM on the solution behavior was not detected. The way of arms grafting to the core, at the lower or upper rim of C8A, significantly affected the behavior of star-shaped PAlOx in aqueous solutions. First of all, it is manifested in a decrease of the phase separation temperatures for stars with C8A, functionalized along the upper rim. The reason for the observed phenomenon is the different core configuration. When functionalization is along the lower rim, the arms tended to shield the hydrophobic C8A, enveloped it, and slightly constricted the upper rim of C8A. The arms grafted to the upper rim, on the contrary, increased its radius due to steric interactions. Thus, the shape of the C8A with arms at the lower rim resembled a basket, while the C8A with arms at the upper rim looks like a plate. A change in the core configuration varied its accessibility to the solvent and, consequently, the phase separation temperature.

Supplementary Materials: The following are available online at https://www.mdpi.com/article/10.3390/polym13152507/s1, Figure S1: Plots of $R_{h\text{-}D}(c)$ vs. concentration c for the studied polymer stars in 2-nitropropane. Figure S2: Concentration dependencies of dn for the C8A-PaOx solutions 2-nitropropane. Figure S3: Reduced viscosity η_{sp}/c vs. c for the studied polymer stars in 2-nitropropane. Figure S4: UV–visible spectrum of the star-shaped C8A-(PEtOx-b-PiPrOx)-UR solution in ethanol. Polymer concentration is 0.005 g·cm^{-3}. Figure S5: GPC traces of star copolymers CA8-PAlOx. Figure S6: Time dependencies of light scattering intensity I/I_0 and the transmission I^*/I_0^* for solutions C8A-PAlOx. I_0 and I^*_0 are values of light scattering intensity and the optical transmission at $t = 0$.

Author Contributions: Conceptualization, A.F.; methodology, A.T.; formal analysis, T.K. and A.B.; investigation, T.K.; resources, A.A. and A.B.; data curation, A.A.; writing—original draft preparation T.K. and A.A.; writing—review and editing, A.F. and A.T.; visualization, A.B. and T.K.; supervision T.K.; project administration, A.F. All authors have read and agreed to the published version of the manuscript.

Funding: The study was carried out under the financing of the Ministry of Science and Highest Education of the Russian Federation, State registration No. AAAA-A20-120022090038-1.

Institutional Review Board Statement: Not applicable.

Informed Consent Statement: Not applicable.

Data Availability Statement: Not applicable.

Conflicts of Interest: The authors declare no conflict of interest.

References

1. Inoue, K. Functional dendrimers, hyperbranched and star polymers. *Prog. Polym. Sci.* **2000**, *25*, 453–571. [CrossRef]
2. Lee, H.; Pietrasik, J.; Sheiko, S.S.; Matyjaszewski, K. Stimuli-responsive molecular brushes. *Prog. Polym. Sci.* **2010**, *35*, 24–44 [CrossRef]
3. Voit, B. Hyperbranched polymers—All problems solved after 15 years of research? *J. Polym. Sci. Part A Polym. Chem.* **2005**, *43*, 2679–2699. [CrossRef]
4. Hadjichristidis, N.; Pitsikalis, M.; Pispas, S.; Iatrou, H. Polymers with complex architecture by living anionic polymerization *Chem. Rev.* **2001**, *101*, 3747–3792. [CrossRef]
5. Wu, W.; Wang, W.; Li, J. Star polymers: Advances in biomedical applications. *Prog. Polym. Sci.* **2015**, *46*, 55–85. [CrossRef]
6. Plamper, F.A.; Schmalz, A.; Penott-Chang, E.; Drechsler, M.; Jusufi, A.; Ballauff, M.A.; Muller, A.H.E. Synthesis and characterization of star-shaped poly(N,N-dimethylaminoethyl methacrylate) and its quaternized ammonium salts. *Macromolecules* **2007**, *40*, 5689–5697. [CrossRef]
7. Rudolph, T.; Crotty, S.; von der Lühe, M.; Pretzel, D.; Schubert, U.S.; Schacher, F.H. Synthesis and solution properties of double hydrophilic poly(ethylene oxide)-block-poly(2-ethyl-2-oxazoline) (PEO-b-PEtOx) star block copolymers. *Polymers* **2013**, *5*, 1081–1101. [CrossRef]
8. Kowalczuk, A.; Kronek, J.; Bosowska, K.; Trzebicka, B.; Dworak, A. Star poly(2-ethyl-2-oxazoline)s—Synthesis and thermosensitivity. *Polym. Int.* **2011**, *60*, 1001–1009. [CrossRef]
9. Kuckling, D.; Wycisk, A. Stimuli-responsive star polymers. *J. Polym. Sci. Part A Polym. Chem.* **2013**, *51*, 2980–2994. [CrossRef]
10. Felberg, L.E.; Brookes, D.H.; Head-Gordon, T.; Rice, J.E.; Swope, W.C. Role of hydrophilicity and length of diblock arms for determining star polymer physical properties. *J. Phys. Chem. B* **2015**, *119*, 944–957. [CrossRef]
11. Li, J.; Kikuchi, S.; Sato, S.; Chen, Y.; Xu, L.; Song, B.; Duan, Q.; Wang, Y.; Kakuchi, T.; Shen, X. Core-first synthesis and thermoresponsive property of three-, four-, and six-arm star-shaped poly(N,N-diethylacrylamide)s and their block copolymers with poly(N,N-dimethylacrylamide). *Macromolecules* **2019**, *52*, 7207–7217. [CrossRef]
12. Zhu, W.; Nese, A.; Matyjaszewsk, K. Thermoresponsive star triblock copolymers by combination of ROP and ATRP: From micelles to hydrogels. *J. Polym. Sci. Part A Polym. Chem.* **2011**, *49*, 1942–1952. [CrossRef]

13. Sezonenko, T.; Qiu, X.-P.; Winnik, F.M.; Sato, T. Dehydration, micellization, and phase separation of thermosensitive polyoxazoline star block copolymers in aqueous solution. *Macromolecules* **2019**, *52*, 935–944. [CrossRef]
14. Zhang, Y.; Guan, T.; Han, G.; Guo, T.; Zhang, W. Star block copolymer nanoassemblies: Block sequence is all-important. *Macromolecules* **2019**, *52*, 718–728. [CrossRef]
15. Hirao, A.; Inushima, R.; Nakayama, T.; Watanabe, T.; Yoo, H.-S.; Ishizone, T.; Sugiyama, K.; Kakuchi, T.; Carlotti, S.; Deffieux, A. Precise synthesis of thermo-responsive and water-soluble star-branched polymers and star block copolymers by living anionic polymerization. *Eur. Polym. J.* **2011**, *47*, 713–722.
16. Mori, H.; Ebina, Y.; Kambara, R.; Nakabayashi, K. Temperature-responsive self-assembly of star block copolymers with poly(ionic liquid) segments. *Polym. J.* **2012**, *44*, 550–560. [CrossRef]
17. Mendrek, B.; Fus, A.; Klarzyńska, K.; Sieroń, A.L.; Smet, M.; Kowalczuk, A.; Dworak, A. Synthesis, characterization and cytotoxicity of novel thermoresponsive star copolymers of N,N′-Dimethylaminoethyl methacrylate and hydroxyl-bearing oligo(ethylene glycol) methacrylate. *Polymers* **2018**, *10*, 1255. [CrossRef]
18. Cortez-Lemus, N.A.; Licea-Claverie, A. Preparation of a mini-library of thermo-responsive star (NVCL/NVP-VAc) polymers with tailored properties using a hexafunctional xanthate RAFT agent. *Polymers* **2018**, *10*, 20. [CrossRef]
19. Gitsov, I.; Fréchet, J.M.J. Stimuli-responsive hybrid macromolecules: novel amphiphilic star copolymers with dendritic groups at the periphery. *J. Am. Chem. Soc.* **1996**, *118*, 3785–3786. [CrossRef]
20. Cohen Stuart, M.A.; Huck, W.T.S.; Genzer, J.; Muller, M.; Ober, C.; Stamm, M.; Sukhorukov, G.B.; Szleifer, I.; Tsurkruk, V.V.; Urban, M.; et al. Emerging applications of stimuli-responsive polymer materials. *Nat. Mater.* **2010**, *9*, 101–113. [CrossRef]
21. Borner, H.G.; Kuhnle, H.; Hentschel, J. Making "smart polymers" smarter: Modern concepts to regulate functions in polymer science. *J. Polym. Sci. Part A Polym. Chem.* **2010**, *48*, 1–14. [CrossRef]
22. Bullet, J.R.; Korchagina, E.V.; Winnik, F.M. High-sensitivity microcalorimetry and gel permeation chromatography in tandem reveal the complexity of the synthesis of poly-(2-isopropyl-2-oxazoline) stars. *Macromolecules* **2021**, *54*, 6161–6170. [CrossRef]
23. Trachsel, L.; Romio, M.; Zenobi-Wong, M.; Benetti, E.M. Hydrogels generated from cyclic poly(2-oxazoline)s display unique swelling and mechanical properties. *Macromol. Rapid Commun.* **2021**, *42*, 2000658. [CrossRef]
24. Viegas, T.X.; Bentley, M.D.; Milton Harris, J.; Fang, Z.; Yoon, K.; Dizman, B.; Weimer, R.; Mero, A.; Pasut, G.; Veronese, F.M. Polyoxazoline: Chemistry, Properties, and Applications in Drug Delivery. *Bioconjugate Chem.* **2011**, *22*, 976–986. [CrossRef] [PubMed]
25. Kirila, T.Y.; Kurlykin, M.P.; Tenkovtsev, A.V.; Filippov, A.P. Behavior of aqueous solutions of thermosensitive starlike polyalkyloxazolines with different arm structures. *Polym. Sci. A* **2017**, *59*, 826–838. [CrossRef]
26. Kirila, T.; Smirnova, A.; Kurlykin, M.; Tenkovtsev, A.; Filippov, A. Self-organization in aqueous solutions of thermosensitive star-shaped and linear gradient copolymers of 2-ethyl-2-oxazoline and 2-isopropyl-2-oxazoline. *Colloid. Polym. Sci.* **2020**, *298*, 535–546. [CrossRef]
27. Amirova, A.; Rodchenko, S.; Milenin, S.; Tatarinova, E.; Kurlykin, M.; Tenkovtsev, A.; Filippov, A. Influence of a hydrophobic core on thermoresponsive behavior of dendrimer-based star-shaped poly(2-isopropyl-2-oxazoline) in aqueous solutions. *J. Polym. Res.* **2017**, *24*, 124. [CrossRef]
28. Shinkai, S.; Mori, S.; Tsubaki, T.; Sone, T.; Manabe, O. New water-soluble host molecules derived from calix[6]arene. *Tetrahedron Lett.* **1984**, *25*, 5315–5318. [CrossRef]
29. Schühle, D.T.; Peters, J.A.; Schatz, J. Metal binding calixarenes with potential biomimetic and biomedical applications. *Coord. Chem. Rev.* **2011**, *255*, 2727–2745. [CrossRef]
30. Kiegiel, K.; Steczek, L.; Zakrzewska-Trznadel, G. Application of calixarenes as macrocyclic ligands for Uranium(VI): A Review. *J. Chem.* **2013**, *2013*, 762819. [CrossRef]
31. Chen, M.-X.; Li, T.; Peng, S.; Tao, D. Supramolecular nanocapsules from the self-assembly of amphiphilic calixarene as a carrier for paclitaxel. *New J. Chem.* **2016**, *40*, 9923–9929. [CrossRef]
32. Di Bari, I.; Picciotto, R.; Granata, G.; Blanco, A.R.; Consoli, G.M.L.; Sortino, S. A bactericidal calix[4]arene-based nanoconstruct with amplified NO photorelease. *Org. Biomol. Chem.* **2016**, *14*, 8047–8052. [CrossRef]
33. Morozova, J.E.; Syakaev, V.V.; Kazakova, E.K.; Shalaeva, Y.V.; Nizameev, I.R.; Kadirov, M.K.; Voloshina, A.D.; Zobov, V.V.; Konovalov, A.I. Amphiphilic calixresorcinarene associates as effective solubilizing agents for hydrophobic organic acids: Construction of nano-aggregates. *Soft Matter* **2016**, *12*, 5590–5599. [CrossRef] [PubMed]
34. Khan, K.; Badshah, L.S.; Ahmad, N.; Rashid, H.U.; Mabkhot, Y. Inclusion complexes of a new family of non-ionic amphiphilic dendrocalix[4]arene and poorly water-soluble drugs naproxen and ibuprofen. *Molecules* **2017**, *22*, 783. [CrossRef]
35. Consoli, G.M.L.; Granata, G.; Picciotto, R.; Blanco, A.R.; Marino, A.; Nostro, A. Design, synthesis and antibacterial evaluation of a polycationic calix[4]arene derivative alone and in combination with antibiotics. *Med. Chem. Comm.* **2018**, *9*, 160. [CrossRef] [PubMed]
36. Pisagatti, I.; Barbera, L.; Gattuso, G.; Patanè, S.; Parisi, M.F.; Notti, A. Novel PEGylated calix[5]arenes as carriers for Rose Bengal. *Supramol. Chem.* **2018**, *30*, 658–663. [CrossRef]
37. Tu, C.; Zhu, L.; Li, P.; Chen, Y.; Su, Y.; Yan, D.; Zhu, X.; Zhou, G. Supramolecular polymeric micelles by the host–guest interaction of star-like calix[4]arene and chlorin e6 for photodynamic therapy. *Chem. Commun.* **2011**, *47*, 6063–6065. [CrossRef] [PubMed]
38. Nimsea, S.B.; Kim, T. Biological applications of functionalized calixarenes. *Chem. Soc. Rev.* **2013**, *42*, 366–386. [CrossRef]

39. Amirova, A.; Tobolina, A.; Kirila, T.; Blokhin, A.N.; Razina, A.B.; Tenkovtsev, A.V.; Filippov, A.P. Influence of core configuration and arm structure on solution properties of new thermosensitive star-shaped poly(2-alkyl-2-oxazolines). *Int. J. Polym. Anal. Charact.* **2018**, *23*, 278–285. [CrossRef]
40. Percec, V.; Bera, T.K.; De, B.B.; Sanai, Y.; Smith, J.; Holerca, M.N.; Barboiu, B.; Grubbs, R.B.; Fréchet, J.M.J. Synthesis of functional aromatic multisulfonyl chlorides and their masked precursors. *Org. Chem.* **2001**, *66*, 2104–2117. [CrossRef]
41. Shpyrkov, A.A.; Tarasenko, I.I.; Pankova, G.A.; Il'ina, I.E.; Tarasova, E.V.; Tarabukina, E.B.; Vlasov, G.P.; Filippov, A.P. Molecular mass characteristics and hydrodynamic and conformational properties of hyperbranched poly-l-lysines. *Polym. Sci. A* **2009**, *51*, 250–258. [CrossRef]
42. Rueb, C.J.; Zukoski, C.F. Rheology of suspensions of weakly attractive particles: Approach to gelation. *J. Rheol.* **1998**, *42*, 1451–1476. [CrossRef]
43. Amirova, A.I.; Rodchenko, S.V.; Filippov, A.P. Time dependence of the aggregation of star-shaped poly(2-isopropyl-2-oxazolines) in aqueous solutions. *J. Polym. Res.* **2016**, *23*, 221. [CrossRef]
44. Kurlykin, M.P.; Razina, A.B.; Ten'kovtsev, A.V. The use of sulfonyl halides as initiators of cationic polymerization of oxazolines. *Polym. Sci. B* **2015**, *57*, 395–401. [CrossRef]
45. Blokhin, A.N.; Razina, A.B.; Ten'kovtsev, A.V. Star poly(2-alkyl-2-oxazolines) based on octa-(chlorosulfonyl)-calix[8]arene. *Polym. Sci. B* **2018**, *60*, 307–316.
46. Vergaelen, M.; Verbraeken, B.; Monnery, B.D.; Hoogenboom, R. Sulfolane as common rate accelerating solvent for the cationic ring-opening polymerization of 2-oxazolines. *ACS Macro Lett.* **2015**, *4*, 825–828. [CrossRef]
47. Verbraeken, B.; Monnery, B.D.; Lava, K.; Hoogenboom, R. The chemistry of poly(2-oxazoline)s. *Eur. Polym. J.* **2017**, *88*, 451–469. [CrossRef]
48. Park, J.-S.; Kataoka, K. Precise control of lower critical solution temperature of thermosensitive poly(2-isopropyl-2-oxazoline) via gradient copolymerization with 2-ethyl-2-oxazoline as a hydrophilic comonomers. *Macromolecules* **2006**, *39*, 6622–6630. [CrossRef]
49. Kirila, T.; Smirnova, A.; Razina, A.; Tenkovtsev, A.; Filippov, A. Synthesis and conformational characteristics of thermosensitive star-shaped six-arm polypeptoids. *Polymers* **2020**, *12*, 800. [CrossRef]
50. Gubarev, A.S.; Monnery, B.D.; Lezov, A.A.; Sedlacek, O.; Tsvetkov, N.V.; Hoogenboom, R.; Filippov, S.K. Conformational properties of biocompatible poly(2-ethyl-2-oxazoline)s in phosphate buffered saline. *Polym. Chem.* **2018**, *9*, 2232–2237. [CrossRef]
51. Tsvetkov, V.N. *Rigid-Chain Polymers*; Plenum: New York, NY, USA, 1989.
52. Sung, J.H.; Lee, D.C. Molecular shape of poly(2-ethyl-2-oxazoline) chains in THF. *Polymer* **2001**, *42*, 5771–5779. [CrossRef]
53. Grube, M.; Leiske, M.N.; Schubert, U.S.; Nischang, I. POx as an alternative to PEG? A hydrodynamic and light scattering study. *Macromolecules* **2018**, *51*, 1905–1916. [CrossRef]
54. Smirnova, A.; Kirila, T.; Blokhin, A.; Kozina, N.; Kurlykin, M.; Tenkovtsev, A.; Filippov, A. Linear and star-shaped poly(2-ethyl-2-oxazine)s. Synthesis, characterization and conformation in solution. *Eur. Polym. J.* **2021**, *156*, 110637. [CrossRef]
55. Zimm, B.H.; Stockmayer, W.H. The dimensions of chain molecules containing branches and rings. *J. Chem. Phys.* **1949**, *17*, 1301–1314. [CrossRef]
56. Burchard, W. Particle scattering factors of some branched polymers. *Macromolecules* **1977**, *10*, 919–927. [CrossRef]
57. Burchard, W. Static and dynamic light scattering from branched polymers and biopolymers. *Adv. Polym. Sci.* **1983**, *48*, 1–124.
58. Daoud, M.; Cotton, J.P. Star shaped polymers: A model for the conformation and its concentration dependence. *J. Phys.* **1982**, *43*, 531–538. [CrossRef]
59. Weissmuller, M.; Burchard, W. Molar mass distributions of end-linked polystyrene star macromolecules. *Polym. Int.* **1997**, *44*, 380–390. [CrossRef]
60. Tsvetkov, V.N.; Lavrenko, P.N.; Bushin, S.V. A hydrodynamic invariant of polymeric molecules. *Russ. Chem. Rev.* **1982**, *51*, 975–993. [CrossRef]
61. Tsvetkov, V.N.; Lavrenko, P.N.; Bushin, S.V. Hydrodynamic invariant of polymer-molecules. *J. Polym. Sci. Polym. Chem. Ed.* **1984**, *22*, 3447–3486. [CrossRef]
62. Filippov, A.P.; Belyaeva, E.V.; Tarabukina, E.B.; Amirova, A.I. Behavior of hyperbranched polymers in solutions. *Polym. Sci. C* **2011**, *53*, 107–117. [CrossRef]
63. Pavlov, G.M.; Korneeva, E.V.; Roy, R.; Michailova, N.A.; Ortega, P.C.; Perez, M.A. Sedimentation, translational diffusion, and viscosity of lactosylated polyamidoamine dendrimers. *Progr. Colloid Polym. Sci.* **1999**, *113*, 150–157.
64. Pavlov, G.M.; Korneeva, E.V.; Nepogod'ev, S.A.; Jumel, K.; Harding, S.E. Translational and rotational of lactodendrimer molecules in solution. *Polym. Sci. A* **1998**, *40*, 12821289.
65. Obeid, R.; Maltseva, E.; Thünemann, A.F.; Tanaka, F.; Winnik, F.M. Temperature response of self-assembled micelles of telechelic hydrophobically modified poly(2-alkyl-2-oxazoline)s in water. *Macromolecules* **2009**, *42*, 2204–2214. [CrossRef]
66. Hruby, M.; Filippov, S.K.; Panek, J.; Novakova, M.; Mackova, H.; Kucka, J.; Ulbrich, K. Polyoxazoline thermoresponsive micelles as radionuclide delivery systems. *Macromol. Biosci.* **2010**, *10*, 916–924. [CrossRef] [PubMed]
67. Caponi, P.F.; Qiu, X.P.; Vilela, F.; Winnik, F.M.; Ulijn, R.V. Phosphatase/temperature responsive poly(2-isopropyl-2-oxazoline). *Polym. Chem.* **2011**, *2*, 306–308. [CrossRef]
68. Takahashi, R.; Sato, T.; Terao, K.; Qiu, X.P.; Winnik, F.M. Self-association of a thermosensitive poly(alkyl-2-oxazoline) block copolymer in aqueous solution. *Macromolecules* **2012**, *45*, 6111–6119. [CrossRef]

9. Witte, H.; Seeliger, W. Cyclische imidsäureester aus nitrilen und aminoalkoholen. *Just. Lieb. Annal. Chem.* **1974**, *6*, 996–1009. [CrossRef]
0. Dimitrov, I.; Trzebicka, B.; Müller, A.H.E.; Dworak, A.; Tsvetanov, C.B. Thermosensitive watersoluble copolymers with doubly responsive reversibly interacting entities. *Prog. Polym. Sci.* **2007**, *32*, 1275–1343. [CrossRef]
1. Rossegger, E.; Schenk, V.; Wiesbrock, F. Design strategies for functionalized poly(2-oxazoline)s and derived materials. *Polymers* **2013**, *5*, 956–1011. [CrossRef]
2. Weber, C.; Hoogenboom, R.; Schubert, U.S. Temperature responsive bio-compatible polymers based on poly(ethylene oxide) and poly(2-oxazoline)s. *Prog. Polym. Sci.* **2012**, *37*, 686–714. [CrossRef]
3. Trinh, L.T.T.; Lambermont-Thijs, H.M.L.; Schubert, U.S.; Hoogenboom, R.; Kjøniksen, A.L. Thermoresponsive poly(2-oxazoline) block copolymers exhibiting two cloud points: Complex multistep assembly behavior. *Macromolecules* **2012**, *45*, 4337–4345. [CrossRef]
4. Steinschulte, A.A.; Schulte, B.; Rütten, S.; Eckert, T.; Okuda, J.; Möller, M.; Schneider, S.; Borisov, O.V.; Plamper, F.A. Effects of architecture on the stability of thermosensitive unimolecular micelles. *Phys. Chem. Chem. Phys.* **2014**, *16*, 4917–4932. [CrossRef]
5. Steinschulte, A.A.; Schulte, B.; Erberich, M.; Borisov, O.V.; Plamper, F.A. Unimolecular Janus micelles by microenvironment-induced, internal complexation. *ACS Macro Lett.* **2012**, *1*, 504–507. [CrossRef]
6. Kirila, T.U.; Smirnova, A.V.; Filippov, A.S.; Razina, A.B.; Tenkovtsev, A.V.; Filippov, A.P. Thermosensitive star-shaped poly-2-ethyl-2-oxazine. Synthesis, structure characterization, conformation, and self-organization in aqueous solutions. *Eur. Polym. J.* **2019**, *120*, 109215. [CrossRef]
7. Kratochvil, P. *Classical Light Scattering from Polymer Solution*; Elsevier: Amsterdam, The Netherlands, 1987.
8. Schärtl, W. *Light Scattering from Polymer Solutions and Nanoparticle Dispersions*; Springer: Berlin, Germany, 2007.
9. Amirova, A.I.; Golub, O.V.; Kirila, T.U.; Razina, A.B.; Tenkovtsev, A.V.; Filippov, A.P. Influence of arm length and number on star-shaped poly(2-isopropyl-2-oxazoline) aggregation in aqueous solutions near cloud point. *Soft Mater.* **2016**, *14*, 15–26. [CrossRef]
0. Simonova, M.A.; Tarasova, E.V.; Dudkina, M.M.; Tenkovtsev, A.V.; Filippov, A.P. Synthesis and hydrodynamic and conformation properties of star-shaped polystyrene with calix[8]arene core. *Int. J. Polym. Anal. Charact.* **2019**, *24*, 87–95. [CrossRef]
1. Smirnova, A.V.; Kirila, T.U.; Kurlykin, M.P.; Tenkovtsev, A.V.; Filippov, A.P. Behavior of aqueous solutions of polymer star with block copolymer poly(2-ethyl-2-oxazoline) and poly(2-isopropyl-2-oxazoline) arms. *Int. J. Polym. Anal. Charact.* **2017**, *22*, 677–684. [CrossRef]
2. Rodchenko, S.; Amirova, A.; Milenin, S.; Kurlykin, M.; Tenkovtsev, A.; Filippov, A. Self-organization of thermosensitive star-shaped poly(2-isopropyl-2-oxazolines) influenced by arm number and generation of carbosilane dendrimer core in aqueous solutions. *Colloid Polym. Sci.* **2020**, *298*, 355–363. [CrossRef]
3. Kirile, T.Y.; Tobolina, A.I.; Elkina, A.A.; Kurlykin, M.P.; Tenkovtsev, A.V.; Filippov, A.P. Self-assembly processes in aqueous solutions of heat-sensitive star-shaped poly-2-ethyl-2-oxazoline. *Fiber. Chem.* **2018**, *50*, 248–251. [CrossRef]
4. Hoogenboom, R.; Schlaad, H. Thermoresponsive poly(2-oxazoline)s, polypeptoids, and polypeptides. *Polym. Chem.* **2017**, *8*, 24–40. [CrossRef]
5. De la Rosa, V.R.; Nau, W.M.; Hoogenboom, R. Tuning temperature responsive poly(2-alkyl-2-oxazoline)s by supramolecular host–guest interactions. *Org. Biomol. Chem.* **2015**, *13*, 3048–3057. [CrossRef] [PubMed]
6. Zaccone, A.; Crassous, J.J.; Béri, B.; Ballauff, M. Quantifying the reversible association of thermosensitive nanoparticles. *Phys. Rev. Lett.* **2011**, *107*, 168303. [CrossRef]
7. Ye, J.; Xu, J.; Hu, J.; Wang, X.; Zhang, G.; Liu, S.; Wu, C. Comparative study of temperature-induced association of cyclic and linear poly(N-isopropylacrylamide) chains in dilute solutions by laser light scattering and stopped-flow temperature jump. *Macromolecules* **2008**, *41*, 4416–4422. [CrossRef]
8. Zhao, J.; Hoogenboom, R.; Van Assche, G.; Van Mele, B. Demixing and remixing kinetics of poly(2-isopropyl-2-oxazoline) (PIPOZ) aqueous solutions studied by modulated temperature differential scanning calorimetry. *Macromolecules* **2010**, *43*, 6853–6860. [CrossRef]
9. Han, X.; Zhang, X.; Zhu, H.; Yin, Q.; Liu, H.L.; Hu, Y. Effect of composition of PDMAEMA-b-PAA block copolymers on their pH and temperature-responsive behaviors. *Langmuir* **2013**, *29*, 1024–1034. [CrossRef]
0. Adelsberger, J.; Grillo, I.; Kulkarni, A.; Sharp, M.; Bivigou-Koumba, A.M.; Laschewsky, A.; Müller-Buschbaum, P.; Papadakis, C.M. Kinetics of aggregation in micellar solutions of thermoresponsive triblock copolymers—Influence of concentration, start and target temperatures. *Soft Matter* **2013**, *9*, 1685–1699. [CrossRef]
1. Meyer, M.; Antonietti, M.; Schlaad, H. Unexpected thermal characteristics of aqueous solutions of poly(2-isopropyl-2-oxazoline). *Soft Matter* **2007**, *3*, 430–431. [CrossRef]
2. Bühler, J.; Muth, S.; Fischer, K.; Schmidt, M. Collapse of cylindrical brushes with 2-isopropyloxazoline side chains close to the phase boundary. *Macromol. Rapid Commun.* **2013**, *34*, 588–594. [CrossRef]
3. Kudryavtseva, A.A.; Kurlykin, M.P.; Tarabukina, E.B.; Tenkovtsev, A.V.; Filippov, A.P. Behavior of thermosensitive graft copolymer with aromatic polyester backbone and poly-2-ethyl-2-oxazoline side chains in aqueous solutions. *Int. J. Polym. Anal. Charact.* **2017**, *22*, 526–533. [CrossRef]

polymers

MDPI

Article

Self-Organization in Dilute Aqueous Solutions of Thermoresponsive Star-Shaped Six-Arm Poly-2-Alkyl-2-Oxazines and Poly-2-Alkyl-2-Oxazolines

Tatyana Kirila [1,*], Anna Smirnova [1], Vladimir Aseyev [2], Andrey Tenkovtsev [1], Heikki Tenhu [2] and Alexander Filippov [1]

[1] Institute of Macromolecular Compounds of the Russian Academy of Sciences, Bolshoy pr., 31, 199004 Saint Petersburg, Russia; av.smirnova536@gmail.com (A.S.); avt@hq.macro.ru (A.T.); afil@imc.macro.ru (A.F.)
[2] Department of Chemistry, University of Helsinki, 00014 Helsinki, Finland; vladimir.aseyev@helsinki.fi (V.A.); heikki.tenhu@helsinki.fi (H.T.)
* Correspondence: tatyana_pyx@mail.ru; Tel.: +7-812-328-4102

Abstract: The behavior of star-shaped six-arm poly-2-alkyl-2-oxazines and poly-2-alkyl-2-oxazolines in aqueous solutions on heating was studied by light scattering, turbidimetry and microcalorimetry. The core of stars was hexaaza [2_6] orthoparacyclophane and the arms were poly-2-ethyl-2-oxazine, poly-2-isopropyl-2-oxazine, poly-2-ethyl-2-oxazoline, and poly-2-isopropyl-2-oxazoline. The arm structure affects the properties of polymers already at low temperatures. Molecules and aggregates were present in solutions of poly-2-alkyl-2-oxazines, while aggregates of two types were observed in the case of poly-2-alkyl-2-oxazolines. On heating below the phase separation temperature, the characteristics of the investigated solutions did not depend practically on temperature. An increase in the dehydration degree of poly-2-alkyl-2-oxazines and poly-2-alkyl-2-oxazolines led to the formation of intermolecular hydrogen bonds, and aggregation was the dominant process near the phase separation temperature. It was shown that the characteristics of the phase transition in solutions of the studied polymer stars are determined primarily by the arm structure, while the influence of the molar mass is not so significant. In comparison with literature data, the role of the hydrophobic core structure in the formation of the properties of star-shaped polymers was analyzed.

Keywords: thermoresponsive star-shaped polymers; poly-2-alkyl-2-oxazines and poly-2-alkyl-2-oxazolines; aqueous solutions; light scattering; turbidimetry; microcalorimetry; phase separation; aggregation

Citation: Kirila, T.; Smirnova, A.; Aseyev, V.; Tenkovtsev, A.; Tenhu, H.; Filippov, A. Self-Organization in Dilute Aqueous Solutions of Thermoresponsive Star-Shaped Six-Arm Poly-2-Alkyl-2-Oxazines and Poly-2-Alkyl-2-Oxazolines. *Polymers* **2021**, *13*, 1429. https://doi.org/10.3390/polym13091429

Academic Editors: Alexey Bubnov and Mattia Sponchioni

Received: 30 March 2021
Accepted: 26 April 2021
Published: 29 April 2021

1. Introduction

Thermoresponsive pseudo-polypeptoids have attracted great interest of researchers in recent years due to the wide potential of their application, for example, in medicine as nanocontainers for targeted delivery of drugs [1–3]. For them, the synthesis conditions were determined, allowing obtaining polymers with a given structure and molar mass characteristics. The linear pseudo-polypeptoids have been studied in sufficient detail, and influence of their chemical structure and molar mass on the physico-chemical properties, in particular, on self-organization of macromolecules and thermoresponsiveness in aqueous solutions was established [4–10].

One of the most well-studied classes of thermosensitive pseudo-polypeptoids is poly-2-alkyl-2-oxazolines (PAlOx). They demonstrate LSCT behavior in water-salt solutions, and the phase separation temperature depends on the length of the side radical and can range from practically zero to 100 °C [11]. Due to their biocompatibility and stability in biological media, they are widely used in medical applications and biotechnology [2,12,13]. In particular, complexes of linear PAlOx with low molecular weight compounds were obtained, which are used as delivery systems for drugs, DNA, as well as materials for creating biocompatible composite structures [14].

Poly-2-alkyl-2-oxazines (PAlOz) are homologs of poly-2-alkyl-2-oxazolines. The presence of an additional methylene group in the monomer unit makes them more hydrophobic than PAlOx [15–17]. In contrast to poly-2-alkyl-2-oxazines, PAlOz have been studied in much less detail [18], although the synthesis of these polymers was described at about the same time [19]. This situation is associated with the difficulties of synthesis of PAlOz, namely, the reaction for the preparation of PAlOz is characterized by low polymerization rate constants and a high chain transfer rate. This makes it difficult to obtain the high molar mass samples [20,21]. It should be noted that PAlOz has a number of advantages over PAlOx. For PAlOz, no irreversible crystallization in water was found upon prolonged heating above the phase separation temperature, which is characteristic of poly-2-isopropyl-2-oxazoline [22,23]. It seems even more significant that for PAlOz good binding of water-insoluble medicinal compounds was found [24].

In connection with biomedical applications, special attention is paid to the study of pseudo-polypeptoids with complex architecture, in particular, polymer stars. For star-shaped PAlOx, the effect of the structure of arms, their number and length, as well as the solvent composition on the thermosensitivity was analyzed [25–27]. The possibilities of using star-shaped polymers as carriers of drugs to a great extent depend on the structure and properties of the branching center. For example, calix[*n*]arenes and resorcinarenes are prone to self-organization and the formation of nanoscale assemblies [28,29]; therefore, their use as the core of star molecules significantly increases the binding efficiency of low-molecular-weight drugs.

Star-shaped polymers based on aza [1_n] cyclophans have practically not been described so far, although the aza [1_n] cyclophans have been known of since 1963 [30]. Hexaase [2_6] metacyclophane and hexaase [2_6] orthoparacyclophane can be obtained in high yield by reduction in the corresponding macrocyclic Schiff bases [31].

In our previous work [25], for the first time, star-shaped six-arm pseudo-polypeptoids with hexaaza [2_6] orthoparacyclophane core were synthesized using cationic ring-opening polymerization. Four polymers were obtained, namely, star-shaped poly-2-ethyl-2-oxazine (CPh6-PEtOz), poly-2-isopropyl-2-oxazine (CPh6-PiPrOz), poly-2-ethyl-2-oxazoline (CPh6-PEtOx), and poly-2-isopropyl-2-oxazoline (CPh6-PiPrOx) (Figure 1). Conformational behavior of star-shaped poly-2-alkyl-2-oxazines (CPh6-PAlOz) and poly-2-alkyl-2-oxazolines (CPh6-PAlOx) were investigated by the methods of molecular hydrodynamics and optics in molecular dispersed solutions. It was established that conformation characteristics of CPh6-PAlOz and CPh6-PAlOx depended on arm length, while the chemical structure weakly affected the solution behavior of the star-shaped pseudo-polypeptoids. The star-shaped CPh6-PAlOz and CPh6-PAlOx are characterized by higher intramolecular density in comparison with their linear analogs. Taking into account the prospects of practical application, the influence of salt on the self-organization in CPh6-PAlOz and CPh6-PAlOx solutions was studied [32]. NaCl and N-methylpyridinium p-toluenesulfonate (*N*-PTS) were used as salts. It was found that the effect of salt on the thermosensitivity of the discussed stars depends on the structure of the salt and polymer and on the salt content in the solution. For NaCl solutions, the phase separation temperature monotonically decreased with the growth of salt concentration. In *N*-PTS solutions, the dependence of the phase separation temperature on the salt concentration was non-monotonic with minimum at salt concentration corresponding to one salt molecule per one arm of a polymer star.

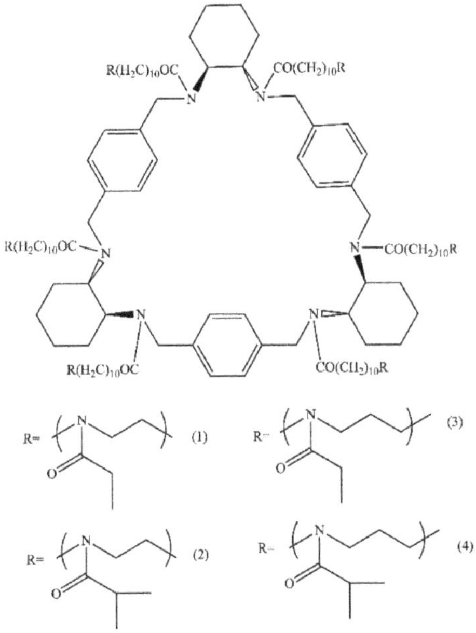

Figure 1. Structure of star-shaped CPh6-PEtOx (**1**), CPh6-PiPrOx (**2**), CPh6-PEtOz (**3**), and CPh6-PiPrOz (**4**).

The aim of the present work is to investigate the effect of the arm structure and, accordingly, the hydrophobicity of the molecules on self-organization and aggregation in water solutions of thermoresponsive star-shaped CPh6-PAlOz and CPh6-PAlOx on heating.

2. Materials and Methods

2.1. Polymer Star Synthesis

The synthesis of star-shaped six-arm thermosensitive poly-2-alkyl-2-oxazolines and poly-2-alkyl-2-oxazines with a hexaase [2_6] orthoparacyclophane core has been described in detail earlier [25]. The molar mass characteristics of the samples were determined in chloroform. The molar masses (MM) of CPh6-PEtOz and CPh6-PiPrOz are higher than MM of CPh6-PEtOx and CPh6-PiPrOx (Table 1). Accordingly, the molar fraction ω of hydrophobic groups in CPh6-PEtOz (7.2 mol %) and CPh6-PiPrOz (8.3 mol %) is almost one and a half times lower than in CPh6-PEtOx (11.0 mol %) and CPh6-PiPrOx (11.8 mol %). The values of dispersity factor $Đ = M_w/M_n$ of investigated samples were calculated using GPC curves which were obtained in [25]. However, it should be taken into account that linear standards were used in the GPC analysis. Accordingly, the obtained $Đ$ values should be considered as an estimate of the polydispersion.

Table 1. Molar masses and average values of hydrodynamic radii of scattering object in aqueous solutions of investigated stars-shaped CPh6-PAlOx and CPh6-PAlOz.

Polymer	M_{sD}, (1) g mol^{-1}	$Đ$ (2)	$R_{h\text{-}D}$, (1) nm	<R_f>, nm	<R_m>, nm	<R_s>, nm	<S_f>	<S_m>	<S_s>	C_s
CPh6-PEtOz	23,000	1.27	3.5	3.9 ± 0.3	-	89 ± 8	0.09	-	0.91	0.07
CPh6-PiPrOz	20,000	1.24	3.3	4.1 ± 0.3	-	89 ± 9	0.10	-	0.92	0.07
CPh6-PEtOx	15,000	1.29	3.0	-	6.3 ± 0.7	84 ± 7	-	0.35	0.65	0.06
CPh6-PiPrOx	14,000	1.19	2.6	-	15 ± 1	86 ± 12	-	0.84	0.16	0.08

(1) the values of M_{sD} and $R_{h\text{-}D}$ were obtained in [25]. (2) the values of $Đ$ were calculated using GPC curves which were obtained in [25].

2.2. Solution Investigation

The thermosensitive behavior of star-shaped pseudo-polypeptoids was studied by light scattering and turbidimetry on a PhotoCor Complex (Photocor Instruments Inc., Moscow, Russia) setup with a diode laser and a sensor for measuring optical transmission. Wavelength is $\lambda = 659.1$ nm. The studies were carried out in the temperature range T from 5 to 75 °C. The value of T was changed discretely with a step from 0.5 to 5 °C and was regulated with an accuracy of 0.1 °C.

The measurement procedure was described in detail [33]. For all solutions, the values of the scattered light intensity I, optical transmission I^*, hydrodynamic radii R_h of scattering objects and their S_i contribution to the total light scattering were determined. The S_i values were obtained from the area of the corresponding peak in the intensity distribution I over R_h.

After the given temperature was established, only the dependences of the light scattering intensity I and optical transmission I^* on time t were recorded. These measurements were carried out at a scattering angle of 90°. When the changes in I were no more than 1% over the time required for the accumulation and processing of the autocorrelation function, the distribution of I over the hydrodynamic radii R_h was obtained, and the angular dependences of the values of I, R_h, and S_i were analyzed in the range of light scattering angles from 45° to 135° in order to prove the diffusion nature of the modes, as well as to obtain extrapolated values of R_h (Figure 2). In all experiments, the time t_{eq} required to achieve time constant values of I, R_h, and S_i was recorded. Since the phase separation temperatures for CPh6-PEtOx and CPh6-PEtOz at $c < 0.01$ g cm^{-3} had values above 75 °C, measurements for these solutions were carried out on a Zetasizer Nano ZS (Malvern Instruments Limited, Worcestershire, UK) particle analyzer in the temperature range from 15 to 100 °C.

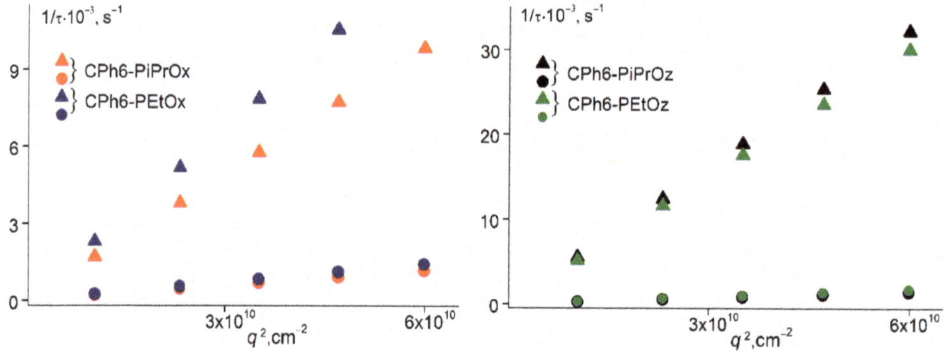

Figure 2. Relaxation time $1/\tau$ on squared wave vector q^2 for CPh6-PAlOz and CPh6-PAlOx at 21 °C.

The polymer concentrations varied from 0.0002 to 0.015 g cm^{-3} for CPh6-PiPrOx and CPh6-PiPrOz and from 0.005 to 0.025 g·cm^{-3} for CPh6-PEtOx and CPh6-PEtOz. The solutions were filtered into dust-free vials using Millipore filters with hydrophilic PTFE membrane and pore size of 0.45 μm (Merck KGaA, Darmstadt, Germany).

Microcalorimetric studies of aqueous solutions of star-shaped CPh6-PAlOz and CPh6-PAlOx were carried out on a MicroCal PEAQ-DSC microcalorimeter (Malvern Instruments Limited, Worcestershire, UK) with a capillary cell with a volume of 0.507 mL at a heating rate of 1 °C/min and a pressure of 50 kPa. The temperature range of measurements was chosen depending on the characteristics of the polymer (from 5 to 60 °C for CPh6-PiPrOz and from 15 to 100 °C for CPh6-PiPrOx, CPh6-PEtOx, and CPh6-PEtOz). The solution concentrations varied from 0.001 to 0.015 g cm^{-3} for CPh6-PiPrOx and CPh6-PiPrOz, and from 0.005 to 0.025 g cm^{-3} for CPh6-PEtOx and CPh6-PEtOz. The phase transition heat ΔH was calculated using the OriginLab taking into account the polymer concentration.

3. Results and Discussion

3.1. Behavior of Star-Shaped Six-Arm Pseudo-Polypeptoids in Aqueous Solutions at Low Temperatures

At low temperatures, the aqueous solutions of the studied polymer stars were not molecularly dispersed; there were two types of scattering objects at all concentrations (Figure 3). However, the behavior of CPh6-PAlOz and CPh6-PAlOx was markedly different. In the case of CPh6-PEtOz and CPh6-PiPrOz, the radii R_f of the particles responsible for the fast mode (Table 1) were close to the hydrodynamic radii $R_{h\text{-}D}$ of the macromolecules of the stars under consideration, determined in an organic solvent in which there were no associative phenomena [25]. Therefore, particles with radius R_f could be considered as individual star-shaped macromolecules. Taking into account the hydrophobicity of the CPh6 core and the hydrophilicity of the PEtOz arms, it can be assumed that the structure of CPh6-PAlOz molecules in aqueous solutions is similar to that of unimolecular micelles of the core-shell type [34–37]. For both PAlOz stars, the R_f values did not depend on the polymer concentration (Figure 4).

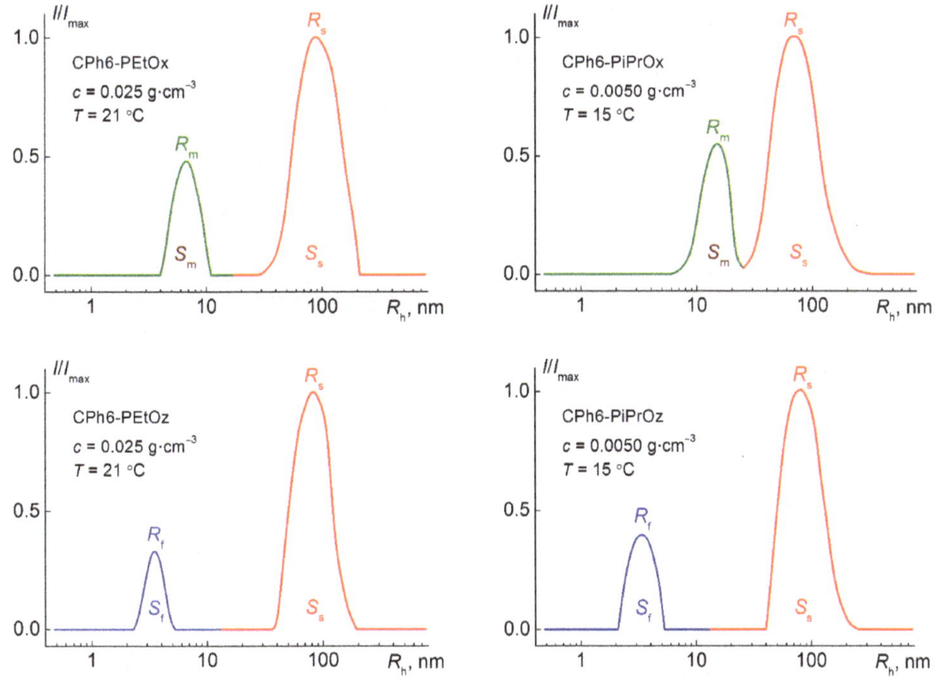

Figure 3. The distributions of light scattering intensity over hydrodynamic radii of scattering species for solutions of star-shaped pseudo-polypeptoids. R_f, R_m, and Rs, are the hydrodynamic radii of macromolecules (fast mode), small aggregates (middle mode) and large aggregates (slow mode), respectively. S_f, S_m, and S_s are contributions of the corresponding modes to the total light scattering.

In exactly the same way, no concentration dependence was found for the species responsible for the faster mode in CPh6-PEtOx and CPh6-PiPrOx solutions. However, the radii R_m of these particles differed markedly from $R_{h\text{-}D}$ for CPh6-PAlOx. For CPh6-PEtOx, the concentration-averaged value $\langle R_m \rangle = (6.3 \pm 0.7)$ nm is 2.1 times higher than $R_{h\text{-}D} = 3.0$ nm. For CPh6-PiPrOx, the discussed parameters differed more strongly: $\langle R_m \rangle = (15 \pm 1)$ nm and $R_{h\text{-}D} = 2.6$ nm. Therefore, the species under discussion are supramolecular structures. Similar aggregates, which are usually called micelle-like structures, have been repeatedly observed for stimulus-sensitive polymers [6,35,38–42], including the star-shaped PAlOx and PAlOz [43–46]. The formation of these supramolecular structures in solutions of the studied CPh6-PAlOx is mainly caused by interaction of hydrophobic CPh6 core. This explains the difference in the behavior of CPh6-PAlOx and CPh6-PAlOz. In the CPh6-PEtOz and CPh6-PiPrOz molecules, the hexaaza [2_6] orthoparacyclophane core is well shielded from water molecules by the hydrophilic corona formed by the arms, that prevents the formation of aggregates. The lengths L_{tsc} of the chains of PEtOz and PiPrOz are 15.9 and 12.1 nm [25]. These values are almost two times higher than L_{tsc} for CPh6-PEtOx (L_{tsc} = 8.5 nm) and CPh6-PiPrOx (L_{tsc} = 6.9 nm). Accordingly, in the case of CPh6-PAlOx, the arm length is not enough for reliable screening of branching centers, and the interaction of the latter leads to the aggregate formation.

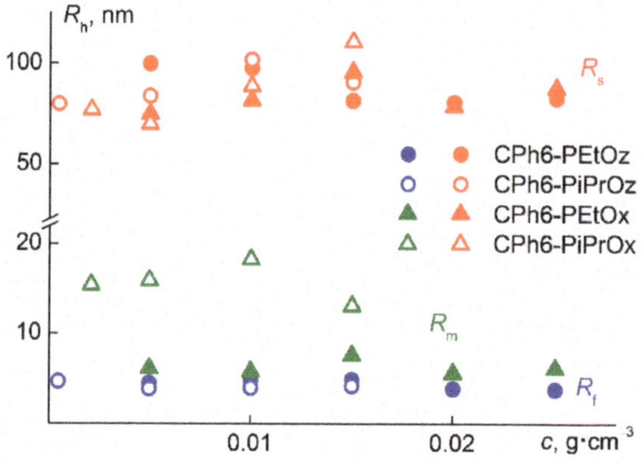

Figure 4. The dependences of the hydrodynamic radii R_h of scattering species on concentration c for aqueous solutions of star-shaped CPh6-PEtOz and CPh6-PEtOx 15 °C and CPh6-PiPrOz and CPh6-PiPrOx at 21 °C.

As was shown earlier, the shape of micelle-like aggregates is close to spherical [45]. The spherical form of molecules of polymer stars with a large number of relatively short arms was proved theoretically [47] and experimentally [48]. These facts allow the estimation of the aggregation degree m_{agg} as the ratio of volumes of the aggregates and the macromolecules. For spherical particles, the volume is proportional to third power of their radius and for investigated CPh6-PAlOx

$$m_{agg} \approx (R_m / R_{h\text{-}D})^3 \tag{1}$$

Thus, $m_{agg} \approx 9$ and 200 for CPh6-PEtOx and CPh6-PiPrOx, respectively. Note that such an approach is rough. First, the hydrodynamic radius is not the real dimension of dissolved species. Second, the densities of aggregates and macromolecules are different. A significant distinction in the m_{agg} values for CPh6-PEtOx and CPh6-PiPrOx may be caused by the different hydrophobicity of their arms. Indeed, as shown by Winnik et al., PEtOx dehydration begins at about 50 °C, while in the case of PiPrOx, this can occur already at 20 °C [7]. Consequently, in CPh6-PiPrOx molecules at 21 °C, there can be a noticeable number of dehydrated units, the intermolecular interactions of which promote aggregation and, in particular, an increase in the size of micelle-like structures.

In addition to unimolecular micelles or small aggregates, large aggregates with a hydrodynamic radius R_s from 70 to 110 nm were detected in CPh6-PAlOz and CPh6-PAlOx solutions. No systematic change in R_s with concentration was found. The average values of the hydrodynamic radii $<R_s>$ of these supramolecular structures for the studied polymers are practically the same (Table 1). It can be assumed that these aggregates were formed mainly by "defective" molecules, the number of arms which was less than six, and the arm length was less than the average for the sample. The relative weight concentration c_s of large aggregates is low. The c_s value can be roughly estimated using the models of sphere for macromolecules and micelle-like structures and coil for large aggregates. As is known, the intensity I_i of ith specie is proportional to both the molar mass M_i and concentration c of particles [49–51]

$$I_i \sim c_i M_i \quad (2)$$

where $I_i = S_i I$, I is integral light scattering intensity of solution, and S_i is relative contribution of ith specie in the I value. $S_i = S_f$ and $c_i = c_f$ for macromolecules, $S_i = S_m$ and $c_i = c_m$ for micelle-like structures, and $S_i = S_s$ for large supramolecular structures (see Figure 3). The particle radius R_i is related to its molar mass as $M_i \sim R_i^x$. Parameter x depends on the particle shape, for example, $x = 3$ for spherical particles, $x = 2$ for coil structures, and $x = 1$ for rigid rods. Within the described approximations, we obtain for CPh6-PAlOz

$$c_s/c_f = (S_f/S_s)\,(R_s^2/R_f^3) \quad (3)$$

and CPh6-PAlOx

$$c_m/c_f = (S_m/S_s)\,(R_s^2/R_m^3) \quad (4)$$

The contributions S_i of different particles for solutions of investigated stars did not depend on polymer concentration. The average values of S_f, S_m, and S_s ($<S_f>$, $<S_m>$, and $<S_s>$, respectively) are shown in Table 1. Substitution of $<S_f>$, $<S_m>$, and $<S_s>$ into Equations (3) and (4) gives values of relative concentration c_s of large aggregates for each polymer studied (Table 1). It is clearly seen that c_s did not exceed 8%.

In order to estimate the aggregation degree m_{agg} for large aggregates, one can use the previously proposed approach [52] based on comparing the translational friction coefficients for macromolecules and aggregates. By modeling unimolecular micelles by spheres, and large aggregates by ellipsoids of revolution, it is possible to show that the aggregation degree associated with the parameters of the ellipsoid by the equation [52]:

$$m_{agg} = V_{ell}/V_{sph} = p^2 R_f^3/8a^3 \quad (5)$$

where V_{ell} is the volume of the modeling ellipsoid of revolution, V_{sph} is the volume of the modeling sphere, $p = a/b$, a and b are the major and minor axes of ellipsoid. Herein, the dependences of a and b on p for a model ellipsoid of revolution with a translational friction coefficient f are described by the formulas

$$a = f/(6\pi\eta_0 F(p)) \quad (6)$$

$$b = f/(6\pi\eta_0 p F(p)) \quad (7)$$

where

$$aF(p) = (p^2-1)^{1/2}/(p\ln((p + (p^2-1)^{1/2})/(p-(p^2-1)^{1/2})) \quad (8)$$

For a given value of the translational friction coefficient of the ellipsoid, the m_{agg} value turns out to be a rather weak function of the ellipsoid parameters. In the region of "reasonable" values of the asymmetry factor $1 < p < 3$, the change in m_{agg} is about 30%. In particular, taking into account the values of $R_{h\text{-}D}$ and R_s from Table 1, at $p = 2$, the values of m_{agg} are from 10,000 to 20,000 for CPh6-PEtOz and CPh6-PiPrOz and from 15,000 to 25,000 for CPh6-PEtOx and CPh6-PiPrOx. Therefore, for all investigated stars at low temperatures, the large aggregates contain from one to two tens of thousands of macromolecules.

Note that when estimating the relative weight concentration c_s of large aggregates and aggregation degree m_{agg}, we used the hydrodynamic radii of the particles and their contributions to the integral value of light scattering, which were determined using a PhotoCor Complex setup. Similar information for CPh6-PEtOz and CPh6-PEtOx solutions was obtained using a Zetasizer Nano ZS particle analyzer. Figure S1 in Supplementary Materials shows the corresponding distributions for solutions of these polymers. As might be expected, they are qualitatively similar to the dependencies shown in Figure 3. The average values of hydrodynamic radii R_f, R_m and R_s obtained with different instruments coincide within the experimental error. A similar situation takes place for the contributions of particles to the integral light scattering intensity. However, one must keep in mind that the accuracy of determining these parameters is significantly lower, and the experimental error can reach 15%.

3.2. *Behavior of Star-Shaped Six-Arm Pseudo-Polypeptoids in Aqueous Solutions on Heating*

A phase transition was detected in aqueous solutions of the studied polymers on heating by methods of light scattering, turbidimetry, and microcalorimetry. The temperature T_1 of the onset of phase separation was determined from the beginning of a sharp decrease in optical transmission I^* and the beginning of a strong increase in the light scattering intensity I (Figure 5). Below T_1, the characteristics of most of the studied solutions did not change with temperature (Figures 5 and 6). Only in some solutions with PAlOx arms, when approaching T_1, a slight increase in intensity I was observed, caused by an increase in the size R_s of large aggregates and their contribution S_s to the total intensity of light scattering. This behavior distinguishes the studied polymers from the star-shaped PAlOz and PAlOx, whose cores were calix [*n*] arenes (*n* = 4, 8) [46,53]. In solutions of the mentioned polymers, the processes of aggregation and self-organization at the molecular level began long before reaching T_1. Probably, this is the manifestation of the role of the branching center in the formation of the properties of thermosensitive stars in aqueous solutions.

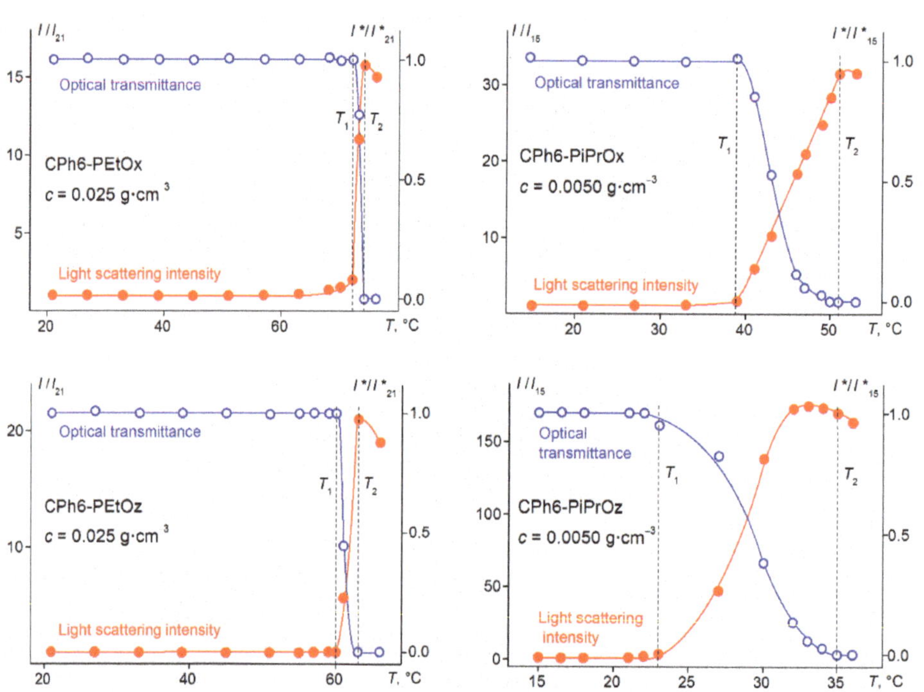

Figure 5. Temperature dependences of relative light scattering intensity I/I_{15} and I/I_{21} and relative transmission I^*/I^*_{15} and I^*/I^*_{21} for solutions of investigated star polymers. I_{15} and I_{21} are light scattering intensity at 15 °C and 21 °C, respectively. I^*_{15} and I^*_{21} are optical transmission at 15 °C and 21 °C, respectively.

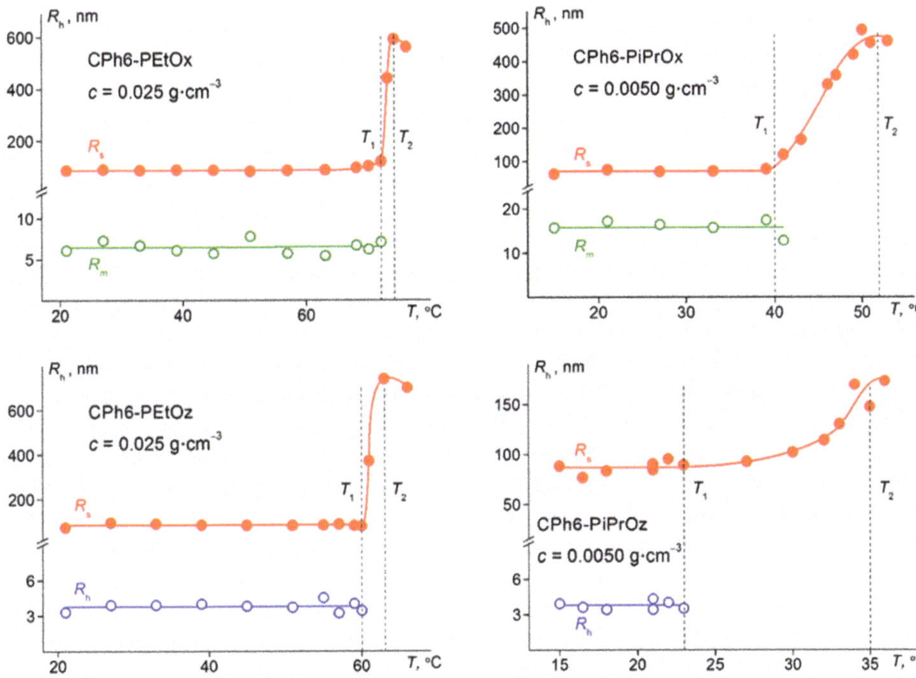

Figure 6. Temperature dependences of hydrodynamic radii R_h of scattering objects for aqueous solutions of star-shaped pseudo-polypeptoids.

Above T_1, the light scattering intensity increased very rapidly with temperature, achieving the maximum value near the temperature T_2, at which the optical transmission became zero (Figure 5). T_2 can be considered as the temperature of the finishing of the phase separation according to turbidimetry data. The reason for the observed behavior of light scattering is the growth of radius of the large aggregates (Figure 6) which is often observed for thermosensitive polymer stars [43,46,54–56]. At $T \geq T_1$, macromolecules in solutions of CPh6-PAlOz and small aggregates in solution of CPh6-PAlOx were not detected by the dynamic light scattering. Thus, at $T > T_1$, the dominant process in solutions of the studied star-shaped polymers was aggregation as a result of an increase in dehydration degree on heating and the formation of intermolecular hydrogen bonds. Above T_2, a decrease in the I and R_s values was observed, however, a quantitative analysis of these data is impossible, since in this region the solutions are turbid and light scattering is multiple.

The microcalorimetric endotherms of the aqueous solutions of CPh6-PAlOx and CPh6-PAlOz are shown in Figure 7. The dependences of the phase transition heat on the concentration ΔH for the studied polymer stars had a form typical for thermosensitive polymers (Figure 8). Thus, no qualitative change in thermodynamic behavior was observed on passage from linear polymers to star-shaped polymers. On the other hand, the influence of the arm structure was clearly visible. As can be seen in Figure 8, for polymer stars containing isopropyl groups, the ΔH value was almost an order of magnitude higher than the phase transition heat for CPh6-PEtOz and CPh6-PEtOx, that can explain the lower hydrophobicity of the latter. It should also be noted that ΔH was slightly higher for the star-shaped CPh6-PAlOz in comparison with this characteristic for CPh6-PAlOx.

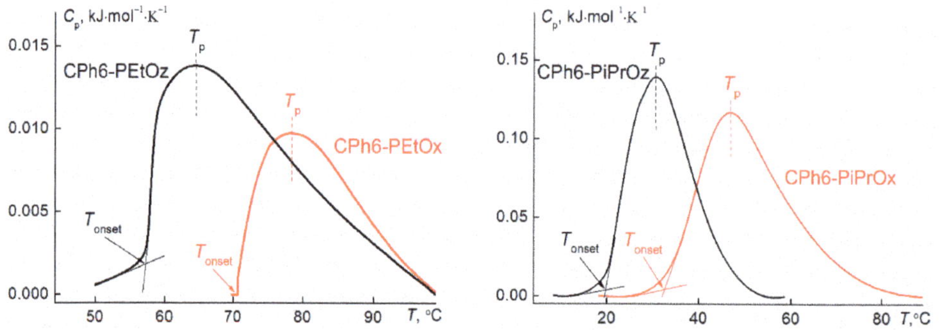

Figure 7. Microcalorimetric endotherms of CPh6-PiPrOz and CPh6-PiPrOx (c = 0.015 g cm^{-3}) and CPh6-PEtOz and CPh6-PEtOx (c = 0.025 g cm^{-3}) in water (rate 1.0 °C/min).

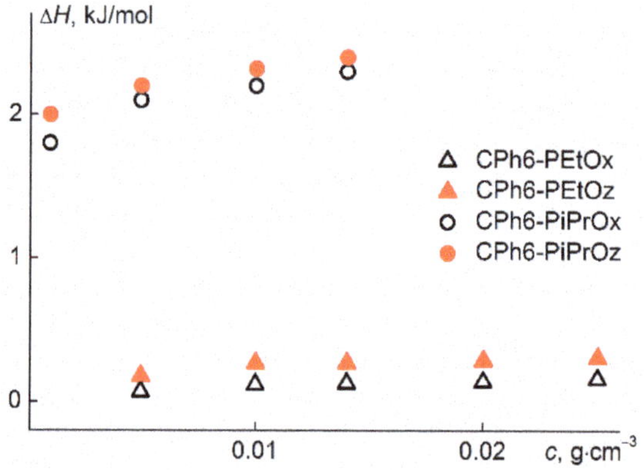

Figure 8. Concentration dependences of enthalpy for the studied aqueous solutions.

From the obtained values of ΔH, the temperatures of the peak maximum and the maximum heat capacities, one can calculate the Van't Hoff enthalpy ΔH^{vH} [57,58]:

$$\Delta H^{vH} = 4RT_p^{2}(C_p^{max}/\Delta H) \quad (9)$$

where T_p is the temperature of the maximum, C_p^{max} is the heat capacity at T_p, and R is universal gas constant. The ratios $n' = \Delta H^{vH}/\Delta H$ gives information about the number of structural units of the polymers that cooperate with each other in the transition [59], i.e., so-called "cooperative units". The values of ΔH^{vH}, ΔH and n' given in Table 2 were obtained for solutions with the maximum polymer concentration: c = 0.015 g cm^{-3} for CPh6-PiPrOz and CPh6-PiPrOx and c = 0.025 g cm^{-3} for CPh6-PEtOz, CPh6-PEtOx. The highest n' value was obtained for CPh6-PEtOx, which is the most hydrophilic, while for the most hydrophobic CPh6-PiPrOz, the n' parameter has the lowest value.

Table 2. Thermodynamic characteristics of solutions of star-shaped pseudo-polypeptoids.

Polymer	ΔH^{vH}, kJ/mol	ΔH, kJ/mol	n'
CPh6-PEtOz	6.0	0.3	20
CPh6-PiPrOz	1.8	2.4	0.75
CPh6-PEtOx	10.1	0.2	50
CPh6-PiPrOx	3.4	2.2	1.5

3.3. Concentration Dependences of Phase Separation Temperatures for Aqueous Solutions of Investigated Polymer Stars

Figure 9 shows the concentration dependences of temperature T_1 obtained for the studied polymers. The phase separation temperatures increased with dilution. This is typical for dilute solutions of thermosensitive polymers. Table 3 compares the phase separation temperatures according to turbidimetry (T_1 and T_2) and microcalorimetry (T_{onset} and T_p) data (T_{onset} is temperature of the onset of phase separation (see Figure 7)). For all polymers, the temperatures of the onset of phase separation, determined by discussed methods, differed insignificantly. On the other hand, the values of T_2 and T_p for stars with arms containing isopropyl groups differed by 8 °C. The observed distinctions can be associated with both different physical bases of the methods and with different experimental procedures (discrete temperature variation and constant heating).

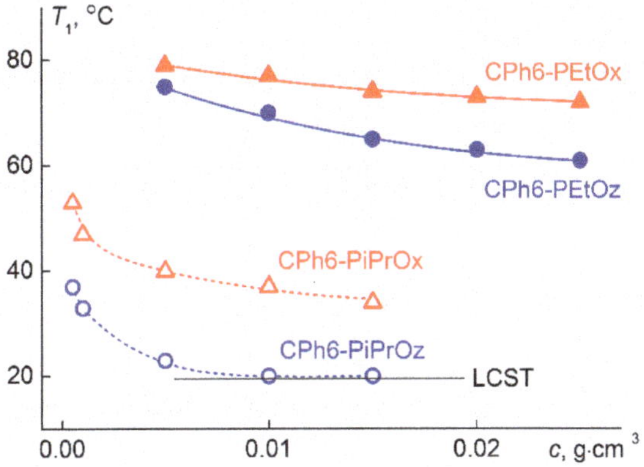

Figure 9. Concentration dependencies of the phase separation temperature T_1 for star-shaped CPh6-PAlOz and CPh6-PAlOx.

Table 3. Phase separation temperatures of solutions of star-shaped pseudo-polypeptoids at concentration $c = 0.015$ g cm^{-3}.

Sample	T_{onset}, °C	T_p, °C	T_1, °C	T_2, °C
CPh6-PEtOz	63	70	65	70
CPh6-PiPrOz	20	31	20	23
CPh6-PEtOx	75	83	74	-
CPh6-PiPrOx	33	47	34	39

The lower critical solution temperature (LCST) was determined only for CPh6-PiPrOz. Taking into account the character of the dependences of T_1 on c, it can be assumed that for other samples LCST slightly differs from the T_1 value obtained at the maximum studied concentration. As for the absolute values of the phase transition temperatures, for stars with oxazoline arms, they do not differ very much from the cloud points for linear [6,20,33,60] and star-shaped [26,27,33,53,61] PEtOx and PiPrOx. For PAlOz, the data are much less even for linear polymers [20,24], and for star-shaped CPh6-PAlOz they are absent at all. Besides, when analyzing the phase separation temperatures, it is necessary to take into account the influence of at least the molar mass of the polymer and the hydrophilic-hydrophobic balance [33], and it is rather difficult to draw reliable conclusions about the "chemical structure—macromolecule architecture—phase separation temperature" correlations.

As expected, for stars with more hydrophilic arms CPh6-PEtOz and CPh6-PEtOx, higher phase separation temperatures were obtained in comparison with T_1 for the CPh6-PiPrOz and CPh6-PiPrOx solutions. This difference is quite large, at a given concentration it is 40–45 °C and correlates with the data for linear PEtOx and PiPrOx [20,33,60]. At passage from PAlOx to PAlOz, the hydrophobicity of the polymer increases as a result of the elongation of the monomer unit by one $-CH_2-$ group. This leads to a decrease in the phase separation temperature. In the investigated concentration range, in the pair of CPh6-PEtOz and CPh6-PEtOx, the T_1 values differ by 5–10 °C, and for CPh6-PiPrOz and CPh6-PiPrOx the distinction is more (from 14 to 17 °C). It should be borne in mind that CPh6-PEtOz and CPh6-PiPrOz are characterized by higher molar mass. This should have led to a decrease in the phase separation temperatures for stars with PAlOz arms. Therefore, the discussed difference in the T_1 values for CPh6-PAlOz and CPh6-PAlOx could be slightly smaller if the compared samples had similar molar masses and hydrophilic-hydrophobic balance.

Note that the values of the phase separation temperatures and LCST can be significantly influenced by the polydispersion of the samples [62]. Therefore, the dispersion should be taken into account when analyzing the T_1 values. However, as can be seen from Table 1, the values of *Đ* for the studied CPh6-PAlOz and CPh6-PAlOx are close. Accordingly, it can be assumed that the effect of polydispersion will be minimal.

The effect of the structure of the studied stars is most pronounced when comparing CPh6-PEtOz and CPh6-PiPrOx. The monomer units of the arms of these polymers have the same set of atoms and groups, but for CPh6-PEtOz the phase separation temperatures are about 40 °C higher than T_1 for CPh6-PiPrOx. Consequently, the position of the $-CH_2-$ group, namely in the main chain or in the side fragments, is a decisive factor determining the phase separation temperature. Note also that the difference in the T_1 values for CPh6-PEtOz and CPh6-PEtOx is significantly less than the corresponding difference for CPh6-PiPrOz and CPh6-PiPrOx, i.e., the higher the hydrophobicity of the homologues, the greater the difference in the phase separation temperatures of their solutions.

4. Conclusions

Aqueous solutions of star-shaped six-arm pseudo-polypeptoids on heating were investigated within a wide concentration and temperature ranges. Poly-2-ethyl-2-oxazine, poly-2-isopropyl-2-oxazine, poly-2-ethyl-2-oxazoline, and poly-2-isopropyl-2-oxazoline were arms and hexaaza [2_6] orthoparacyclophane was core.

At low temperatures, the behavior of the solutions depended on the structure and size of arms. Macromolecules, more precisely unimolecular micelles, and large aggregates were observed in solutions of poly-2-alkyl-2-oxazines, while in solutions of poly-2-alkyl-2-oxazolines there were two types of aggregates. The formation of the smaller ones is caused by interaction of hydrophobic CPh6 core, because the short arms do not sufficiently shield them from the solvent. In addition to unimolecular micelles and small aggregates, in aqueous solutions of the studied polymer stars, large loose aggregates were present, which contained from 10,000 to 25,000 macromolecules. These supramolecular structures were formed mainly from molecules, the arm number in which was less than six, and the arm length was less than the average for the sample. The weight fraction of large aggregates did not exceed 8%.

On heating below the phase separation temperature, in contrast to the previously studied pseudo-polypeptoids with a calix[*n*]arene core, the characteristics of the CPh6-PAlOz and CPh6-PAlOx solutions were practically independent of temperature. This behavior can be explained by the influence of the structure of the branching center. Within the phase separation interval, the prevailing process was aggregation due to the formation of intermolecular hydrogen bonds. This led to the fact that unimolecular micelles and small aggregates attached to large aggregates or formed new supramolecular structures, the size of which increased on heating.

For all investigated solutions, the phase separation temperatures increased with dilution, but LCST was determined reliably only for star with poly-2-isopropyl-2-oxazine arms. At a given concentration, the phase separation temperatures for poly-2-alkyl-2-oxazoline stars with CPh6 core differed from their values for linear analogs and stars with branch centers of a another structure. These differences are insignificant and can be caused both by the difference in the structure and architecture of the molecules of the compared samples, and by the influence of the molar mass of the polymers. It was shown that an increase in the arm hydrophobicity leads to a decrease in the phase separation temperature of the aqueous solutions of studied stars. In particular, their values for poly-2-alkyl-2-oxazolines are higher than for poly-2-alkyl-2-oxazines. On passage from CPh6-PEtOz to CPh6-PiPrOz and from CPh6-PEtOx to CPh6-PiPrOx, the phase separation temperatures decreased. Besides, for polymer stars containing isopropyl groups, the phase transition heat was almost an order of magnitude higher than the ΔH value for star with ethyl groups in side fragment of arms.

Supplementary Materials: The following are available online at https://www.mdpi.com/article/10.3390/polym13091429/s1, Figure S1: The distributions of scattering species over hydrodynamic radii for solutions of star-shaped CPh6-PEtOx and CPh6-PEtOz. Rf, Rm, and Rs, are the hydrodynamic radii of macromolecules (fast mode), small aggregates (middle mode) and large aggregates (slow mode), respectively.

Author Contributions: Conceptualization, A.F.; methodology, H.T.; writing—original draft preparation, A.T., A.F. and T.K.; formal analysis V.A., A.T., and T.K.; investigation, T.K., and A.S.; resources V.A., and A.S.; data curation, A.S.; writing—review and editing, A.S. and V.A.; visualization, T.K.; supervision, A.F.; project administration, A.F. and H.T.; funding acquisition, T.K. All authors have read and agreed to the published version of the manuscript.

Funding: This research was funded by the Russian Science Foundation, grant number 19-73-00175.

Institutional Review Board Statement: Not applicable.

Informed Consent Statement: Not applicable.

Data Availability Statement: Not applicable.

Acknowledgments: Authors acknowledge the support for microcalorimetric studies EC H2020-MSCA-RISE-2018/823883: Soft Biocompatible Polymeric NANOstructures: A Toolbox for Novel Generation of Nano Pharmaceuticals in Ophthalmology (NanoPol).

Conflicts of Interest: The authors declare no conflict of interest.

References

1. Glassner, M.; Vergaelen, M.; Hoogenboom, R. Poly(2-oxazoline)s: A comprehensive overview of polymer structures and their physical properties. *Polym. Int.* **2017**, *67*, 32–45. [CrossRef]
2. Wu, W.; Wang, W.; Li, J. Star polymers: Advances in biomedical applications. *Prog. Polym. Sci.* **2015**, *46*, 55–85. [CrossRef]
3. Zahoranova, A.; Luxenhofer, R. Poly(2-oxazoline)- and poly(2-oxazine)-based self-assemblies, polyplexes, and drug nanoformulations—An update. *Adv. Healthcare Mater.* **2021**, *10*, 2001382. [CrossRef] [PubMed]
4. Pei, Y.; Jarrett, K.; Saunders, M.; Roth, P.J.; Buckley, C.E.; Lowe, A.B. Triply responsive soft matter nanoparticles based on poly[oligo(ethylene glycol) methyl ether methacrylate-block-3-phenylpropyl methacrylate] copolymers. *Polym. Chem.* **2016**, *7*, 2740–2750. [CrossRef]
5. Verbraeken, B.; Monnery, B.D.; Lava, K.; Hoogenboom, R. The chemistry of poly(2-oxazoline)s. *Eur. Polym. J.* **2017**, *88*, 451–469. [CrossRef]
6. Weber, C.; Hoogenboom, R.; Schubert, U.S. Temperature responsive bio-compatible polymers based on poly(ethylene oxide) and poly(2-oxazoline)s. *Prog. Polym. Sci.* **2012**, *37*, 686–714. [CrossRef]
7. Takahashi, R.; Sato, T.; Terao, K.; Qiu, X.-P.; Winnik, F.M. Self-association of a thermosensitive poly(alkyl-2-oxazoline) block copolymer in aqueous solution. *Macromolecules* **2012**, *45*, 6111–6119. [CrossRef]
8. Caponi, P.F.; Qiu, X.P.; Vilela, F.; Winnik, F.M.; Ulijn, R.V. Phosphatase/temperature responsive poly(2-isopropyl-2-oxazoline). *Polym. Chem.* **2011**, *2*, 306–308. [CrossRef]
9. Vlassi, E.; Papagiannopoulos, A.; Pispas, S. Amphiphilic poly(2-oxazoline) copolymers as self-assembled carriers for drug delivery applications. *Eur. Polym. J.* **2017**, *88*, 516–523. [CrossRef]
10. Zhang, N.; Luxenhofer, R.; Jordan, R. Thermoresponsive poly(2-oxazoline) molecular brushes by living ionic polymerization: Modulation of the cloud point by random and block copolymer pendant chains. *Macromol. Chem. Phys.* **2012**, *213*, 1963–1969. [CrossRef]
11. Amirova, A.I.; Blokhin, A.N.; Razina, A.B.; Tenkovtsev, A.V.; Filippov, A.P. The behavior of thermoresponsive star-shaped poly-2-isopropyl-2-oxazoline in saline media. *Mendeleev Commun.* **2019**, *29*, 472–474. [CrossRef]
12. Luxenhofer, R.; Han, Y.; Schulz, A.; Tong, J.; He, Z.; Kabanov, A.V.; Jordan, R. Poly(2-oxazoline)s as polymer therapeutics. *Macromol. Rapid Commun.* **2012**, *33*, 1613–1631. [CrossRef] [PubMed]

13. De la Rosa, V.R. Poly(2-oxazoline)s as materials for biomedical applications. *J. Mater. Sci. Mater.* **2013**, *25*, 1211–1225. [CrossRef]
14. Drakalska, E.; Momekova, D.; Manolova, Y.; Budurova, D.; Momekov, G.; Genova, M.; Antonov, L.; Lambov, N.; Rangelov, S. Hybrid liposomal PEGylated calix[4]arene systems as drug delivery platforms for curcumin. *Int. J. Pharm.* **2014**, *472*, 165–174. [CrossRef]
15. Morgese, G.; Verbraeken, B.; Ramakrishna, S.N.; Gombert, Y.; Cavalli, E.; Rosenboom, J.G.; Zenobi-Wong, M.; Spencer, N.D.; Hoogenboom, R. Chemical design of non-ionic polymer brushes as biointerfaces: Poly(2-oxazine)s outperform both poly(2-oxazoline)s and PEG. *Angew. Chem. Int. Ed.* **2018**, *57*, 11667–11672. [CrossRef] [PubMed]
16. Bloksma, M.M.; Schubert, U.S.; Hoogenboom, R. Poly(cyclic imino ether)s beyond 2-substituted-2-oxazolines. *Macromol. Rapid Commun.* **2011**, *32*, 1419–1441. [CrossRef]
17. Sinnwell, S.; Ritter, H. Microwave accelerated polymerization of 2-phenyl-5,6-dihydro-4H-1,3-oxazine: Kinetics and influence of end-groups on glass transition temperature. *Macromol. Rapid Commun.* **2006**, *27*, 1335–1340. [CrossRef]
18. Terashima, T.; Kojima, H.; Sawamoto, M. Core-imprinted star polymers via living radical polymerization: Precision cavity microgels for selective molecular recognition. *Chem. Lett.* **2014**, *43*, 1690–1692. [CrossRef]
19. Kobayashi, S.; Igarashi, T.; Moriuchi, Y.; Saegusa, T. Block copolymers from cyclic imino ethers: A new class of nonionic polymer surfactant. *Macromolecules* **1986**, *19*, 535–541. [CrossRef]
20. Bloksma, M.M.; Paulus, R.M.; van Kuringen, H.P.C.; van der Woerdt, F.; Lambermont-Thijs, H.M.L.; Schubert, U.S.; Hoogenboom, R. Thermoresponsive poly(2-oxazine)s. *Macromol. Rapid Commun.* **2011**, *33*, 92–96. [CrossRef]
21. Lambermont-Thijs, H.M.L.; Fijten, M.W.M.; van der Linden, A.J.; van Lankvelt, B.M.; Bloksma, M.M.; Schubert, U.S.; Hoogenboom, R. Efficient cationic ring-opening polymerization of diverse cyclic imino ethers: Unexpected copolymerization behavior. *Macromolecules* **2011**, *44*, 4320–4325. [CrossRef]
22. Meyer, M.; Antonietti, M.; Schlaad, H. Unexpected thermal characteristics of aqueous solutions of poly(2-isopropyl-2-oxazoline). *Soft Matter* **2007**, *3*, 430–431. [CrossRef] [PubMed]
23. Demirel, A.L.; Meyer, M.; Schlaad, H. Formation of polyamide nanofibers by directional crystallization in aqueous solution. *Angew. Chem. Int. Ed.* **2007**, *46*, 8622–8778. [CrossRef] [PubMed]
24. Lübtow, M.M.; Hahn, L.; Haider, M.S.; Luxenhofer, R. Drug specificity, synergy and antagonism in ultrahigh capacity poly(2-oxazoline)/poly(2-oxazine) based formulations. *J. Am. Chem. Soc.* **2017**, *139*, 10980–10983. [CrossRef] [PubMed]
25. Kirila, T.; Smirnova, A.; Razina, A.; Tenkovtsev, A.; Filippov, A. Synthesis and conformational characteristics of thermosensitive star-shaped six-arm polypeptoids. *Polymers* **2020**, *12*, 800. [CrossRef]
26. Sezonenko, T.; Qiu, X.P.; Winnik, F.M.; Sato, T. Dehydration, micellization, and phase separation of thermosensitive polyoxazoline star block copolymers in aqueous solution. *Macromolecules* **2019**, *52*, 935–944. [CrossRef]
27. Kowalczuk, A.; Kronek, J.; Bosowska, K.; Trzebicka, B.; Dworak, A. Star poly(2-ethyl-2-oxazoline)s-synthesis and thermosensitivity. *Polym. Int.* **2011**, *60*, 1001–1009. [CrossRef]
28. Angot, S.; Murthy, K.S.; Taton, D.; Gnanou, Y. Scope of the copper halide/bipyridyl system associated with calixarene-based multihalides for the synthesis of well-defined polystyrene and poly(meth)acrylate stars. *Macromolecules* **2000**, *33*, 7261–7274. [CrossRef]
29. Strandman, S.; Hietala, S.; Aseyev, V.; Koli, B.; Butcher, S.J.; Tenhu, H. Supramolecular assemblies of amphiphilic PMMA-block-PAA stars in aqueous solutions. *Polymer* **2006**, *47*, 6524–6535. [CrossRef]
30. Smith, G.W. Crystal Structure of a Nitrogen Isostere of Pentacyclo-Octacosadodecaene. *Nature* **1963**, *198*, 879. [CrossRef]
31. Gawroński, J.; Kołbon, H.; Kwit, M.; Katrusiak, A. Designing large triangular chiral macrocycles: Efficient [3 + 3] diamine−dialdehyde condensations based on conformational bias. *J. Org. Chem.* **2000**, *65*, 5768–5773. [CrossRef] [PubMed]
32. Kirila, T.; Smirnova, A.; Razina, A.; Tenkovtsev, A.; Filippov, A. Influence of salt on the self-organization in solutions of star-shaped poly-2-alkyl-2-oxazoline and poly-2-alkyl-2-oxazine on heating. *Polymers* **2021**, *13*, 1152. [CrossRef]
33. Amirova, A.; Rodchenko, S.; Milenin, S.; Tatarinova, E.; Kurlykin, M.; Tenkovtsev, A.; Filippov, A. Influence of a hydrophobic core on thermoresponsive behavior of dendrimer-based star-shaped poly(2-isopropyl-2-oxazoline) in aqueous solutions. *J. Polym. Res.* **2017**, *24*, 124. [CrossRef]
34. Salzinger, S.; Huber, S.; Jaksch, S.; Busch, P.; Jordan, R.; Papadakis, C.M. Aggregation behavior of thermo-responsive poly(2-oxazoline)s at the cloud point investigated by FCS and SANS. *Colloid. Polym. Sci.* **2012**, *290*, 385–400. [CrossRef]
35. Krumm, C.; Fik, C.P.; Meuris, M.; Dropalla, G.J.; Geltenpoth, H.; Sickmann, A.; Tiller, J.C. Well-defined amphiphilic poly(2-oxazoline) ABA-triblock copolymers and their aggregation behavior in aqueous solution. *Macromol. Rapid Commun.* **2012**, *33*, 1677–1682. [CrossRef] [PubMed]
36. Xu, J.; Luo, S.; Shi, W.; Liu, S. Two-stage collapse of unimolecular micelles with double thermoresponsive coronas. *Langmuir* **2006**, *22*, 989–997. [CrossRef]
37. Kyriakos, K.; Aravopoulou, D.; Augsbach, L.; Sapper, J.; Ottinger, S.; Psylla, C.; Aghebat Rafat, A.; Benitez-Montoya, C.A.; Miasnikova, A.; Di, Z.; et al. Novel thermoresponsive block copolymers having different architectures—Structural, rheological, thermal, and dielectric investigations. *Colloid. Polym. Sci.* **2014**, *292*, 1757–1774. [CrossRef]
38. Witte, H.; Seeliger, W. Cyclische imidsäureester aus nitrilen und aminoalkoholen. *Lieb. Ann.* **1974**, *6*, 996–1009. [CrossRef]
39. Trinh, L.T.T.; Lambermont-Thijs, H.M.L.; Schubert, U.S.; Hoogenboom, R.; Kjoniksen, A.L. Thermoresponsive poly(2-oxazoline) block copolymers exhibiting two cloud points: Complex multistep assembly behavior. *Macromolecules* **2012**, *45*, 4337–4345. [CrossRef]

0. Steinschulte, A.A.; Schulte, B.; Rutten, S.; Eckert, T.; Okuda, J.; Moller, M.; Schneider, S.; Borisov, O.V.; Plamper, F.A. Effects of architecture on the stability of thermosensitive unimolecular micelles. *Phys. Chem. Chem. Phys.* **2014**, *16*, 4917–4932. [CrossRef] [PubMed]
1. Dimitrov, I.; Trzebicka, B.; Müller, A.H.E.; Dworak, A.; Tsvetanov, C.B. Thermosensitive water-soluble copolymers with doubly responsive reversibly interacting entities. *Prog. Polym. Sci.* **2007**, *32*, 1275–1343. [CrossRef]
2. Rossegger, E.; Schenk, V.; Wiesbrock, F. Design strategies for functionalized poly(2-oxazoline)s and derived materials. *Polymers* **2013**, *5*, 956–1011. [CrossRef]
3. Amirova, A.; Golub, O.; Kirila, T.; Razina, A.; Tenkovtsev, A.; Filippov, A. Influence of arm length on aqueous solution behavior of thermosensitive poly(2-isopropyl-2-oxazoline) stars. *Colloid Polym. Sci.* **2017**, *295*, 117–124. [CrossRef]
4. Kirila, T.; Smirnova, A.; Kurlykin, M.; Tenkovtsev, A.; Filippov, A. Self-organization in aqueous solutions of thermosensitive star-shaped and linear gradient copolymers of 2-ethyl-2-oxazoline and 2-isopropyl-2-oxazoline. *Colloid Polym. Sci.* **2020**, *298*, 535–546. [CrossRef]
5. Smirnova, A.V.; Kirila, T.U.; Kurlykin, M.P.; Tenkovtsev, A.V.; Filippov, A.P. Behavior of aqueous solutions of polymer star with block copolymer poly(2-ethyl-2-oxazoline) and poly(2-isopropyl- 2-oxazoline) arms. *Int. J. Polym. Anal. Charact.* **2017**, *22*, 677–684. [CrossRef]
6. Kirila, T.U.; Smirnova, A.V.; Filippov, A.S.; Razina, A.B.; Tenkovtsev, A.V.; Filippov, A.P. Thermosensitive star-shaped poly-2-ethyl-2-oxazine. Synthesis, structure characterization, conformation, and self-organization in aqueous solutions. *Eur. Polym. J.* **2019**, *120*, 109215. [CrossRef]
7. Daoud, M.; Cotton, J.P. Star shaped polymers: A model for the conformation and its concentration dependence. *J. Phys.* **1982**, *43*, 531–538. [CrossRef]
8. Simonova, M.A.; Tarasova, E.V.; Dudkina, M.M.; Tenkovtsev, A.V.; Filippov, A.P. Synthesis and hydrodynamic and conformation properties of star-shaped polystyrene with calix [8] arene core. *Int. J. Polym. Anal. Charact.* **2019**, *24*, 87–95. [CrossRef]
9. Kratochvil, P. *Classical Light Scattering from Polymer Solution*, 1st ed.; Elsevier: Amsterdam, The Netherlands, 1987.
0. Schärtl, W. *Light Scattering from Polymer Solutions and Nanoparticle Dispersions*, 1st ed.; Springer: Berlin, Germany, 2007.
1. Øgendal, L.H. *Light Scattering Demystified Theory and Practice*; University of Copenhagen: Copenhagen, Danmark, 2017.
2. Rodchenko, S.; Amirova, A.; Kurlykin, M.; Tenkovtsev, A.; Milenin, S.; Filippov, A. Amphiphilic molecular brushes with regular polydimethylsiloxane backbone and poly-2-isopropyl-2-oxazoline side chains. 2. Self-organization in aqueous solutions on heating. *Polymers* **2021**, *13*, 31. [CrossRef] [PubMed]
3. Kirile, T.Y.; Tobolina, A.I.; Elkina, A.A.; Kurlykin, M.P.; Ten'kovtsev, A.V.; Filippov, A.P. Self-assembly processes in aqueous solutions of heat-sensitive star-shaped poly-2-ethyl-2-oxazoline. *Fibre Chem.* **2018**, *50*, 248–251. [CrossRef]
4. Kowalczuk, A.; Mendrek, B.; Żymełka-Miara, I.; Libera, M.; Marcinkowski, A.; Trzebicka, B.; Smet, M.; Dworak, A. Solution behavior of star polymers with oligo(ethylene glycol) methyl ether methacrylate arms. *Polymer* **2012**, *53*, 5619–5631. [CrossRef]
5. Qiu, F.; Wang, D.; Wang, R.; Huan, X.; Tong, G.; Zhu, Q.; Yan, D.; Zhu, X. Temperature-induced emission enhancement of star conjugated copolymers with poly(2-(dimethylamino) ethyl methacrylate) coronas for detection of bacteria. *Biomacromolecules* **2013**, *14*, 1678–1686. [CrossRef] [PubMed]
6. Le Dévédec, F.; Strandman, S.; Baille, W.E.; Zhu, X.X. Functional star block copolymers with a cholane core: Thermo-responsiveness and aggregation behavior. *Polymer* **2013**, *54*, 3898–3903. [CrossRef]
7. Privalov, P.L. Physical basis of the stability of the folded conformations of proteins. In *Protein Folding*; Creighton, T.E., Ed.; W.H. Freeman and Co.: New York, NY, USA, 1992; pp. 83–126.
8. Tiktopulo, E.I.; Bychkova, V.E.; Ricka, J.; Ptitsyn, O.B. Cooperativity of the coil-globule transition in a homopolymer: Microcalorimetric study of Poly(N-sopropylacrylamide). *Macromolecules* **1994**, *27*, 2879–2882. [CrossRef]
9. Ladbury, J.E.; Chowdhry, B.Z. *Biocalorimetry: Applications of Calorimetry in the Biological Sciences*; John Wiley and Sons: Chichester, UK, 1998.
0. Hoogenboom, R.; Schlaad, H. Thermoresponsive poly(2-oxazoline)s, polypeptoids, and polypeptides. *Polym. Chem.* **2017**, *8*, 24–40. [CrossRef]
1. Rodchenko, S.; Amirova, A.; Milenin, S.; Kurlykin, M.; Tenkovtsev, A.; Filippov, A. Self-organization of thermosensitive star-shaped poly(2-isopropyl-2-oxazolines) influenced by arm number and generation of carbosilane dendrimer core in aqueous solutions. *Colloid Polym. Sci.* **2020**, *298*, 355–363. [CrossRef]
2. Atanase, L.I.; Riess, G. Thermal cloud point fractionation of poly(vinyl alcohol-co-vinyl acetate): Partition of nanogels in the fractions. *Polymers* **2011**, *3*, 1065–1075. [CrossRef]

MDPI
St. Alban-Anlage 66
4052 Basel
Switzerland
Tel. +41 61 683 77 34
Fax +41 61 302 89 18
www.mdpi.com

Polymers Editorial Office
E-mail: polymers@mdpi.com
www.mdpi.com/journal/polymers